The Biological Basis of Radiation Protection Practice

The Biological Basis of Radiation Protection Practice

EDITED BY

KENNETH L. MOSSMAN, Ph.D., M.Ed.

WILLIAM A. MILLS, Ph.D.

WILLIAMS & WILKINS
BALTIMORE • HONG KONG • LONDON • MUNICH
PHILADELPHIA • SYDNEY • TOKYO

Publisher: Deanna Gemmill
Associate Publisher: Carole Pippin
Copy Editor: Rebecca Marnhout
Designer: Norman Och
Illustration Planner: Ray Lowman
Production Coordinator: Sandy Renovetz

Printed in the United States of America

Library of Congress Cataloging-in-Publication Data
The Biological basis of radiation protection practice / edited by Kenneth L. Mossman, William A. Mills.
p. cm.
Includes bibliographical references and index.
ISBN 0-683-06140-2
1. Radiation—Safety measures. I. Mossman, Kenneth L., 1946– II. Mills, William A., 1929– .
[DNLM: 1. Radiation Genetics. 2. Pregnancy—radiation effects. 3. Neoplasms, Radiation-Induced. 4. Radiation Injuries. WN 610 B61535]
RA1231.R2B44 1992
616.9′89705—dc20
DNLM/DLC
for Library of Congress 92-14976
CIP

Foreword

This monograph is based on the Health Physics Society Summer School "The Biological Basis of Radiation Protection Practice," held at Georgetown University in Washington, DC, July 14–19, 1991. This course was designed to address the biological effects of ionizing radiation exposure of practical concern in radiation protection.

A distinguished faculty was assembled to present lectures on a wide spectrum of acute and late effects. This monograph has allowed for expansion of the topics covered during the week-long course, resulting in more extensive reviews and more detailed analyses. Each paper was carefully reviewed by other authorities in the field. We hope this monograph will be of value to the practicing health physicist as well as to other professionals involved with uses of radiation in medicine and industry (e.g., medical physicists, radiologists, industrial hygiene professionals).

The editors wish to express their appreciation to the distinguished contributors to this monograph, to the manuscript referees for their many hours of effort and valuable suggestions, and to Ms. Diane S. Flack of Oak Ridge Associated Universities, Inc., for her efforts in initially assembling manuscripts. Ms. Michelle Johnson provided invaluable editorial assistance in the final preparation of papers for publication. The editors are grateful to Ms. Deanna Gemmill and Ms. Carole Pippin, of Williams & Wilkins, for their valuable support and cooperation in the publication of this monograph.

The editors are particularly grateful to the Committee on Interagency Radiation Research and Policy Coordination (CIRRPC), Office of Science and Technology Policy, Executive Office of the President, for their grant (administered through Oak Ridge Associated Universities, Inc.) to the Health Physics Society to support publication of this book. Without CIRRPC support, publication would not have been possible.

KENNETH L. MOSSMAN, PH.D., M.ED.
WILLIAM A. MILLS, PH.D.
Editors

Introduction

As a public health profession, health physics is concerned with the protection of man from the harmful effects of exposure to radiation. This book reviews the major biological effects of ionizing radiation exposure of interest in radiation protection practice. Chapters have been written by an outstanding group of contributors who bring a wide range of perspectives in health physics and radiobiology.

This book is organized into four sections. The first two sections deal with low dose radiation effects including genetic effects, developmental effects, and cancer. Discussions include reviews of the National Academy of Sciences' BEIR IV and BEIR V reports and ICRP 60. Specific papers, or portions of papers, in these sections focus on underlying biology and radiobiology principles to provide the reader with a comprehensive review of these important public health effects. The third section explores acute effects of radiation which occur following whole-body or partial body exposure to high dose. An historical review of radiation accidents and discussions of the radiation syndromes and radiation responses in specific tissues of interest in radiation protection, including skin, gonads and lens of the eye, are provided. The final section explores special topics in radiobiology of interest in health physics including radiation protection of human subjects in research studies, contemporary issues in radiation health, and the role of radiobiology in operational health physics training.

This text should be particularly valuable to the practicing health physicist. In many areas of health physics, such as operational radiation protection, little knowledge of radiobiology is necessary to carry out various technical operations. However, a knowledge of radiobiology and understanding of the public health consequences of radiation protection practices, enhances the professionalism of the health physics practitioner. In other areas of health physics practice, knowledge of specific health effects is necessary. For instance, in providing guidance to pregnant or potentially pregnant workers, the health physicist should have a fundamental understanding of effects of radiation on human development and reproduction. Finally, health physicists should have a basic understanding of the radiobiology principles of radiation protection practice because they are probably asked more questions about health effects of radiation exposure by professional co-workers and members of the general public than about any other aspect of radiation protection.

For those readers concerned with radiation risk assessment and management, the papers by Abrahamson, Boice, Meinhold, Mills, and Upton should be of particular interest. These contributors provide an overview of the extensive human data base, and its limitations and uncertainties, which has been used in the development of radiation risk estimates. The National Academy of Sciences' BEIR IV and V reports and the ICRP 60 report are also discussed in detail in the context of how the human radiation exposure experience has been used to develop risk estimates and the uncertainties that must be considered in the use of these estimates in risk management practice and in developing standards.

This book should also be of value to medical physicists, nuclear physicians, radiologists, and radiation oncologists who use radiation producing equipment and radioactive materials in the diagnosis and treatment of disease. The sections of the book dealing with carcinogenesis, genetic effects, and exposure during pregnancy should be particularly useful in explaining radiation risks to patients and in evaluating radiation risks following medical exposure.

KENNETH L. MOSSMAN, PH.D., M.ED.
WILLIAM A. MILLS, PH.D.
Editors

Contributors

SEYMOUR ABRAHAMSON, PH.D.
Professor, Department of Zoology, University of Wisconsin, Madison, WI

ROBERT C. BAUMILLER, PH.D.
Dean, School of Health Science, University of Detroit Mercy, Detroit, MI

MARY ELLEN BERGER
Oak Ridge Associated Universities, Inc., REAC/TX, Oak Ridge, TN

JOHN D. BOICE, JR., PH.D.
Chief, Radiation Epidemiology Branch, National Cancer Institute, Bethesda, MD

ROBERT L. BRENT, M.D., PH.D.
Professor and Chairman, Department of Pediatrics, Jefferson Medical College, Philadelphia, PA

ROGER J. CLOUTIER, M.S.
Oak Ridge Associated Universities, Inc., Oak Ridge, TN

ANATOLY DRITSCHILO, M.D.
Professor and Chairman, Department of Radiation Medicine, Georgetown University Medical Center, Washington, DC

THOMAS B. ELLIOTT, PH.D.
Armed Forces Radiobiology Research Institute, Defense Nuclear Agency, Bethesda, MD

FUN H. FONG, M.D., ACEP
Oak Ridge Associated Universities, Inc., REAC/TX, Oak Ridge, TN

SHIRLEY A. FRY, M.D., BCH, MPH
Oak Ridge Associated Universities, Inc., REAC/TX, Oak Ridge, TN

RALPH H. JOHNSON, J.D.
Schmeltzer, Aptaler, and Shepard, Washington, DC

TIMOTHY J. JORGENSEN, PH.D.
Assistant Professor, Department of Radiation Medicine, Georgetown University Medical Center, Washington, DC

ERIC KEARSLEY, PH.D.
Chairman, Radiation Biophysics Department, Armed Forces Radiobiology Research Institute, Defense Nuclear Agency, Bethesda, MD

G. DAVID LEDNEY, PH.D.
Department of Experimental Hematology, Armed Forces Radiobiology Research Institute, Defense Nuclear Agency, Bethesda, MD

CLARENCE C. LUSHBAUGH, M.D., PH.D.
Oak Ridge Associated Universities, Inc., REAC/TX, Oak Ridge, TN
CHARLES B. MEINHOLD
President, National Council on Radiation Protection and Measurements, Bethesda, MD
WILLIAM A. MILLS, PH.D.
Senior Technical Advisor, Oak Ridge Associated Universities, Inc., Washington, DC
MARCUS M. MOORE
Armed Forces Radiobiology Research Institute, Defense Nuclear Agency, Bethesda, MD
KENNETH L. MOSSMAN, PH.D., M.ED.
Assistant Vice President for Research, Arizona State University, Tempe, AZ
ROBERT C. RICKS, PH.D.
Oak Ridge Associated Universities, Inc., REAC/TX, Oak Ridge, TN
ANN H. SIPE
Oak Ridge Associated Universities, Inc., REAC/TX, Oak Ridge, TN
ARTHUR C. UPTON, M.D.
Director, Institute of Environmental Medicine, New York University Medical Center, New York, NY
NIEL WALD, M.D.
Professor, Graduate School of Public Health, Department of Environmental and Occupational Health, University of Pittsburgh, Pittsburgh, PA

Contents

Foreword by *Kenneth L. Mossman* and *William A. Mills* ... v
Introduction by *Kenneth L. Mossman* and *William A. Mills* ... vii
Contributors ... ix

Section I: Genetic and Developmental Consequences of Radiation Exposure

1. Principles of Genetics and Mutagenesis ... *Robert C. Baumiller* 3
2. Genetic Effects of Irradiation ... *Seymour Abrahamson* 12
3. The Effects of Embryonic and Fetal Exposure to Ionizing Radiation: Counseling the Patient and Worker about These Risks *Robert L. Brent* 23
4. Radiation Exposure in Pregnancy: Health Physics Considerations *Kenneth L. Mossman* 63

Section II: Radiation and Cancer

5. Cancer and Carcinogenesis ... *Anatoly Dritschilo* 77
6. Radiation Carcinogenesis-Human Epidemiology .. *John D. Boice, Jr.* 89
7. Cancer Risk Estimates for External Radiation Exposure (BEIR V) *Arthur C. Upton* 121
8. Estimates of Cancer Risk from Internally Deposited Radionuclides, with Emphasis on Alpha-Emitters: BEIR IV ... *William A. Mills* 138
9. Risk Assessment in Radiation Protection ... *Charles B. Meinhold* 154
10. The Radioepidemiological Tables in Radiation Litigation *Ralph H. Johnson* 158

Section III: Whole Body Radiation Exposure and Effects on Specific Tissues and Organs

11. Biology of Radiosensitive Tissues and Organs *Timothy J. Jorgensen* 169
12. Acute Radiation Injuries and Their Medical Management *Niel Wald* 184
13. Modulation of Mortality by Tissue Trauma and Sepsis in Mice after Radiation Injury ... *G. David Ledney, Thomas B. Elliott, and Marcus M. Moore* 202
14. History of Radiation Accidents ... *Robert C. Ricks, Shirley A. Fry, Ann H. Sipe, Mary Ellen Berger, Fun H. Fong, Jr. and Clarence C. Lushbaugh* 218

Section IV: Special Topics

15. Contemporary Issues in Radiation Health .... *Eric Kearsley* 229
16. Radiation Biology and Operational Health Physics *Roger J. Cloutier* 240
17. Radiation Protection of Human Subjects in Research Protocols *Kenneth L. Mossman* 255

Index 263

Section I

Genetic and Developmental Consequences of Radiation Exposure

Chapter 1

Principles of Genetics and Mutagenesis

ROBERT C. BAUMILLER

ABSTRACT

Genetics has become the dominant biological science. Mendelian principles still are accepted, with certain modifications: (*a*) Genes occur in pairs (except for sex-linked genes in animals), (*b*) the paired genes segregate and recombine in the offspring, and (*c*) different pairs of genes segregate independently of other pairs (unless they are in relatively close approximation on the same chromosome). Cytogenetics has supplied the microscopically visible basis for these principles and their exceptions. Mutation at the chromosome level was described by cytogeneticists, but gene mutation was not fully understood until the chemical makeup of the gene was elucidated. DNA and its coded message, and the variations in that code caused by mutations, have become the work of the 1990s under the research endeavor termed "The Human Genome Project." Radiation, long known to be mutagenic, produces damage by production of chemically reactive radicals and direct ionization of atoms. Relevant genetic lesions can now be investigated at the molecular level. Radiation-induced malignancy can now be understood as specific genetic changes in cell regulation due to random mutations induced by radiation.

PRINCIPLES OF GENETICS AND MUTAGENESIS

Gregor Mendel was an Augustinian monk in Brno, which now is in Czechoslovakia. As a botanist he was deeply interested in plant breeding and impressed by the variants within single species available from the traveling seed vendors. Mendel was by profession a man of prayer and reflection. Certainly the latter quality of his personality led him to observe nature and choose his experimental system with extraordinary perspicacity and great good luck (or maybe the prayerfulness paid off!).

Mendel chose a closed flowering species that self-fertilized unless human intervention brought about controlled cross-fertilization. Mendel chose variants that always bred true, generation after generation, and he chose variants that were always distinguishable from one another; i.e., his plants with green seeds never shaded into a yellowish level that was indistinguishable from the yellow-seeded variants, which never produced seeds that shaded toward green and crossed in appearance with the yellowest green seeds.

Mendel then was ready to cross-fertilize lines. The first generation of offspring was like one of the two variants under test, not a blend. The seeds produced were, in turn, planted and allowed to self-fertilize and produce the second generation. The results were an average of three seeds like the parent plant to one like the original parent used in cross-fertilization whose appearance was not seen in the first generation.

P_1 green × yellow
F_1 all green
P_2 F_1 self-fertilizing
F_2 3 green: 1 yellow

If then the second generation were allowed to self-fertilize, the one previously suppressed variant bred true. One of the three with the dominating variant trait bred true, but the other two produced a third-generation split 3 to 1.

P_3 F_2 self-fertilizing

	green	green	green	yellow
F_3	all green	3:1	3:1	all yellow

Mendel's greatest feat was examining his data with a statistical mind and proposing a hypothesis to explain what he recorded. He suggested that factors for the variant type being followed occurred in pairs within an individual. In reproduction each parent passed on one of their paired factors to the offspring. Factors in the hybrid were dominant or recessive (green dominant to yellow). The nonexpressing factor (recessive) maintained its full integrity although not expressed. The self-fertilized hybrid recombined factors, randomly producing the 3:1 ratio of dominant homozygote to hybrid heterozygote to recessive homozygote. (In a homozygote both members of the gene pair are alike–AA or aa; in a heterozygote the paired genes differ–Aa.)

Statistical methods for examining data (Student's t or Chi square) were not available to Mendel, but the data were so close to the predicted 3:1 and 1:2:1 ratios that no one of good will could deny the fit (power of prayer or Mendelian selection–i.e., was Mendel fortunate to have such a good fit, or did he fudge the data?) (Mendel 1865).

Mendel was so practical and reasonable that he was made an administrator, and as abbot could no longer get into his garden or find the time to think through experiments to generate data to test hypotheses or create new ones.

For almost 40 years Mendel's discovery was lost and/or ignored, which may realistically have been a blessing. Today we relate genetics and evolution and genetic recombination and mutation to natural selection. If Darwin had known the Mendelian principles, perhaps his view would have been so narrowed that he might not have made the dramatic and sweeping conclusions that we now call the theory of evolution. Mendelian principles evoke permanence rather than change.

The time was right for the rediscovery of Mendelism at the dawn of the twentieth century. Discovery in genetics exploded as the Western world came to think of science and technology as the future of mankind. To geneticists, the real work began at Columbia University with Thomas Hunt Morgan and his collaborators and students. The Fly Room (*Drosophila melanogaster*) was the center of activity, with the professors and students using this marvelous organism to deepen and expand the understanding of genetics by rapid breeding experiments. The insights of the Morgan group were fueled by the work of cytologists and particularly the 1898 second edition of E.B. Wilson's *The Cell in Development and Heredity* (Wilson was also at Columbia). Wilson was convinced that the nucleus and especially the chromatin were involved with heredity. Boveri and Sutton, 3 years later, separately enunciated the chromosome theory of inheritance (Boveri 1902, Sutton 1902).

Chromosomes occur in pairs just like genes. They duplicate, and by mitotic division each daughter cell receives a copy of each of the chromosomes found in the parent cell. Germ cells (sperm and mature egg) contain one half the amount of genetic material. The process of meiosis is a special double cell division with only a single duplication of genetic material; most important, the homologous chromosomes pair along their common length in order to separate the paternally and maternally derived genes into separate germ cells to be reunited again in the next generation of individuals.

Thus genes existed, and each gene could have an alternative form. One student of Morgan's demonstrated that the genes were in linear order along the chromosome (Sturtevant 1913). A gene usually remained the same generation after generation, but occasionally a gene could change from one form to another, changing the genotypic makeup and, when paired with another appropriate gene, yielding a phenotype that was different.

Mendel noted the difference between dominance and recessiveness in genes, but since most plants are both male and female, he could not have discovered sex-linkage as a type of inheritance. Animals have, in general, one sex that produces two types of gametes and the other that produces gametes of one type. In the human, males have an X and a Y chromosome. The Y chromosome has few genes but does, under normal circumstances, determine a male phenotype. Two X chromosomes yield a normal female phenotype. Since the Y chromosome has almost no genes in common with those on the X, males manifest any trait carried on the X they inherit.

P_1	XY	× XXc
F_1	XY	(normal male)
	XX	(normal female)
	XXc	(carrier female)
	XcY	(affected male)

Thus males are affected by an array of X-linked diseases or conditions (e.g., color blindness in humans) that appear to be recessive in females. If a gene is X-linked and dominant in the female, it is often lethal in the male, just as dominant abnormal genes in the non-sex-determining chromosomes are often lethal if homozygous.

Mendel was not the only scientist of his time interested in genetics. He was bright enough, or fortunate enough, to select clear single-gene controlled traits for study. The well-funded and prestigious Galton in England, some years after Mendel, was fascinated with traits that existed in a graded fashion through the population, traits like height or intelligence (Galton 1869). Matings did not produce Mendelian ratios but normal curves around the parental mean. Knowing Mendelian principles, we can conceptualize traits dependent on many gene pairs, all contributing to the end phenotype, and even include the variation caused by environment. However, ignorant of Mendelian principles, one could not unequivocally arrive at such a theory since many other hypotheses would be just as likely to explain the results.

The genius of Mendel came in good part because the seeds he used were true breeders. The distinguishing trait was present in homozygous condition; i.e., both members of the gene pair were identical. These are now called pure lines. The fly breeders worked with randomly mating populations, requiring them to selectively pair and mate them. They discovered heterozygotes in nature by brother/sister mating the offspring of a single-pair mating. Pure lines breed true generation after generation, demonstrating that genes are stable, though not completely. Genes also change rarely. These rare mutations are then equally stable in pure lines if they are viable. Mutants almost always are less viable than the normal, i.e., predominant gene in the population. Dominant mutations are often very serious in effect, changing physical structure. Recessive mutants usually cause only slight alterations in viability when heterozygous but are frequently lethal when homozygous. Recessive genes often are either enzyme producers or regulatory genes.

Genetic change occurs by altering the makeup of a gene, taking away all or part of a gene or adding genetic material or rearranging the genetic material.

The grossest change in amount of material is loss or gain of a whole genome. Reduction to haploid state in the human occurs briefly in the lifetime of each individual sperm cell. Polyploidy does occur, viz., triploidy or tetraploidy, the latter being lethal in the human, almost always early in pregnancy. Triploidy is less lethal, and newborns with triploidy have been reported. They have a short life span. An extra single chromosome, trisomy, or loss of a single chromosome, monosomy, is frequent at conception in humans. It has been suggested that 50% or more of human concepti are trisomic or monosomic. Sex chromosome abnormalities, as a group, are most frequent at birth because of the relative genetic unimportance of the Y and the fact that all X chromosomes over one are almost completely shut down early in development. Next in frequency is trisomy 21, the cause of Down syndrome. Risk for Down syndrome is related to maternal age. Trisomy 13 and 18 occur rarely; in fact these trisomies get to birth only if a portion of the placenta has reverted to totally normal diploidy.

Chromosome breakage can also lead to changes in expected reproductive outcome. A chromosome is made up of many genes and has three necessary components for its continued functioning from mitosis to mitosis to mitosis. The ends of the chromosome have an entity termed a telomere that effectively closes the DNA, and between the telomeres is a centromere, which is the spindle fiber attachment site during the process of cell division. If a centromere does not exist, the chromosome will not be divided and drawn into the daughter nuclei in the process of cell division.

Single chromosome breakage may heal with no perceptible change in the resultant chromosome or its activities. If break points do not heal, the acentric fragment will be lost and the centric fragment upon the next round of synthesis will join broken ends, forming a dicentric chromosome. Unless the dicentric fragment is small and has the two centromeres close together, the cell containing it will die.

Two breaks in a single chromosome can result in a loss of the interstitial piece, which may be cell lethal or, if small, causative of a single syndrome or of several simultaneous syndromes. Another option is that the piece may be inverted, changing the order of the genes but not the total genetic content. Inversions can cause major problems in pairing and exchange during meiosis. The whole human race is amazingly almost free of inversions.

One break in each of two chromosomes may result in a translocation of pieces, yielding functional chromosomes but new gene positions. If there is no loss of material, this usually results in normal growth and development but presents problems in meiosis. Individuals carrying a translocation often suffer repeat miscarriages. Other yet more complicated events can and do occur but are lost from the population rapidly.

Gene mutation was thought to be caused by a mistake of nature and was totally unpredictable and very rare until, in 1927 H.J. Muller, another collaborator of Morgan's, discovered the mutagenic quality of ionizing radiation (Muller 1927). Radiation was found to cause submicroscopic changes in the gene and also to cause chromosome breakage leading to changes in structure of chromosomes and to loss or gain of genetic material.

In 1941 Charlotte Auerbach discovered chemical mutagenesis. The discovery was withheld from the public because stockpiled gas weapons were involved (Auerbach and Robson 1947).

A huge literature has accumulated on the effects of radiation. Much of this work is of great importance in understanding the types of mutation that result from different types of radiation, the effects of dose in relationship to time, and the effect of environment on the mutation process. The ingenious experiments designed to probe the nature of the genetic material and to understand the induction of change all were left to the history of science when the actual composition of the genetic material became known and biochemical technology breakthroughs allowed more and more sensitive probing.

In 1944 Avery, MacLeod, and McCarty showed that deoxyribonucleic acid (DNA) was the molecule that contained the genetic information (Avery, MacLeod, and McCarty 1944). In 1953 Watson and Crick published the structure of DNA, and the code leading to all basic biological activity was ready to be broken (Watson and Crick 1953). Today the Human Genome Project under the direction of Dr. Watson is pointed at unraveling the code for the entire human genome as well as the genome of other important genetic organisms including our old friend *Drosophila melanogaster*.

Cytogenetics, which began before the rediscovery of Mendel, revealed many secrets, but always in special organisms. Genetic statistical inference promoted by breeding experiments was verified by the physical evidence provided by the cytogeneticist. But cytogenetic techniques were crude, and the chromosome number of higher organisms was only guessed at and often wrongly. In 1956 Tjio and Levan were able to prepare human chromosomes with accuracy and show that we contain not 48 but 46 chromosomes (Tjio and Levan 1956). Today chromosome preparation is a routine diagnostic procedure for pregnant women at risk because of age or drug or radiation exposure. It is also routine in the diagnosis and staging of cancer.

Genetic variation is common. We are each of us different, and yet in essential genes, genes necessary for metabolism, movement, nervous reactions, etc., we are amazingly alike, in fact, almost identical, and we are amazingly like other living organisms. The essential nature of a gene is demonstrated by its conservation between species and, yes, between genera. Variation is important for adaptability; continuous mutation is important for maintaining variation and continuing nature's experimentation, with all of life plodding ever so slowly along the evolutionary process.

In general, genes persist in populations at a frequency that is constant through the generations. Dominant genes that are not passed on exist with the frequency of spontaneous mutation. Recessive genes exist at a frequency balanced by new mutation and selection against the homozygous recessive individual. For a recessive gene the frequency of affected people is the random possibility of two genes' coming together. If the gene frequency is called q, then the frequency of people affected should be $q \times q$ or q^2. If there is only one alternate gene p, then $p + q = 1$. By counting individuals who are affected with cystic fibrosis and knowing the size of the population that they come from, one finds one in 2500 Caucasians are affected, therefore, $1/2500 = q^2$ and $q = 1/50$ and $p = 49/50$. The whole population genotypically are p^2 homozygous unaffected, 2pq heterozygotes and q^2 homozygous affected where $p^2 + 2pq + q^2 = 1$. In a population where selection and mutation are in equilibrium and where size of the population is very large (infinite) and where mating is random generation after generation, these frequencies will endure. Nature is in balance. Such a population is said to be in Hardy-Weinberg equilibrium.

But we humans have taken nature into our own hands and by our many manipulations have changed the mutation and selection process. We produce high quantities of chemicals rarely found in nature and synthesize others never found in nature. We have learned how to use radiation in many of its forms to save life and to destroy life. We seek to save and serve life, individual by individual. We also cause genetic changes that if inherited by individuals lead to death postconception, postimplantation, during embryogeny, prenatally and postnatally, and at various other stages of our lives. Even the aging process and death at 100 plus is genetic in causation. The fear is that by our environmental pollution, chemically and radiologically, we can destroy life in a brief period of time and throughout our whole planet. Understanding mutagenesis is, therefore, a key to survival.

Mutation, in general, is a change in DNA, either in the amount of DNA, the relationship of one segment of DNA to another, or an altered coding subunit.

DNA is the genetic material. It is a double-stranded helix whose distinguishing feature is four organic bases who pair by hydrogen bonds. Adenine pairs with thymine, and cytosine pairs with guanine. The double strands separate to reproduce themselves or to produce a copy of messenger RNA, which is similar to DNA, varying by its sugar moiety and replacing thymine with uracil. The message ultimately is translated into a product protein that can be structural or metabolic in its function.

The meaningful reading frame for DNA is three nucleotides in length. These in proper order ultimately translate into an amino acid. Amino acids are the building blocks of protein. Different triplets designate each of the 20 essential amino acids, some amino acids having more than one code word. There is a single start triplet and three stop triplets. The code is universal to all forms of life on our planet.

Each human cell has 3 billion pairs of DNA bases and, we think, about 100,000 genes. But 80% of our DNA does not code for protein. The coding DNA is of obvious importance, but the rest is not unimportant but shows the lowest conservation rate between individuals and between species.

Changes in the amount of DNA passed from cell to cell or parent to offspring occur on the gross chromosomal level because of errors in cell division, spontaneous and induced. Such errors are frequent in humans at conception–in fact, more frequent than the correct division. However, 99% of such errors never come to birth, and 98% never generate enough growth to be recovered. Extra material (e.g., trisomy 21, Down syndrome) is tolerated better than the loss of material. The result of most extra material or loss is cell or embryo death. The exception to this rule is found in cancer cells, which no longer function under normal control. The evolution of cancer cell lines is dependent on a series of gene changes and, while thought of as out of control, is really in an altered but narrowly limited state.

Agents such as radiation do cause chromosome sets to be increased (polyploidy) and single chromosome additions or loss (trisomy or monosomy). These changes usually occur because of an altered cell cycle or disruption of the normal process of meiosis or mitosis. Radiation alters chromosomes by physical breakage of the DNA by ionization or by the creation of chemical complexes, such as free radicals or peroxidases, which cause breakage or change in the DNA that is not perfectly repaired by the various systems available to the nuclear material. Ionizing radiation or radioactive decay can and most frequently does cause breakage or disruption that demands healing or repair of DNA. This repair may lead to loss of nucleotides, gain of nucleotides, or change from one base to another. Base substitution, be it only one base, can change the code so that a different amino acid is placed in a protein; the sickle cell mutation represents such a change. The loss or gain of three nucleotides would subtract or add an amino acid (e.g., the cystic fibrosis gene is frequently [70%] caused by the loss of three nucleotides.) Loss or gain of one or more nucleotides but not a number divisible by three would lead to a shifting of the reading frame and usually leads to early termination of the messenger RNA produced and the loss of normal protein.

Geneticists think transgenerationally. But as human beings we think of our own well-being and the well-being of those we love. A threat to that well-being is age and disease, with vascular problems and malignancy our greatest concerns. A particular type of cancer has intrigued human geneticists for years. Retinoblastoma (Rb) is a cancer of the retina that appears in late fetal life or in the 1st year or so of life. Rb is inherited as a dominant, but not every retinal cell becomes cancerous; in fact, the carrier can escape the malignancy entirely but then produce a child or children who again manifest the condition. Sporadic, nontransmissible cases do appear, but these are almost always unilateral as opposed to bilateral in the inherited type. They also usually have one focus where the nonsporadic type average six foci per eye. These bits of information culled from many family studies led to the two-hit hypothesis (Knudson 1971). Knudson proposed that two separate changes must occur in a cell for it to become malignant for Rb. Some individuals inherit a genome that has already undergone step 1 and now needs only step 2. Step 2 happens randomly. In sporadic cases 1 and 2 have to occur in the same cell or its descendants. Rb was shown to be cytologically visible as a small deletion on the long arm of chromosome 13. In cancerous cells it was demonstrated that both 13 chromosomes had the same deletion. Thus step 1 and step 2 could be demonstrated. The Rb gene produces a product that exercises positive control over cell division in retinal cells; loss of control leads to cancer.

Other cancers, especially leukemias, have been known to be radiation induced. Many leukemias have chromosome changes, usually translocations in the cancer cell lines where one or both break points are often diagnostic of the type of leukemia. The break point is near what is now termed an oncogene. An oncogene is a decommissioned retroviral bit of DNA that, when broken or joined with another bit of DNA, interferes with the normal feedback regulation of cell division. No longer negatively regulated, the cell divides out of control.

Since we now consider cancer to be genetic, we now understand how radiation used to preferentially kill cancerous cells can at the same time and at a low frequency be inducing a new cancer to begin or to take further steps toward malignancy.

The process of mutation entails, then, any change on the DNA level that results in the change of hereditary information in its structure or in the quantity of meaningful information produced. While radiation mutagenesis is of prime concern for this audience, chemical and virally induced mutagenesis is equally or perhaps more important. Combined with spontaneous mutation, such changes in our DNA can lead to cancer, aging, and birth defects, all problems we seek to lessen if not eliminate.

REFERENCES

Auerbach, C., and Robson, J.M. Tests of chemical substances for mutagenicity. *Proc. R. Soc. Edinburgh B62*:284–291, 1947.

Avery, O.T., MacLeod, C.M., and McCarty, M. Studies on the chemical nature of the substance inducing transformation in pneumococcal types. *J. Exp. Med. 79*:137–158, 1944.

Boveri, T. On multipolar mitoses as a means for the analysis of the cell nucleus. *Verh. Phys. Med. Gesellch. Wurzburg 35:*67–90, 1902.

Galton, F. *Hereditary Genius.* London, MacMillan & Co., 1869.

Knudson, A.G. Mutation and cancer: Statistical study of retinoblastoma. *Proc. Natl. Acad. Sci. 68:*820–823, 1971.

Mendel, G. Versuche uber pflanzen hybriden. *Verh. Naturf. Ver. Brunn 4:*3–47, 1865. Translated and reprinted in Peters, T.A. *Classic Papers in Genetics.* Englewood Cliffs, NJ, Prentice-Hall, Inc., 1959.

Muller, H.J. Artificial transmutation of the gene. *Science 66:*84–87, 1927.

Sturtevant, A.H. The linear arrangement of six sex-linked factors in Drosophila as shown by their mode of association. *J. Exp. Zool. 14:*43–59, 1913.

Sutton, W.S. On the morphology of the chromosome group in Brachystola magna. *Biol. Bull. 4:*124–139, 1902.

Tjio, J.H., and Levan, H. The chromosome number in man. *Hereditas 42:*1–6, 1956.

Watson, J.D., and Crick, F.H.C. Molecular structure of nucleic acids. *Nature 171:*737–738, 1953.

Chapter 2

Genetic Effects of Irradiation

SEYMOUR ABRAHAMSON

ABSTRACT

The studies of several exposed populations of humans have provided the data from which various committees have been able to develop somatic risk estimates for radiation-induced cancers and possibly noncancer diseases. However, even the largest genetic study ever undertaken, on the offspring of the A-bomb survivors, has failed to reveal a significant increase in any of the endpoints studied to date. As will be discussed, these genetic data do provide a lower baseline against which other risk estimates can be compared. With few exceptions most of the risk estimates for human genetic effects are derived from well-controlled animal experiments with accurate dosimetry. They also involve a variety of assumptions in the extrapolation process to man. It is an equally important fact that for every animal, plant, and microorganism system studied, high-energy radiation can produce a variety of genetic changes, and thus there is no reason to believe that humans are different. The lack of responses seen in the Japanese A-bomb studies may reflect a different sensitivity to irradiation than assumed from the prediction made from animal study extrapolation (as some believe), or they may simply reflect the lack of statistical power based on a small sample size, particularly in the higher dose ranges (as others believe). We shall examine this issue in some detail. In the material that follows, an overview of genetic diseases, their natural incidence, radiation-induced frequencies as determined by major national commissions, and the uncertainties associated with them and their social implications is provided.

INTRODUCTION: NATURE OF GENETIC DAMAGE

Genetic damage can be subdivided into three categories.

1. Change in gene structure, which includes changes in a single DNA nucleotide base or the deletion of one, many, or all the nucleotides making up that gene, including neighboring genes, or a rearrangement of that DNA sequence, all of which lead to an alteration of protein structure and function. At this level the damage would be submicroscopic.
2. Changes in the structure of chromosomes. This usually includes microscopically observable changes and deletions or rearrangements of blocks of

material within a chromosome (intrachromosome changes) or between two or more chromosomes (interchromosomal rearrangements).

3. Changes in the appropriate number of chromosomes transmitted to the next generation. Aneuploidy is the general name for this phenomenon.

NATURE OF RADIATION-INDUCED GENETIC DAMAGE

As molecular capabilities for measuring genetic changes become more sophisticated, including precise sequencing of a gene's nucleotide coding structure, it has been possible to differentiate at the qualitative level the kinds of mutational events brought about by high-energy radiation versus specific chemical mutagens and perhaps even the so-called "spontaneous" mutations. The majority of radiation-induced mutations appear to result from deletions or rearrangements of blocks of the DNA sequence of genes rather than specific nucleotide substitutions. Many (not all) chemical mutagens, on the other hand, act almost exclusively by affecting a single nucleotide.

From a purely physical standpoint, target theory proposes the larger the target the more frequently it will be traversed by random ionization tracks. Thus larger genes will be mutated more frequently than smaller genes and we know that mammalian genes are much larger (i.e., total amount of DNA per gene) than the genes found in lower organisms. The lower organisms use their genic DNA to code for the amino acid sequence of proteins. In the mammal the coding sequence is interrupted in multiple sites by large sequences of noncoding DNA. This probably explains the observation made 20 years ago (Abrahamson et al. 1973) which indicated that the mutation induction rate/cSv/per locus differed by three orders of magnitude among species that differed proportionately in DNA content by three orders of magnitude; i.e., those species with the smallest DNA content had the smallest average gene mutation rates per unit dose, and as DNA levels increased so too did mutation rates.

Factors other than effective target size will also influence the mutation rate. Repair systems appear to operate more effectively on regions of DNA (genes) that are actively transcribing their message, and the efficiency of the enzymatic repair systems may vary considerably within the cell cycle and in different cell types. Chromosomes undergoing cell division are particularly sensitive to irradiation; they are highly condensed, and repair systems are less functional.

EXPRESSION OF MUTANT TYPES

Genetic changes can be expressed in a number of ways. A *dominant mutant* is a single gene change that produces an observable effect in the presence of a normal gene. Examples include Huntington disease and neurofibromatosis. In a *recessive mutant* the expression of a defect requires that both parents contribute an identically defective gene to their offspring. Examples include sickle cell anemia, Tay Sachs disease, and cystic fibrosis. *Sex-linked mutants* involve recessive genes on the X chromosome; they affect males most frequently, since they have a single X chromosome. Females have two X chromosomes and require two identical recessive mutant genes to express a defect. Examples include hemophilia and muscular dystrophy (Duchenne type). In *unbalanced chromosome rearrangements,* the unbalanced translocation, and interchromosomal rearrangements, provide too much material from one chromosome and too little from the other chromosome; this imbalance can lead to prenatal death or severe abnormalities and death within 1–2 years after birth. *Aneuploidy* is an abnormal chromosome number; this phenomenon occurs during the meiotic stages of germ cell formation. Failure to properly segregate members of a chromosome pair from each other can lead to a mature germ cell with either one too many or one too few chromosomes. Most of these aneuploid conditions are prenatal lethals as a result of the severe imbalance of genetic material. The most common and viable type is trisomy 21 (Down syndrome), but the sex chromosomes X and Y are also frequently involved, and they cause physical and physiological disturbances.

To date the least understood form of genetic disease but yet the largest class of disease does not result from a single mutant gene or a specific type of chromosome abnormality. It is called the *complexly, irregularly inherited, or multifactorial disease class,* and as the names imply, more than one gene may be required to interact with other genes and with appropriate environmental factors (e.g., dietary factors such as lipid, sugar, and salt intakes). Such diseases as diabetes, psychosis (manic depressive), schizophrenia, asthma, epilepsy, a large number and variety of cardiovascular disorders, and hereditary forms of cancer are part of this group.

GERM CELL STAGES OF IMPORTANCE

Because the accumulated radiation exposure that a population receives over their entire reproductive lifetime (30–35 years) is of primary concern, interest is directed to two germ cell populations in humans, spermatogonial cells and immature dictyate oocytes. The spermatogonial cells of the males are a stem cell population that continues to replenish itself by cell division and produces a population of cells destined to differentiate into mature sperm. The stem cells continue to accumulate the environmental and spontaneous insults leading to mutation. Most of the experimental data come from studies on male mice. Unlike the male, the female is born with a large but finite number of immature egg cells (nonreplaceable). It is these premature resting oocytes that will be both aging and accumulating potential mutagenic damage prior to maturation, which occurs shortly before ovulation. It is the radiosensitivity of these two germ cell stages that are of primary concern.

CURRENT INCIDENCE OF GENETIC DISEASE IN THE HUMAN POPULATION

Over the past 40 years, understanding of the natural prevalence of genetically determined diseases in a population has undergone continued evolution, primarily because more and more diseases are recognized to have a genetic basis. Forty years ago, the correct chromosome number for humans (46) was not known, and we had such poor cytogenetic techniques that we did not know that there were chromosomal abnormalities associated with Down, Klinefelter, and Turner syndromes and other diseases as well. Today each week seems to bring the finding of a new gene association with an old disease–Alzheimer's, breast cancer, colon cancer.

Even given all this, there have only been a few major epidemiological studies attempting to estimate the prevalence of genetic disease, and none of these studies involve the U.S. population. A British Columbia study now has followed a large population through the 1st 25 years of life and categorized the recognized disorders into their respective bins on a case-by-case basis. A major problem lies in the fact that so many dominantly inherited diseases have a much later onset of development than 25 years of age, so that they have been missed by this study. It would also appear that the incidence of complexly inherited diseases was seriously underestimated in this data base, for the same reasons.

A second major study in Hungary has suggested that nearly an order of magnitude increase in complexly inherited diseases is warranted, since the disease prevalence was about 60% in the Hungarian population over age 20 whereas in the British Columbia study it was 9% for those under 20. This figure was adopted in the most recent United Nations Scientific Committee on the Effects of Atomic Radiation (UNSCEAR) report (1988). However, the BEIR V committee of the National Academy of Sciences suggested that this category as presented by UNSCEAR was still too limiting and did not include cancer and cardiovascular diseases, which in their view had a genetic component of unknown size. Their estimate of multifactorial diseases jumped from about 90,000 cases in the BEIR 1980 report (National Research Council 1980) to over 1,200,000 cases per million liveborn in the BEIR 1990 report (National Research Council 1990). Table 2.1 summarizes the BEIR V estimates of genetic disease incidence and implies that on average each individual will suffer from one or more serious disorders with an

TABLE 2.1. THE CURRENT INCIDENCE OF GENETIC DISEASE IN A POPULATION OF 1 MILLION PERSONS

Diseases[a]	Incidence/10^6 liveborn[b]
Autosomal dominants	10,000
X-linked	400
Trisomies	3,800
Unbalanced translocations	600
Complex etiology[c]	
Congenital abnormalities	20,000–30,000
Heart	600,000
Cancer	300,000
selected others	300,000
TOTAL (rounded off)	1,245,000

[a] The genetic basis of the diseases are described in more detail in the text.

[b] BEIR V estimates of the current incidence of genetic disease (National Research Council 1990). They provide the baseline from which the doubling dose risk estimates in the subsequent tables are derived. These values are used by regulatory agencies to establish protection guidelines.

[c] Diseases of complex etiology may have an inherited genetic susceptibility component in some proportion of the cases. As BEIR V states, "From a genetic point of view, the mutational component of diseases with complex etiologies results from a number of genes, usually with small individual effects, that in combination determine susceptibility to environmental factors causing the disease" (National Research Council 1990).

underlying genetic basis over his or her lifetime, a large fraction of which will be of late onset.

METHODS OF RISK ESTIMATION

There are two major methods of estimating genetic risk to a population. The direct method, provides an estimate after parental exposure of first-generation effects for dominant or chromosomal disorders; the indirect or doubling dose method (D.D.; or relative mutation risk 1/D.D.) estimates the risk from a single exposure summed over all time and extrapolates back to the first generation.

In the first case, dominant effects in both skeletal and cataract disorders have been scored in mice, and breeding experiments subsequently demonstrated their dominant nature. All these experiments were done at a high total dose or radiation acutely delivered in the range of 5 Gy or more. Extrapolation to low-dose, low dose rate conditions is required. The estimates are then adjusted to account for the proportion and seriousness of all dominant human diseases that those specific diseases represent. The validity of the method of course depends on how accurate these adjustments are. They do lead to similar estimates for male germ cells. The different committees' estimates by this method range between 10 and 30 cases expected in the first generation.

The indirect method uses the mouse-specific locus rates to estimate that dose required to double the spontaneous mutation rate (of mice or humans). The specific locus tests were the largest scale mouse genetics experiments ever performed. They were being undertaken primarily at the Oak Ridge National Laboratories under the direction of Dr. William Russell; smaller but similar experiments were being conducted at Harwell, England, by Dr. Mary Lyon and in the GSF Laboratories (Germany) of Dr. Ehling. Extensive dose responses for chronic gamma exposures are linear in kinetics; and a D.D. of 1 Sv has been accepted. As used by BEIR and UNSCEAR, the estimate of first-generation cases ranges between 6 and 35 (Table 2.2).

The current incidence of dominant diseases × 1/D.D. projects the total genetic damage over all time from a single unit exposure to a million parents per 10^6 offspring. Since on the average, dominant mutations will persist for no more than 5–6 generations because of their disease seriousness, 1/5–1/6 would be expected in the first generation. For recessive mutations the persistence or equilibrium time would be much longer, about 40–50 generations, with most elimination resulting from a slight 2½% dominance effect.

We have again used the direct method in the Nuclear Regulatory Commission report (NUREG/CR-4214) for sex-linked mutations (NUREG 1989). This involves multiplying the total number of genes (estimated as 250) on the X chromosome associated with genetic disease by the average mouse-induced specific locus (gene) rate per unit dose. Only maternal exposure is involved in this estimate of eight first-generation cases. The indirect method results in less than one additional case per 10^6 offspring.

To compute the risk of unbalanced rearrangements all the committees used as a basis for their calculations the data of Brewen et al. (1975) on human and marmoset x-ray-induced translocations in spermatogonia scored in the sperma-

TABLE 2.2. SIMPLE MENDELIAN TRAIT INCREASE AFTER 0.01-GY PARENTAL EXPOSURE AND 10^6 LIVE BIRTHS[a]

	First-Generation Effects[e]				
Type	UNSCEAR	BEIR V	NUREG		All Generations[f]
Dominants[b]	15	6–35			
[c]	(10–30)		30	→	140
X-linked[c]		<1			
			8	→	40
Trisomies[b]	—	<1			
			8	→	10
Unbalanced[b]	2.4				
Translocations[c]	(2–20)	<5	12	→	16
Congenital	—	30[d]			30
TOTAL	20	35–70	60		240

[a] Adapted from United Nations Scientific Committee on the Effects of Atomic Radiation. Report of the United Nations Scientific Committee on the Effects of Atomic Radiation. Sources, effects and risks of ionizing radiation. General Assembly Official Records, 43 Session, Suppl. 45. New York: United Nations, 1988; NUREG/CR-4214. Health effects model for nuclear power plant accident consequence analyses. Washington, DC: U.S. Nuclear Regulatory Commission, 1989; National Research Council. Health effects of exposure to low levels of ionizing radiation (BEIR V). Washington, DC: National Academy Press, 1990.

[b] Estimation based on doubling dose method.

[c] Estimation based on direct method.

[d] The report shows 10 cases, but there is a calculation error.

[e] The NUREG report assumes that the first-generation congenital effects would be included in the dominant disease category.

[f] The number of genetic disorders accumulating over all generations from the single exposure to both parents.

tocytes. The observed rate of 7.4×10^{-4} per .01 Gy was corrected for dose rate, segregation, and viability factors to yield a rate of 7.4×10^{-6} per .01 Gy for paternal exposure and 5.6×10^{-6} per .01 Gy for maternal exposure (NUREG 1989). The range of values seen in Table 2.2 is from 2 to 20 cases per 10^6 liveborn.

The NUREG (1991) report uses the calculations of Abrahamson et al. (1989, 1990) to derive estimates of aneuploidy induction. This is based on human spermatogonial irradiation producing aneuploid sperm and mouse oocyte irradiations resulting in aneuploid eggs. The observed data were corrected for dose rate and viability factors and yielded estimates of eight cases per million per 0.01 Gy. This estimate is larger than that developed by the BEIR V committee, but they provide no calculations for their estimate (National Research Council 1990).

Irregularly inherited disease would have long persistence times–10–50 generations–and only a portion of the disease is due to mutation-dependent factors. That portion, the mutation component (MC) modifies the doubling dose in a proportionate manner.

It is immediately obvious that the incidence of irregularly inherited disease is vastly greater than that in all other categories. Before both BEIR and UNSCEAR increased their estimates of the incidence by a factor of 10, they had provided risk estimates for this class of disease, albeit estimates that were always expressed with great uncertainty. This was because neither an accurate measure of the MC nor the average persistence time of the diseases (i.e., their extinction curves) was known. BEIR 1980 used an MC range of 0.05–0.5; UNSCEAR used 0.05 (National Research Council 1980, UNSCEAR 1986). However, in their most recent reports the BEIR and UNSCEAR committees declined to provide numerical estimates (UNSCEAR 1988, National Research Council 1990). BEIR V states: "The risks may be negligibly small, or they may be as large or larger than the risks for all other traits combined" (National Research Council 1990).

The NUREG report (1991) attempts to quantify these risks (see Table 2.3). The approach for the cancer issue was to assume that humans possess about 50–100 tumor suppressor genes (see also Knudson 1989 and Cavenee 1989). The recessive form of the gene when homozygous causes loss of suppression, but one doesn't have to be completely homozygous. Inheritance of a recessive gene from one parent preconditions the individual for a second somatic cell mutation, making it a virtual certainty that some cells will be homozygous for the *lack* of tumor suppression. Thus this direct approach provided us with one estimate (10–20 cases/10^6/0.01 Sv in the first generation). An independent approach was based on the observation that between 5% and 10% of all cancers are of familial origin (i.e., resulting from mutant forms of the tumor suppressor genes). This led to overlapping estimates (15–30 cases) with the first approach. Note that triggering of the inherited cancer requires two mutational steps, one inherited germinally, the other occurring somatically, but this is not sufficient–other mutational steps in these somatic cell lines and promotional steps are also needed. Apparently, however, the mutant clone is at a selective growth advantage, probably facilitating the occurrence of the additional steps. Thus while we are dealing with a multifactorial disease, the problem is becoming more tractable since a relatively few genes need be involved in the hereditary contribution. It may be premature to expect that the familial inheritance of cardiovascular disease may follow similar

TABLE 2.3. GENETIC RISK ESTIMATES FOR SELECTED IRREGULARLY INHERITED DISEASES IN A POPULATION OF 1 MILLION BASED UPON BEIR V ESTIMATE OF NATURAL INCIDENCE OF THESE DISEASES ASSUMING A 0.01-GY PARENTAL DOSE[a, b]

Type	First Generation	All Generations
Cancer	15[c]	400[c]
Heart	35[d]	780[d]
Others	20[d]	400[d]
TOTAL[e]	70	1,580

[a] NUREG/CR-4214. Health effects models for nuclear power plant accident consequence analysis. Modifications of models resulting from recent reports on health effects of ionizing radiation. Low LET radiation. Part II. Scientific bases for health effects models. Washington, DC: U.S. Nuclear Regulatory Commission, 1991.

[b] The extremely tenuous nature of these numerical estimates in view of the very large uncertainties involved is emphasized.

[c] Assumes there are between 50 and 100 tumor suppressor genes that might be mutated.

[d] Assumes a mutational component of 0.13 and the time to equilibrium of between 10 and 50 generations. If the mutational component were an order of magnitude lower, the risk estimates (number of cases) would also be an order of magnitude lower.

[e] The numbers have been rounded to avoid impression of great precision.

simple genetic patterns, although there is a strong suggestion that atherosclerotic plaques result from mutant cell lines.

Lacking more discrete information, NUREG (1991) has attempted to derive risk estimates for these other multifactorial diseases described in the BEIR 1990 report, assuming a mutational component range of 0.05–0.35, with a central estimate, the geometric mean, of 0.13. As can be seen from the numbers projected in Table 2.3, albeit with great uncertainty, the irregularly inherited disorders may be as large or much larger than the simpler Mendelian traits, even though they represent a very small increment of the normal incidence.

DOSE-RESPONSE MODELS USED IN RISK PROJECTION

Two dose-response models are used to extrapolate risk. For low dose, or low dose rate exposures, all the experimental data in many experimental systems show a linear nonthreshold response. For moderate to high doses, acutely delivered (i.e., within a few hours), the dose-response curve is described as linear-quadratic. The induced yield is proportional to $\alpha D + \beta D^2$, where D is dose and α and β are coefficients. We believe that the densely ionizing portion of the single track is responsible for the two-break gene deletion in the linear portion of the curve and that two independent tracks interact to produce the dose squared component. The β coefficient is dependent on dose rate; the more protracted the dose, the smaller the probability of interaction, the more linear the curve, because rejoining of the broken ends by repair enzymes removes them from the interacting pool before the second break event is introduced. This dose-response curve was first demonstrated for the induction of chromosome aberrations and later for gene mutations.

HIROSHIMA-NAGASAKI F_1 STUDIES

As pointed out in the introduction, the Hiroshima-Nagasaki ongoing studies on the children of the A-bomb survivors have failed to reveal a significant increase in any of the genetic endpoints studied by Neel and Schull and their colleagues at the Radiation Effects Research Foundation as reviewed in their most recent summary paper (1990). These authors conclude that man may be less sensitive than the estimates derived from mouse mutation studies suggest. This may be true; we (including BEIR and UNSCEAR committees) have tended to use the largest scale mouse studies based on the seven specific locus method developed by W. Russell to develop our doubling dose estimates, and these loci may be less than representative of the total genome. On the other hand, what would be expected using the risk estimates that were developed in the recently completed NUREG report (NUREG 1991), and are these values in significant excess for the endpoints considered? As shown in Table 2.4, we have calculated expected numbers for the five endpoints given the number of children whose parents received specific mean DS86 doses. For each endpoint, we have used equations shown in the Table 2.4 footnotes to derive the number of expected cases over the lifetime of the approximately 27,000 children (or other sample sizes indicated). I think it should be clear that none of the endpoints would ever be detected in the noise of the current incidence of disease. Thus the sample size

TABLE 2.4. EXPECTED YIELD OF DIFFERENT CLASSES OF GENETIC DISORDERS AMONG CHILDREN OF ATOMIC BOMB SURVIVORS

Genetic Classes	Predicted Excess cases	No. F_1 Studied[e]	Number of Controls
Dominants[a]	30	27,000	41,000
Multifactorial[b]	56	27,000	41,000
Protein variants[c]	<2.4	13,052	10,609
Sex trisomies	2	8,000	8,000
Cancer[d]	8	27,000	41,000

[a] Only a portion of the dominants would have been expressed as early mortality, mortality to age 26, and congenital defects.

[b] Most multifactorial diseases become evident after the age of 20.

[c] Protein variants: The majority of deletional mutations will not be observable in one-dimensional gel electrophoresis; the number in the table assumes all cases will be observed.

[d] Cancer would be expressed over the lifetime of the population.

[e] Numbers of F_1 studied represent those for whom parental DS86 dosimetry has been applied.

Mean parental doses and number of children are as follows:

Dose (Sv)	.02	.12	.35	.80	1.9
No. F_1	10,648	9,864	3,500	2,300	600

Dose response equations:

Dominants = $Y = (3 \times 10^{-3})d + (3 \times 10^{-3})d^2$ includes both parents. d = dose in Sv or Gy

Multifactorial = $Y = (5.6 \times 10^{-3})d + (5.6 \times 10^{-3})d^2$ includes both parents

Protein = $Y = (4.8 \times 10^{-4})d + (4.8 \times 10^{-4})d^2$ includes per locus mutation rate, 8×10^{-6}/Sv × 30 loci/parent × 2 parents

Sex trisomies = $Y = (7.4 \times 10^{-4})d + (7.4 \times 10^{-4})d^2$

Cancer = $Y = (8 \times 10^{-4})d + (8 \times 10^{-4})d^2$: includes per locus rate of 8×10^{-6}/Sv × 50 loci/parent × 2 parents

Expected number of cases = Y × No. F_1 = (mut. rate/locus) × (# loci) × (2 parental genomes) = (mutation rate/genome) × 2 (genomes)

and particularly the number of children born to parents in the 1–2 Sv range may have been too small to show an effect even when the risk estimate from the most sensitive system was used for three of the five calculations.

SOCIETAL COST OF GENETIC DISEASES

UNSCEAR and NUREG have provided some rough estimates of the detriment associated with genetic disease, which includes life shortening and effective years of impaired life (degree of impairment × years impaired). These values can be equated to cancer mortality and cancer morbidity, although they may have quite different socioeconomic impacts. For example, some 25% of all sex-linked diseases are associated with a lifetime of mental retardation (Arena and Lubs 1991), and in New York the average cost of caring for such an individual is $2 million (Nolin et al. 1991). The societal cost of such a disorder is much greater than that of the child who is suffering an unbalanced translocation destined to survive for only a year or two. Based on the BEIR estimates of genetic disease incidence and the mortality-morbidity burden of the disease categories weighted by their incidence and average time of onset, I would estimate that on average, 20 years of life are lost and a minimum 5–10 years of effective life are lost by illness. Finally, in a report to the EPA in 1974, the Arthur D. Little research group estimated that the cost in dollars per cSv was about four times larger (range 2–7) for excess genetic disease than for excess cancer. This also illustrates that risk estimates provide only a very limited dimension to the impact of either induced somatic or genetic damage.

SUMMARY

In the preceding sections I have outlined the nature of genetic diseases with which humankind is afflicted, the current incidence of such diseases, and how that incidence will be affected by the low-dose or low-dose-rate ionizing radiation to which populations are usually exposed. I reemphasize here that there are great uncertainties in our knowledge of the true incidence of genetic disease and in the procedures required to extrapolate from animal experiments to the human situation. Whereas data on male germ cells are extensive, those for female germ cells are not. It is assumed that they are about equal in induced sensitivity and in response to the influence of dose rate. We have poor knowledge of the mutational components of the irregularly inherited diseases and the persistence time (in generations) of such diseases. If medical progress continues at the pace of the last decades, the issue of genetic diseases 10–20 generations from now may be a relatively insignificant one. For this reason, risk estimates have been based where possible on the direct first-generation approach. Such an analysis is not inconsistent with the observations of the A-bomb F_1 studies, and while this does not validate the methods used, it does show that they are not unreasonable.

REFERENCES

Abrahamson, S., Awa, A.A., and Nakamura, N. New risk estimates for trisomy induction by radiation exposure. *RERF Update 1(4):*34, 1989.

Abrahamson, S., Awa, A.A., and Nakamura, N. Induced trisomies "revisited." *RERF Update 2(2):*3, 1990.

Abrahamson, S., Bender, M.A., Conger, A.D., and Wolff, S. Uniformity of radiation-induced mutation rates among different species. *Nature 245:*460–462, 1973.

Arena, J.F., and Lubs, H.A. Computerized approach to X-linked mental retardation syndromes. *Am. J. Med. Genet. 38:*190–199, 1991.

Brewen, J.G., Preston, R.J., and Gengozian, N. Analysis of x-ray induced chromosomal translocation in human and marmoset spermatogonial stem cells. *Nature 253:*468–470, 1975.

Cavenee, W.K. Current knowledge of heritable tumors. In: *Genetic Susceptibility to Environmental Mutagens and Carcinogens,* Monogr. 2, edited by A.D. Bloom, L. Spatz, and N.W. Paul. Pittsfield, MA, Environmental Health Institutes, 1989.

Knudson, A.G. The genetic predisposition to cancer. In: *Genetic Susceptibility to Environmental Mutagens and Carcinogens,* Monogr. 2, edited by A.D. Bloom, L. Spatz, and N.W. Paul. Pittsfield, MA, Environmental Health Institutes, 1989.

National Research Council. *The Effects on Populations of Exposure to Low Levels of Ionizing Radiation (BEIR III).* Washington, DC, National Academy Press, 1980.

National Research Council. *Health Effects of Exposure to Low Levels of Ionizing Radiation (BEIR V).* Washington, DC, National Academy Press, 1990.

Neel, J.V., Schull, W.J., Awa, A.A., Satoh, C., et al. The children of parent exposure to atomic bombs: estimates of the gametic doubling dose of radiation for humans. *Am. J. Hum. Genet. 46:*1053–1072, 1990.

Nolin, S.L., Snider, D.A., et al. Fragile X screening program in New York State. *Am. J. Med. Genet. 38:*251–255, 1991.

Nuclear Regulatory Commission. *Health Effects Model for Nuclear Power Plant Accident Consequence Analyses*, NUREG/CR-4214. . Edited by J.S. Evans, D.W. Moeller, and D.W. Cooper. Washington, DC, U.S. Nuclear Regulatory Commission, 1989.

Nuclear Regulatory Commission. *Health Effects Models for Nuclear Power Plant Accident Consequence Analysis. Modifications of Models Resulting from Recent Reports on Health Effects of Ionizing Radiation. Low LET Radiation. Part II. Scientific Bases for Health Effects Models*, NUREG/CR-4214. Edited by S. Abrahamson, M.A. Bender, B.B. Boecker, et al. Washington, DC, U.S. Nuclear Regulatory Commission, 1991.

United Nations Scientific Committee on the Effects of Atomic Radiation. *Report of the United Nations Scientific Committee on the Effects of Atomic Radiation. Genetic and Somatic Effects of Ionizing Radiation.* General Assembly Official Records, 41st Session, Suppl. 16. New York, United Nations, 1986.

United Nations Scientific Committee on the Effects of Atomic Radiation. *Report of the United Nations Scientific Committee on the Effects of Atomic Radiation. Sources, Effects and Risks of Ionizing Radiation.* General Assembly Official Records, 43 Session, Suppl. 45. New York, United Nations, 1988.

Chapter 3

The Effects of Embryonic and Fetal Exposure to Ionizing Radiation: Counseling the Patient and Worker about These Risks

ROBERT L. BRENT

ABSTRACT

The effects of ionizing radiation from external sources and internally deposited radionuclides on the developing embryo and fetus are discussed. An extensive review of animal and human data that have been used to develop risk estimates for the teratogenic and carcinogenic effects of in utero radiation exposure is provided. Case histories of radiation exposures in medical settings are included as a basis for discussing strategies for counseling patients and workers.

INTRODUCTION

The effects of high-energy radiation (x-rays, gamma rays, particulate radiation) have been studied more extensively than has any other environmental hazard. In spite of the vast amount of knowledge available about ionizing radiation, many physicians are unfamiliar with the quantitative and qualitative effects of ionizing radiation. In this respect my laboratory frequently is consulted when women of reproductive age are exposed to diagnostic x-rays. More often than not, the exposed women have been given erroneous information.

Radiation is an anxiety-provoking term because of its association with the effects of ionizing radiation. The associations and effects follow.

1. The term "radiation" is related to the effects of the atomic bomb. In the minds of many, it is impossible to separate the effects of low-dose ionizing radiation from the psychological, physical, and radiation effects of the atomic bomb. The fact that the horrendous effects of the atomic and hydrogen bombs are at one end of a continuum of radiation effects distorts to a greater or lesser extent the risks of all forms of energy labeled *radiation*.
2. Populations that have high exposures to ionizing radiation have increased incidence of cancer. These populations include radium dial workers, uranium miners, patients receiving radiation therapy or isotope therapy for various diseases, and the persons who received the higher exposures in Hiroshima

and Nagasaki following the atomic bomb detonation (National Research Council 1990). "Cancer" itself is also an anxiety-provoking term. In our culture it is a disease with which people are "afflicted," and it is a disease that is dreaded by a large segment of the population. Few persons are aware of the extremely small maximum risk for the occurrence of cancer in populations exposed to much lower doses of radiation.

3. The immense psychological consequences of high-energy radiation exposure are extremely important and cannot be ignored when one considers the deleterious effects of radiation. A review of a 30-year study of the survivors of the Hiroshima and Nagasaki bombings revealed some interesting findings (Okada et al. 1975):

Thirty years of study clearly showed that radiation effects are dose dependent. A survivor aware of the magnitude of the radiation dose can better understand the extent of his own risks. More than 90 percent of survivors received much less than

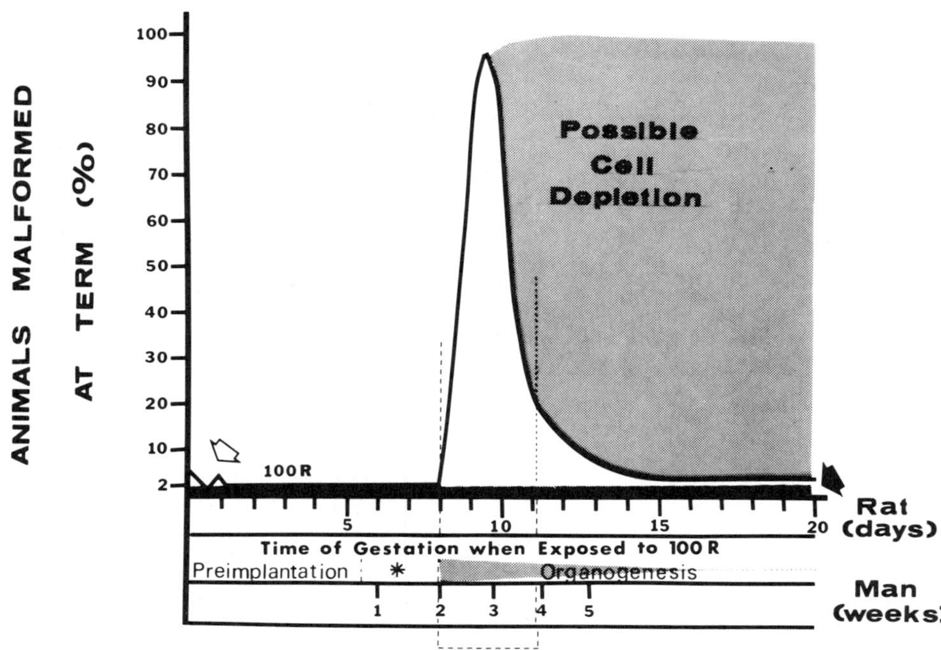

FIG. 3.1. This graph depicts the relative incidence of gross congenital malformations in the rat following an exposure of 1 Gy. The solid arrow on the right side of the ordinate points to the 2% control incidence of congenital malformations seen in this species. The incidence of malformations before the 9th day is the same as in the controls. The open arrow on the left side of the ordinate points to a slight increase in malformations on the 1st day. A very large increase in malformations occurs during the early organogenic period. Note that this corresponds to the 3rd and 4th weeks of human pregnancy. This high incidence of gross malformations falls off rapidly as organogenesis diminishes. Note that organogenesis to some degree (central nervous system) continues to term. Also note that although gross malformations may not be produced during the late fetal stages with 1 Gy, an irreversible cell loss occurs. The significance of these cell losses is under study. The asterisk (*) is placed at the stage of implantation to indicate that although malformations are not readily produced at this stage, growth retardation can be induced.

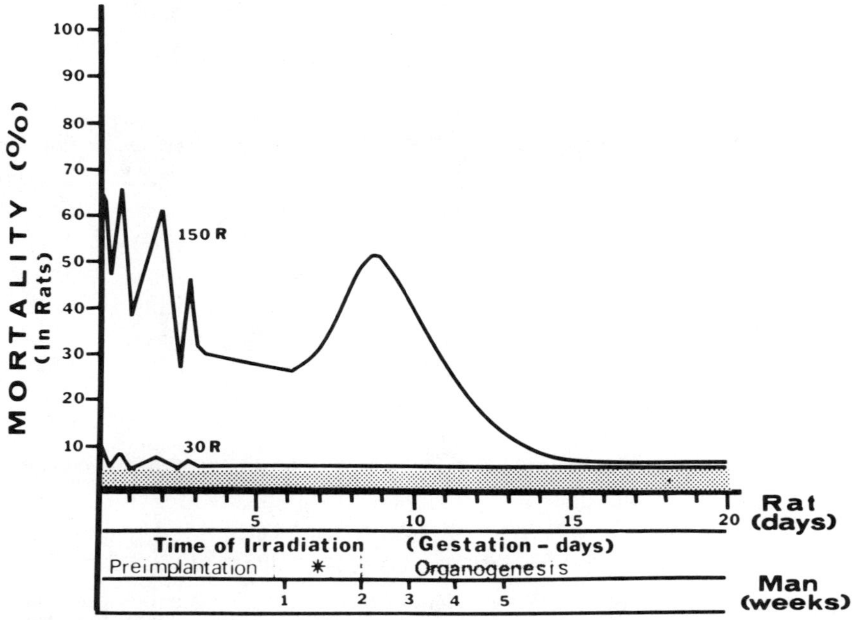

FIG. 3.2. The lethal effect of radiation in rats is greater on the 1st day. It appears that the LD_{50} shifts at different times of the day during early gestation, possibly because the cells are dividing synchronously; therefore, the zygote will vary in its radiosensitivity, depending on the stage of the cell cycle. Note that the embryo becomes somewhat resistant in the implantation stage and then becomes sensitive to the lethal effect of radiation during early organogenesis. A 0.3-Gy dose apparently does increase the resorption rate when radiation takes place on the 1st day. The superimposition of the lethality curve for rats onto the human embryonic development timetable may not be appropriate and is for purposes of comparison only, but it allows one to estimate roughly the stages at which the human embryo might be most readily killed by high doses of radiation.

100 mGy from the A-bombs. With this knowledge, such survivors can realize that their fears of disease from A-bomb exposure may be exaggerated and that the possibility of their developing any such disease is no greater than those of nonexposed individuals. Those who received higher doses have greater risks. It is worth noting that, thus far, the life expectancy of the exposed population under study and of their offspring is equal to that for the population in the rest of Japan. Thirty years of study has yielded reasonable estimates of the risks involved.

In spite of the fact that the results of the Atomic Bomb Casualty Commission data have been widely disseminated in Japan, the psychological burden of radiation exposure is still prominent: "The parents of a man wishing to marry a woman known to be a survivor, or the daughter of a survivor, may object to or even attempt to prevent their marriage, fearing the women's health may be adversely affected or their children might have genetic defects" (Okada et al. 1975).

Granted the risks from high-energy radiation exposure may be exaggerated or not well understood, there is no question that high-energy radiation can damage tissue and produce long-term effects. If the quantitative data pertaining to radiation effects are understood, the effects of radiation can be placed in better

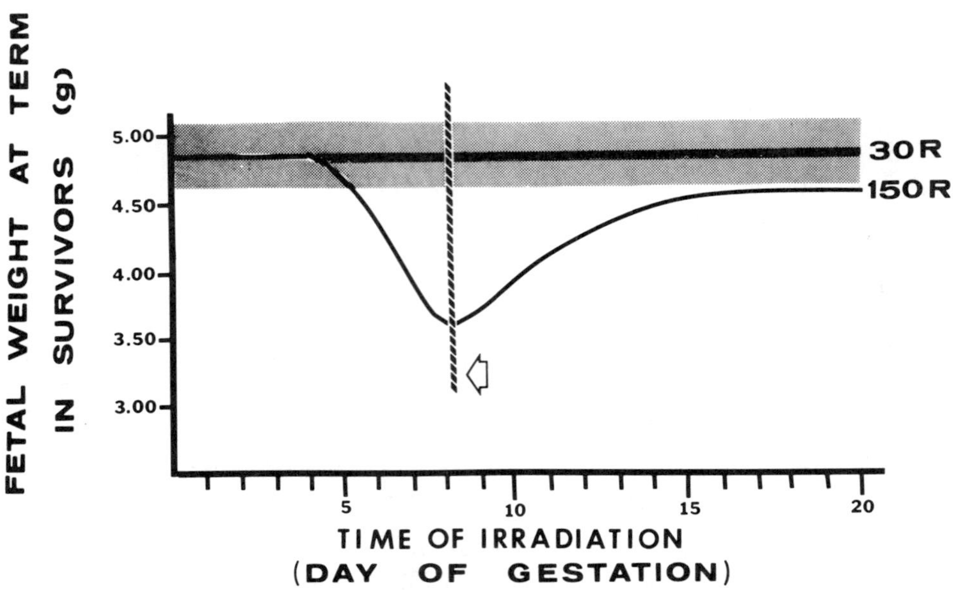

FIG. 3.3. The fetal weight in the Wistar colony of rats in this laboratory averages 4.8 g on the 22nd day when delivered by cesarean section that morning. Note that 1.5 Gy causes growth retardation just after implantation but before the stage is sensitive to teratogenesis (arrow pointing to vertical line). From the time of implantation until term, 1.5 Gy produced some degree of growth retardation. Note that 0.3 Gy does not appear to produce significant deviations in growth, although it may be that all experimental groups are too small to show differences in weight at this low dose.

perspective. Radiation risks should be compared with spontaneous risks and risks produced by exposure to other environmental hazards.

TERATOGENIC AND CARCINOGENIC EFFECTS OF IN UTERO IONIZING RADIATION

There have been many experiments and observations about the effects of radiation on the developing embryo. Nevertheless, we have a great deal to learn about radiation teratogenesis and teratogenesis in general. The subject is summarized briefly here. For further details the reader is referred to more complete reviews (Brent 1970, 1972, 1976, 1977, 1980, Brent and Gorson 1972).

Prior to summarizing the effects of irradiation at different stages of gestation, it is important to emphasize that all the effects described are produced at or above specific doses. Since the vast majority of embryopathological effects are threshold phenomena, subthreshold doses will not produce these effects.

Before implantation, the embryo is a multicellular organism with a decreased sensitivity to the teratogenic and growth-retarding effects of radiation and a greater degree of sensitivity to the lethal effects of irradiation (Figs. 3.1, 3.2, 3.3, Tables 3.1, 3.2, 3.3) (Brent and Bolden 1967a, Russell 1950, Russell and Russell 1950). This does not indicate that malformations cannot be produced, but the cytogenetic abnormalities and malformations, if produced, result in a high incidence of mortality. During early organogenesis the embryo is very sensitive to

TABLE 3.1. A COMPILATION OF THE EFFECTS OF 100 MGY OR LESS OF ACUTE RADIATION IN VARIOUS STAGES OF GESTATION IN RAT AND MOUSE[a]

	Stage of Gestation (Days)				
	Preimplantation	Implantation	Early Organogenesis	Late Organogenesis	Fetus Stages
Mouse	0–4½	4½–6½	6½–8½	8½–12	12–18
Rat	0–5½	5½–8	8–10	10–13	13–22
Corresponding human gestation period	0–9	9–14	15–28	28–50	50–280
Lethality	+[b,c]				
Growth retardation at term	—	—	—	—	—
Growth retardation as adult	—	—	—	—	—
Gross malformations (aplasia, hyperplasia, absence or overgrowth of organs or tissues)	±[d]	—	—	—	—
Cell depletions, minimal but measurable tissue hypoplasia	—	—	—	—	—
Sterility[e]	—	—	—	—	—
Significant increase in germ cell mutations[e]	±	±	±	±	±
Cytogenetic abnormalities	—	—	—	—	—
Neuropathology[e]	—[i]	—	—	—	—
Tumor induction[e,f]	—	—	±	±	±
Behavior disorders[g]	—	—	—	—	—
Reduction in life span[e]	—	—	—	—	—

[a] Dose fractionation or protraction effectively reduces the biological result of all the pathological effects reported to this table. (—) indicates no observed effect; (±) questionable but reported or suggested effect; (+) demonstrated effect; (++) readily apparent effect; (+++) occurs in high incidence.

[b] At this stage the murine embryo is most sensitive to the lethal effects of irradiation. With 100 mGy in the mouse, Rugh reports a slight decrease in litter size in the mouse (Brent 1977b, Rugh 1965).

[c] Recent reevaluation of the atomic bomb victims' data suggests the possibility that mental retardation is a risk in the 100–200 mGy range. This is not supported by most other data.

[d] Rugh reports exencephalia with 150 and 250 mGy in a strain of mice with a 1% incidence of exencephalia (Rugh 1965). Others have not been able to repeat these results (Russell and Saylors 1963).

[e] The potential for mutation induction exists in the embryonic term cells or their precursors. Several long-term studies indicate that considerably greater doses in mice and rats do not affect longevity, tumor incidence, incidence of congenital malformations, litter size, growth rate, or fertility.

[f] Stewart and others have reported that 20 mGy increases the incidence of malignancy by 50% in the offspring. See text for discussion (Laskey et al. 1973, Diamond et al. 1973, Jablon 1973, Miller 1970).

[g] Piontkovskii reports behavioral changes in the rat after 10 mGy daily irradiation (Piontkovskii 1958, 1961). This work has not been reproduced. See text for discussion.

TABLE 3.2. A COMPILATION OF THE EFFECTS OF 1 GY OF ACUTE RADIATION ON EMBRYONIC DEVELOPMENT AT VARIOUS STAGES OF GESTATION IN RAT AND MOUSE[a]

	Stage of Gestation (Days)				
Feature	Preimplantation	Implantation	Early Organogenesis	Late Organogenesis	Fetal Stages
Mouse	0–4½	4½–6½	6½–8½	8½–12	12–18
Rat	0–5½	5½–8	8–10	10–13	13–22
Corresponding human gestation period	0–9	9–14	18–36	36–50	50–280
Lethality	+++[b]	+	++	±	—
Growth retardation at term	—	+	+++	++	+
Growth retardation as adult	—	+	++	+++	++
Gross malformations (aplasia, hypoplasia, absence or overgrowth of organs or tissues)	—	—	+++	±[c,d]	—[c,d]
Cell depletions, minimal but measurable tissue hypoplasia	—	—	±	++	+
Sterility[e]	—	—	±	—	++
Significant increase in germ cell mutations[f]	±	±	±	±	±
Cytogenetic abnormalities[g]	±			+	+
Cataracts	—	—	+	+	+
Neuropathology[d,i]	—	—	+++	+++	++
Tumor induction[h]	—	—	±	±	±
Behavior disorders[d,i]	—	—	±	±	±
Reduction in life span (in nonmalformed embryos)[j]	—	—	—	—	—

[a] Dose fractionation or protraction effectively reduces the biological result of all the pathological effects reported in this table. (—) no observed effect; (±) questionable but reported or suggested effect; (+) demonstrated effect; (++) readily apparent effect; (+++) occurs in high incidence.

[b] Russell reported that 2 Gy increased the incidence of XO aneuploidy in 2%–5% of offspring in mice with a spontaneous incidence of 1%. A dose of 1 Gy kills substantial numbers of mouse and rat embryos at this state, but the survivors appear and develop normally.

[c] One Gy produce changes in the irradiated fetus that are subtle and necessitate detailed examination and comparison with comparable controls.

[d] There is a consensus that the brain maintains a marked sensitivity to radiation throughout all of gestation. Mental retardation is a serious risk at this dose.

[e] The male gonad in the rat can be made extremely hypoplastic by irradiation in the fetal stages with 150 mGy. In the mouse the newborn female is most sensitive to the sterilizing effects of radiation. Much of this research on other animals cannot be applied to the human.

[f] The potential for mutation induction exists in embryonic germ cells or their precursors. The relative sensitivity of the embryonic germ cells when compared to adult germ cells is not known. Several long-term studies in animals do not indicate any exceptional differences.

[g] Bloom has reported a much higher percentage of chromosome breaks in human embryos receiving 1–2 Gy in utero than in adults receiving the same dose of irradiation. As yet there have been no diseases associated with this increase in frequency of chromosome breaks.

[h] Animal experiments and the data from Hiroshima and Nagasaki do not support the concept that in utero irradiation is much more tumorigenic than extrauterine irradiation. On the other hand, Stewart and colleagues and many others report that irradiation from pelvimetry (20 mGy) increases the incidence of leukemia and other tumors (Stewart 1972, 1973, Stewart and Kneale 1970, Stewart et al. 1956, Stewart et al. 1958).

[i] A statistically significant increase in percentage of mental retardation occurs with this dose of radiation. On the other hand, normal intelligence has been found in children receiving much higher doses in utero.

[j] Animal experiments indicate that survivors of in utero irradiation have a life span that is longer than that of groups of animals given the same dose of radiation during their extrauterine life and the same life expectancy as nonirradiated controls.

TABLE 3.3. ESTIMATION OF THE RISKS OF RADIATION IN THE HUMAN EMBRYO BASED ON HUMAN EPIDEMIOLOGICAL STUDIES AND MOUSE AND RAT RADIATION EMBRYOLOGICAL STUDIES

Embryonic Age (Days)	Minimal Lethal Dose (mGy)	Approximate LD_{50} (mGy)	Minimum Dose (mGy) for Permanent Growth Retardation in the Adult	Increased Incidence Mental Retardation	Minimum Dose For Recognized Gross Anatomical Malformation (mGy)	Minimum Dose For Induction of Genetic, Carcinogenic, and Minimal Cell Depletion Phenomena (mGy)
1–5	100	<1000	No effect in survivors			Unknown
18–28	250–500	1400	200–500	20–50 Severe CNS anatomical malformations more likely than mental retardation	200	Unknown
36–50	500	2000	250–500	50[a]	500	Unknown
50–150	>500	>1000	250–500	50[a]	—[b]	Unknown
To Term	>1000	Same as mother	>500	100	—	Unknown

[a] Otake and Schull (1984) suggest an increased risk at lower exposures.

[b] Anatomical malformations of a severe type cannot be produced this late in gestation except in the genitourinary system and tissue hypoplasia in specific organ systems, such as the brain and testes.

the growth-retarding, teratogenic, and lethal effects of irradiation but has the ability to recover from the growth-retarding effects in the postnatal period (Rugh 1962, Russell 1950, Russell 1956, Russell and Russell 1954). During the early fetal period the fetus has diminished sensitivity to multiple organ teratogenesis but retains CNS sensitivity; it is growth-retarded at term and recovers poorly from the growth retardation in the postpartum period. During the later fetal stages, the fetus is not grossly deformed by radiation but can sustain permanent cell depletion of various organs and tissues if the radiation exposure is high enough (Brent 1977b, Brent and Gorson 1972).

One can postulate many mechanisms for the effects of irradiation, including cell death, mitotic delay, disturbances of cell migration, and others. Until we know more about the mechanisms involved in embryopathology, it is difficult to determine which mechanisms are most important in the process of radiation-induced embryopathological disorders. Furthermore, the same pathogenic mechanisms may not be primarily operative at all stages of gestation. Radiation-induced cell death may be minimally important at one stage of gestation because of the embryo's ability to replace the killed cells. At another stage of gestation, cell death may be a primary factor because the embryo has lost the ability to replace dead cells and the fetus may be permanently cell-depleted.

Other radiation-induced effects that have been invoked to explain embryopathological conditions are cytogenetic abnormalities and somatic mutations (Russell and Major 1957, Russell and Saylors 1963). Cytogenetic abnormalities may be responsible for preimplantation death in the irradiated conceptus, but point mutations are less likely to be a contributing factor to abnormal morphogenesis (Brent and Bolden 1967a, Rugh 1965, Russell and Saylors 1963). If one uses the known radiation mutation rate in mammalian cells (as determined by Russell and colleagues [Russell 1950, 1956, Russell et al. 1963)] in the mouse) for estimating potential teratogenicity, it becomes clear that radiation-induced point mutations cannot account for even a small proportion of radiation-induced teratogenicity unless it simply involved cell death. Of course, neither phenomenon has been studied adequately at every stage of mammalian gestation.

Radiation effects may appear at once and result in cell death, embryonic death, or teratogenesis. Other effects may not be obvious immediately and can be measured or ascertained only in the postpartum or adult period. Neuronal depletion, infertility, tissue hypoplasia, neoplasia, or shortening of life span are phenomena the significance of which can be evaluated only in the postpartum or adult organism (Brent 1970, 1980, Brent and Bolden 1961, Cowen and Geller 1960, Hicks and D'Amato 1966, Murphree and Pace 1960, Rugh and Wohlfromm 1964a, 1964b).

The neurophysiological and behavioral changes in adult organisms that have been irradiated in utero are the most difficult to evaluate. The difficulties arise for several reasons: (*a*) It is not easy to eliminate postnatal environmental effects as contributing factors, (*b*) reported findings are frequently not reproducible and are not dose related, (*c*) behavioral changes may not be able to be correlated with the neuroanatomical changes, and (*d*) animal behavior tests may have minimal application to the human situation. This subject has been reviewed (Brent 1976,

Furchgott 1963). Except for the work of Piontkovskii, all behavioral studies of animals or human beings exposed to radiation during gestation do not exhibit any measurable changes if exposure was below the 200–250 mGy range (Piontkovskii 1958, 1961).

HUMAN RADIATION TERATOGENESIS

The classic effects of radiation on the developing mammal are gross congenital malformations, intrauterine growth retardation (IUGR), and embryonic death (Table 3.3). Each of these effects has a dose-response relationship and a threshold exposure below which there is no difference between the irradiated and the control populations. Because human pregnancies predominantly consist of a single embryo, it would be unusual to see the classic triad of effects, although it is conceivable that if the dose and timing were appropriate, a stillborn, growth-retarded, malformed embryo could result.

Growth retardation and CNS effects, such as microcephaly or eye malformations, are the cardinal manifestations of intrauterine radiation effects in humans (Tables 3.4, 3.5). The occurrence of microcephaly is a significant component of the effects of intrauterine radiation (Miller and Mulvihill 1976). A number of independent studies indicate that microcephaly is the most common malformation observed in human beings exposed to high doses of radiation during pregnancy (Zappert 1926; Goldstein and Murphy 1929a, 1929b; Murphy and Goldstein 1930; Dekaban 1968; Miller 1969; Wood et al., 1967a; Wood et al. 1967b; Plummer 1952; Yamazaki et al. 1954). In Goldstein and Murphy's reports (Goldstein and Murphy 1929a, 1929b, Murphy and Goldstein 1930), 19 of 75 irradiated embryos became microcephalic or hydrocephalic children, having received an estimated

TABLE 3.4. FREQUENCY OF SMALL HEAD CIRCUMFERENCE AT AGES 10–19 YEARS ACCORDING TO CITY, GESTATIONAL AGE, AND RADIATION DOSE[a]

	Gestational Age (Wks)			
	Hiroshima		Nagasaki	
Dose (mGy)	0–17	18+	0–17	18+
Not in city or distally exposed	31/764		10/246	
0–90	4(1)/63+	4/65+	0/1	0/9
100–199	6(1)/54	0/44	0/7	0/6
200–299	6/24	1/14	0/5	2/7
300–399	4/8	0/10	2/4	0/6
400–499	3/11	0/6	0/6	0/3
500–599	9(2)/20	2/24	0/9	0/11
1000–1499	2/4	0/10	0/2	1/5
1500+	5(5)/13	1(1)8+	8(3)9	2(1)9
Unknown	1/7	0/3	0/0	0/0
Total (in city)	40(9)/204	8(1)/184	10(3)/43	5(1)/56

[a] From Blot W. Growth and development following prenatal and childhood exposure to atomic radiation. J Radiat Res 1975;16(suppl): 82–88.

[b] Cases in parentheses give number with small head circumference and mental retardation. Plus (+) indicates one person with mental retardation without small head circumference.

TABLE 3.5. NUMBER OF CASES AND RELATIVE RISK OF MENTAL RETARDATION ACCORDING TO DOSE CATEGORY[a]

	Hiroshima				Nagasaki			
Dose (mGy)	Sample Size	Cases[b]	Relative Risk[c]		Sample Size	Cases	Relative Risk	
Not in city or distally exposed	830	5(2)	1.0	(1.0)	246	4(2)	1.0	(1.0)
0–99	145	3(1)	3.4	(3.8)	11	0		—
100–499	189	2(1)	1.8	(1.5)	45	0		—
500–999	47	3(1)	10.6	(11.8)	20	0		—
1000–1999	29	4(1)	22.9	(28.6)	13	0		—
2000–2999	8	3	66.3	(104)	5	1(1)	12.3	
3000	6	2	55.3	(92.2)	7	3	26.4	(52.7)

[a] From Blot W. Growth and development following prenatal and childhood exposure to atomic radiation. J Radiat Res 1975;16(suppl):82–88.

[b] Numbers in parentheses are numbers of cases (included in the totals) whose mental retardation was apparently due to causes other than intrauterine radiation.

[c] Numbers in parentheses are relative risks, excluding cases whose mental retardation was apparently due to nonradiation causes.

dose of greater than 1 Gy as a minimum. Almost all the microcephalic children were mentally retarded. The following malformations also were reported: two children with hypoplastic genitalia, one child with cleft palate, and one child with hypospadias, an abnormality of the large toe, and an abnormality of the ear. Many of the children had intrauterine growth retardation, and three had abnormalities of the eyes, including microphthalmia, cataracts, strabismus, retinal degeneration, and optic atrophy. The unique finding was that no visceral, limb, or other malformations were found unless the child exhibited intrauterine growth retardation, microcephaly (congenital malformations of the brain), or readily apparent eye malformations. Dekaban (1968) reported that 22 infants exposed between the 3rd to 20th weeks of human gestation were microcephalic, mentally retarded, or both (Fig. 3.4). The estimated protracted exposure was 2.5 Gy, and all of the malformed infants exhibited growth retardation. The most frequent abnormalities were (*a*) small size at birth and stunted growth, (*b*) microcephaly, (*c*) mental retardation, (*d*) microphthalmia, (*e*) pigmentary changes in the retina, (*f*) genital and skeletal malformations, and (*g*) cataracts (see Fig. 3.4). All of the embryos reported by Dekaban received several gray. The patients were treated with radiation for various reasons, including dysmenorrhea, menorrhagia, myomata, arthritis or tuberculosis of the sacroiliac joint, and malignant tumor of the uterus or cervix. In several instances, radiation was initiated without knowing the patients were pregnant.

It is of interest to note that in one published case a typical camptomelic dwarf was reported, and the cause was believed to be radiation (Maroteaux et al. 1971). High-dose radiation was delivered from the 2nd to the 5th month. This rare recessive syndrome was not described in 1930, and the authors were not in a position to recognize its possible cause (Maroteaux et al. 1971).

Radiation-induced microcephaly, growth retardation, and microphthalmia are represented in Figures 3.5 and 3.6.

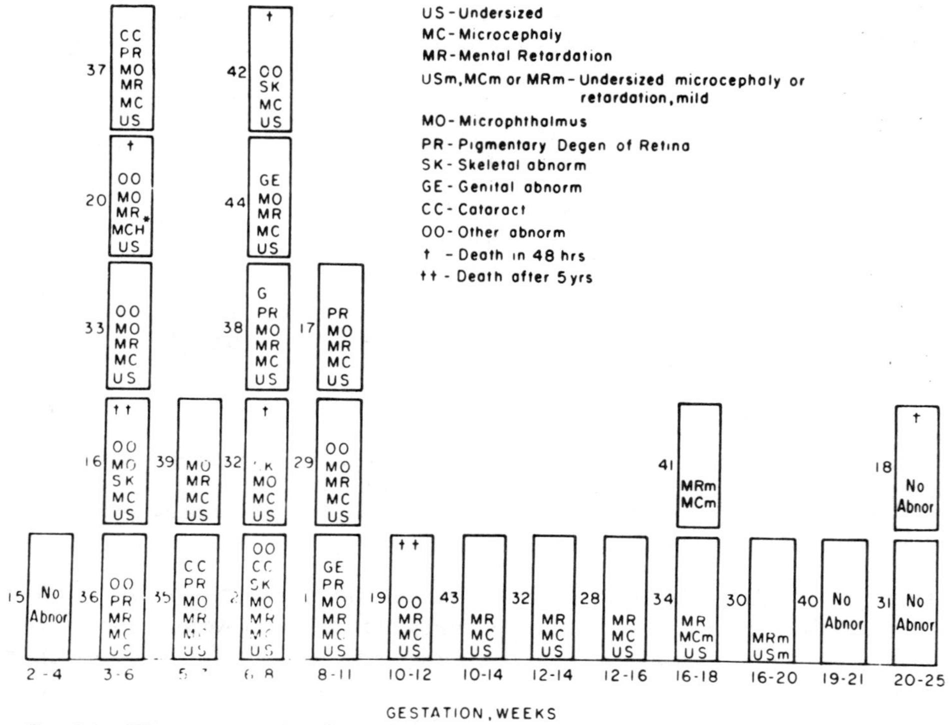

FIG. 3.4. The spectrum of malformations in 26 humans in utero. Note that all infants receiving therapeutic doses of radiation are growth retarded if they were irradiated between 3 and 20 weeks of gestation. Note that in no instance is there an associated malformation in structures other than the nervous system without either microcephaly or mental retardation. (From Dekaban AS. Abnormalities in children exposed to x-irradiation during various stages of gestation: tentative timetable of radiation injury to the human fetus. J Nucl Med 1968;9:471–477.) Reprinted with permission from Society of Nuclear Medicine.

From a sample of 1265 subjects exposed in utero at Hiroshima, 183 were analyzed by Miller (1969) and by Wood and colleagues (Wood et al., 1967a; Wood et al. 1967b). Seventy-eight of these fetuses were younger than 16 weeks of age at the time of irradiation; 105 were 16 weeks or older. Of the 78, 25 were microcephalic (head circumference more than 2 SD below the mean), and 11 were mentally deficient. Of the 105, 7 were microcephalic as defined, and 4 were mentally deficient. The incidence of microcephaly increased with increasing exposure. In 14 of the offspring with smaller than normal head circumference who were also mentally deficient, 10 were greater than 2 SD below the mean height for age. In 16 children with small heads who were not mentally deficient, there was no marked reduction in stature. Tables 3.4 and 3.5 summarize some of the characteristics of the patients exposed in utero before the 16th week of gestation. Severe mental retardation was not observed in any patient given less than 0.5 Gy while in utero. In a summary of the data on microcephaly, Blot (Blot 1975, Blot and Miller 1973) noted that although there was no increased risk for microcephaly in the population exposed to less than 1500 mGy in Nagasaki,

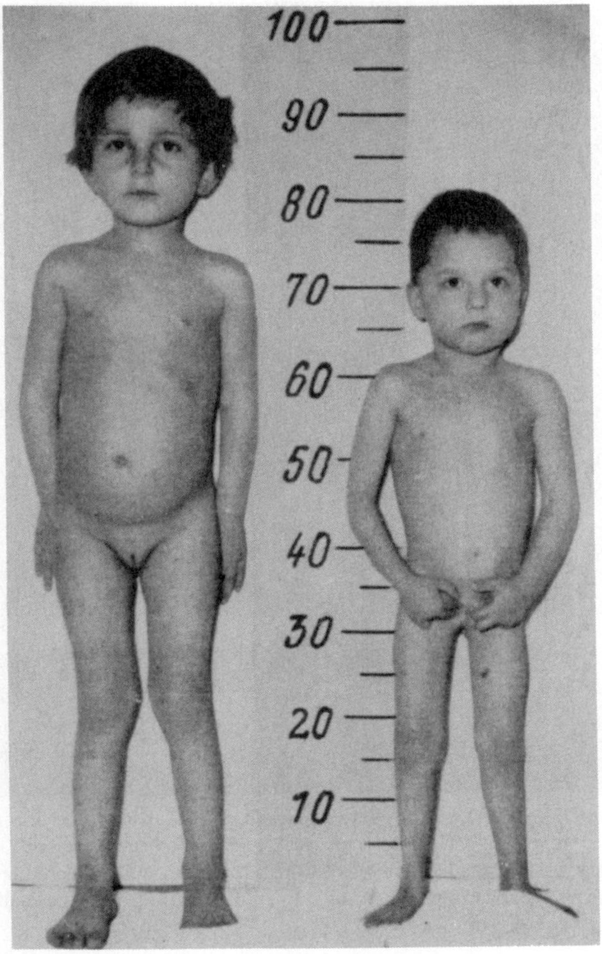

FIG. 3.5. The smaller child received in utero radiation during the 3rd–4th months of gestation. The skin dose to the abdomen was 9 Gy, delivered for the purpose of sterilization of the mother because she had metastatic carcinoma of the breast. The child was full-term but weighed only 1900 g. He had microcephaly and cerebral atrophy. Growth retardation persisted into childhood, as evidenced by comparison with a child the same age who is in the 50th percentile. There was shortening of the left tibia. The eyes had decreased vessel size and pallor of the left disk. There were abnormal motor movements, weakness, and severe mental retardation.

there was an increased risk in the population exposed to much lower doses (see Table 3.5). There is a possibility that the decreased head size observed at the lower doses in Hiroshima was unrelated to the radiation exposure in Hiroshima but was instead due to other causes. It also should be pointed out that 100–200 mGy of low linear energy transfer (LET) radiation will not increase the incidence of microcephaly in experimental animals (Brent 1977b). Otake and Schull (1984) report severe mental retardation at very low exposures; however, their findings are not supported by animal experiments, histopathological effects of radiation, or other human studies.

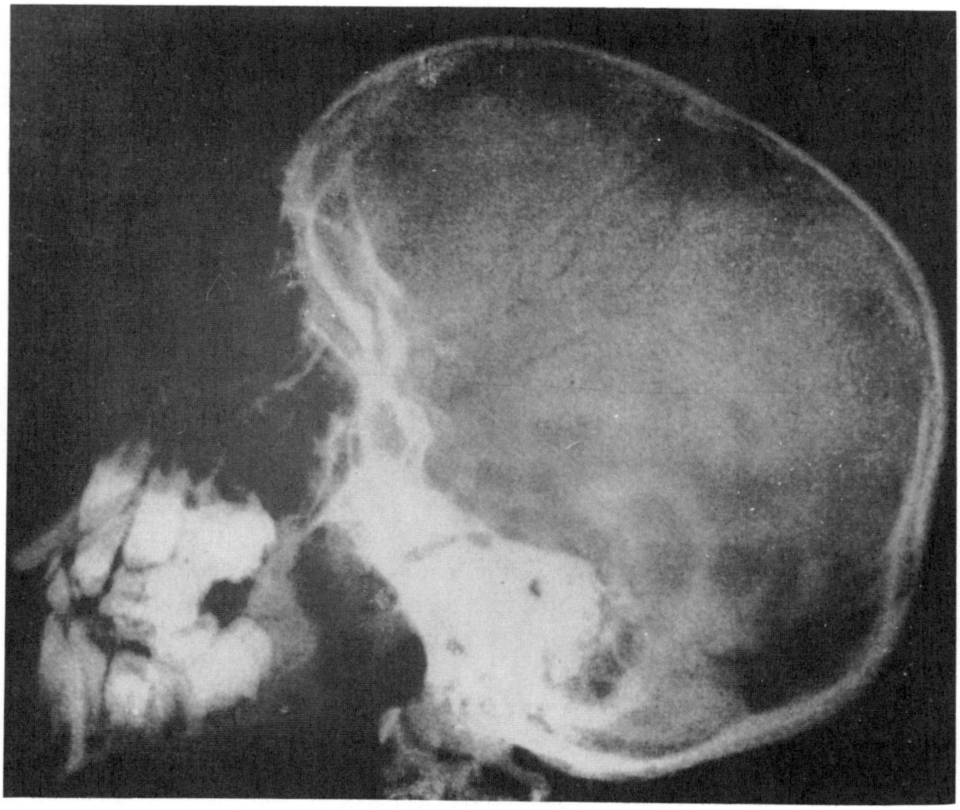

FIG. 3.6. Lateral view of the skull of the patient in Figure 3.5. The encephalogram revealed microcephaly and porencephaly. The mother received 9 Gy to the abdomen. Based on the x-ray exposure factors, the dose to the embryo was estimated to be 2–3 Gy. The embryo could have been as young as 9 weeks or as old as 20 weeks at the time of irradiation, on the basis of the meager history that is available.

Analysis of the human data reveals some interesting and important facts. Growth retardation, microcephaly, and mental retardation are predominant observable effects following acute exposures to greater than 0.5 Gy (low-LET radiation) (Tables 3.2, 3.3). No report of a bona fide, radiation-induced morphologic malformation has been reported in a human being who has not exhibited growth retardation or a CNS abnormality. One might question why the CNS seems to be more sensitive in humans than in experimental animals, in which many organ systems are malformed by irradiation. The explanation for this apparent discrepancy is that, first, the CNS maintains its sensitivity to radiation throughout gestation and into the neonatal period, whereas the other organ systems have much narrower periods during which obvious morphologic alterations can be produced (see Fig. 3.1). Second, the period of exquisite sensitivity to multiple system malformations is proportionately quite small in humans (the 2nd through the 5th week from conception) when compared with the rat (the 9th to the 11th day). In humans, all organ systems can be readily malformed by high doses of radiation during 5% of pregnancy, whereas in the rat radiation can produce gross malformations during 20% of pregnancy. Third, human exposures have occurred at random during pregnancy, and the exposure is controlled for purposes of therapeutic irradiation or not controlled, as at Hiroshima and Nagasaki. Therefore, the exposures that occur in humans during the period of maximum sensitivity will be infrequent and/or will result in a high incidence of abortions, whereas the great majority of exposures will occur during the fetal stages when the CNS is still sensitive to radiation-induced damage. Thus the nature of radiation effects in the human and the apparent discrepancies with animal data can be explained by the manner in which human populations have been exposed.

Irradiation of the human fetus from diagnostic exposures of less than 50 mGy have not been observed to cause congenital malformations or growth retardation (Kinlen and Acheson 1968, Nokkentred 1968, Tabuchi 1964, Tabuchi et al. 1967, Vilumsen 1970), but not all epidemiological studies are negative (Jacobsen and Mellemgaard 1968). These are difficult studies to perform, and it appears that the animal data support the contention that the number of gross congenital malformations will not be increased in a human pregnant population exposed to 200 mGy or less (Table 3.1). This contention is strengthened further by the fact that most human exposures to extensive diagnostic radiation studies are fractionated and/or protracted. This type of exposure is likely to be less effective in producing malformations than is an acute exposure of low-LET radiation (Brent 1971, Brizzee and Brannon 1972).

Although the animal and human data support the contention that exposures in the 50-mGy range would not be expected to increase the incidence of anatomical malformations, growth retardation, mental retardation, or abortion, there is the theoretical possibility that functional or biochemical changes may be produced at low levels with low incidence. Brent and colleagues have investigated a number of these functional and biochemical parameters, such as thyroid function, liver function, and fertility, and have not found them to be more sensitive to radiation than the above-mentioned parameters. That does not mean, however, that some other parameter, still unstudied, would not be permanently altered by low-level radiation.

There are three stochastic phenomena that must be evaluated when one is considering prenatal radiation: (*a*) mutagenesis, (*b*) abortion during the preimplantation period, and (*c*) oncogenesis (see Table 3.6). The risk estimates for the mutagenic effects of radiation for germ cell mutations or somatic mutations listed in Table 3.6 are based on the mutation rates derived from adult animal radiation experiments. Although there is a threshold effect for the induction of abortion during and after early organogenesis (>250 mGy; see Table 3.3), there is a possibility that abortion in the preimplantation period may be a stochastic phenomenon. Doses of 100 mGy on the 1st day of mammalian development produced no apparent increase in abortions, but these experiments involved only hundreds of embryos. It would not be possible to increase the spontaneous risk of abortion by 1.0% or 0.1% unless the size of the experimental groups were considerably increased. Similarly, it would be even more difficult to evaluate the

TABLE 3.6. ESTIMATE OF RISKS OF 10-MGY DOSE (LOW LET) TO THE DEVELOPING HUMAN EMBRYO[a]

Age (Days)	Mutagenic Effect[b,c,d]	Childhood Carcinogenic Effect (Stewart)[e]	Maximum Childhood Carcinogenic Effect (ABCC)[f]	Gross Congenital Malformations, Death, Growth Retardation[g]	Permanent Cell Depletion[g]
1	No data	No data	No data	?	No Effect
18–28	10^{-7} per locus	3.2×10^{-4}	5×10^{-6}	Same as controls	?
50		3.2×10^{-4}	5×10^{-6}		
Late fetus to term		3.2×10^{-4}	5×10^{-6}		

[a] LET, linear energy transfer.

[b] Based on an estimated doubling dose for mutagenesis of 1 Gy, assuming a linear dose-response curve and no threshold for mutagenic effects.

[c] The mutagenic effects have not been studied in the preimplantation period; the surviving embryos are not reduced in size even when the dose is very high, although at this stage the embryo is very sensitive to the lethal effects of radiation.

[d] The estimate is assumed to be adult risk because there was no increased carcinogenic effect in the population of exposed fetuses in Hiroshima and Nagasaki.

[e] The data of Stewart et al. (1956, 1958) would indicate that the embryo is more sensitive to the carcinogenic effect of radiation than the adult. This is a controversial matter and others (Russell 1950, Rugh 1962, Piontkovskii 1958, Plummer 1952, Brizzee and Brannon 1972) feel that this association may be other than a radiation effect.

[f] Atomic Bomb Casualty Commission (ABCC) data on carcinogenesis do not indicate that the embryo and fetus are at increased risk. The risk presented is the same carcinogenic risk attributed to adults, assuming maximal effect at low doses, namely, a linear dose-response curve and no threshold for carcinogenic effects (Brent and Gorson 1972, Hoshino et al. 1965, Wood et al. 1967c).

[g] Radiation-induced embryonic death might possibly be a stochastic effect in the first few days of gestation, although the present data involving hundreds of embryos indicate no effect at 50–100 mGy.

effects of a 10-mGy exposure on the spontaneous abortion rate. There are data neither to support nor to refute the notion that radiation-induced abortion during the preimplantation period is a stochastic phenomenon. Besides the production of congenital malformations, abortion, or intrauterine growth retardation, it has been suggested that prenatal irradiation is more oncogenic than the same exposures in the child or the adult.

ONCOGENIC EFFECT OF PRENATAL RADIATION

Stewart and her colleagues suggested that the human embryo was more sensitive to the leukemogenic effects of radiation, and in later publications they have concluded that other cancers also occur more frequently in persons exposed in utero to diagnostic radiological procedures (primarily pelvimetry) (Stewart 1972, 1973, Stewart and Kneale 1970, Stewart et al. 1956, Stewart et al. 1958). The present estimate is that a 10 to 20 mGy in utero radiation dose increases the chance of leukemia developing in the offspring by a factor of 1.5–2.0 over the natural incidence. This incidence is considerably greater than the increase resulting from 20 mGy delivered to an adult (Table 3.6). In fact, a 20-mGy dose delivered to a large adult population would not make a perceptible change in the incidence of leukemia (Lewis 1957, National Research Council 1972). Lilienfeld (1966) reviewed the epidemiological considerations with respect to leukemogenesis. A number of independent studies support the thesis that in utero radiation exposure is associated with an increased risk of leukemia (Ager et al. 1965, Ford and Patterson 1959, Graham et al. 1966, Lilienfeld 1966, McMahon and Hutchinson 1964, Polhemus and Koch 1969, Yamazaki et al. 1954). Six of nine studies reported in Lilienfeld's paper indicate an increase in leukemia risk of 1.3- to 1.8-fold following diagnostic radiation exposure in utero. Lilienfeld states, "When one considers the variety of control groups used and the sampling variability, the results are remarkably consistent in showing an excess frequency of leukemia among children of radiation-exposed pregnant mothers" (Lilienfeld 1966). Diamond et al. have extended the studies of Lilienfeld and have corroborated their early finding of a higher incidence of leukemia (threefold) in children exposed to diagnostic radiation in utero (Diamond et al. 1973). They also report that this effect did not occur in the black population.

There are a number of interesting associations in these data that should be pointed out. In the studies of Stewart and colleagues there was a higher incidence of previous abortion in the mothers receiving pelvimetry, and also the children in the pelvimetry group had a higher incidence of upper respiratory infections prior to the development of leukemia (Stewart et al. 1956, 1958). Others have reported that infants from families with a strong family history of allergy are also more susceptible to radiation-induced leukemia when exposed to diagnostic radiation in utero (National Research Council 1972). The problem with these data is that in some instances patients with an allergic history and no preconception radiation had a higher frequency of leukemia than did some groups that had received irradiation in utero.

Tabuchi (1964) reported no increase in leukemia following diagnostic radiological procedures. In some of the studies that did not report an increased risk of leukemia, the number of patients was small. Of the 86 persons exposed in utero at Nagasaki, none developed leukemia (Burrow et al. 1964). These persons received considerably higher doses of radiation than did those patients in the previous studies. Kato (1971) studied 1300 people, some of whom were exposed to the atomic bomb while in utero, and observed no increased evidence of malignancy in the 1st 24 years of follow-up, although there was an increased mortality in the 1st year of life and after 10 years of age.

It is of interest that Graham et al. (1966) reported an increased risk of leukemia that was identical whether a mother had received radiation from diagnostic procedures shortly before or after conception. Hoshino et al. (1965) reported no increase in leukemia in a study of 17,000 children of parents who had received radiation before conception. Graham et al. (1966) pointed out that children of mothers with a history of abortion or stillbirth also had children with a higher risk of leukemia. Neutel and Buck (1971) found that childhood malignancy occurred more often in offspring of mothers who smoked. Fasal et al. (1971) reported that infants who were heavier at birth were more likely to get leukemia. It appears that whenever one looks for the association of an event with the occurrence of leukemia, it can be readily found.

At present it is not clear whether radiation exposure during the preconception or postconception period is a causative or associative factor in the increased incidence of leukemia. Miller (1970) and others (Brent and Gorson 1972, Jablon 1973, Wright 1973) disagree with the conclusions of Stewart and colleagues and other reports that support their hypothesis. Miller writes:

> Minimal doses of x-rays are equally oncogenic whether exposure occurred before conception or during pregnancy, whether the neoplasm studied was leukemia or any other major cancer of childhood, and whether the study was based on interviews which may be biased, or on hospital records. Taken in aggregate, the similarity of results, in the absence of a dose-response effect or of supporting data from animal experimentation, raises a question about biologic plausibility of a causal relationship. (Miller 1970)

Furthermore, Miller points out that siblings of leukemic children have an incidence of leukemia of 1 in 720 per 10 years, which is greater than the 1:2000 risk of leukemia following pelvimetry exposure and the 1:3000 probability of leukemia in the general population of children followed for 10 years (see Table 3.7).

Rugh et al. (1966) irradiated mice with 1 Gy on each day of gestation and observed the incidence of tumors in the offspring. There was no statistical increase in the incidence of tumors in adult animals from irradiation in utero on any day. Brent and Bolden (1961, 1968) exposed pregnant mice to doses of 300, 600, and 900 mGy after ½, 7½, 8½, 12½, and 16½ days of gestation. They also did not observe an increase in the incidence of tumors.

At present, a number of investigators believe that in utero exposure to small amounts of radiation increases the risk of leukemia and other malignancies, whereas other investigators seriously question whether the embryo is markedly more sensitive to the leukemogenic effects of irradiation when compared to the

TABLE 3.7. RISK OF LEUKEMIA IN VARIOUS GROUPS WITH SPECIFIC EPIDEMIOLOGICAL AND PATHOLOGICAL CHARACTERISTICS IN POPULATIONS FOLLOWED UP FOR 10 OR 30 YEARS[a]

Group	Approximate Risk	Increased Risk Over Control Population	Occurrence
Identical twin of leukemia twin	1/3	1000	weeks to months
Irradiation-treated polycythemia vera	1/6	500	10–15 years
Bloom syndrome	1/8	375	<30 years old
Hiroshima survivors who were within 1000 meters of the hypocenter	1/60	50	average, 12 years
Down syndrome	1/95	30	<10 years old
Irradiation-treated patients with ankylosing spondylitis	1/270	10	15 years
Siblings of leukemic children	1/720	4	to 10 years
Children exposed to pelvimetry in utero (gestational exposure)	1/2000	1.5	<10 years
U.S. white children <15 years	1/2800	1	to 10 years

[a]Adapted from Miller RW. Epidemiologic conclusions from radiation toxicity studies. In: Fry RJM, Grahn D, Griem ML, eds. Late effects of radiation. London: Taylor and Francis, 1970: 245–256.

TABLE 3.8. RISK OF 5 mSv (MAXIMUM PERMISSIBLE EXPOSURE FOR WOMEN RADIATION WORKERS WITH REPRODUCTIVE POTENTIAL)[a]

	0 mSv		Additional Risk of 5 mSv
Risk of spontaneous abortion	0.15		0
Risk of major congenital malformations	0.03		0
Risk of severe mental retardation	0.005		0
Risk of childhood malignancy/10-year period	0.007		1.6×10^{-4a} or 2.5×10^{-6b}
Risk of early- or late-onset genetic disease	0.10		Risk in next generation is negligible
Total risk (using Stewart data)[a]	0.29		1.6×10^{-4c}
Ratio of total risk to additional risk radiation		1800:1	
Total risk (using ABCC data)[b]	0.29		2.5×10^{-6b}
Ratio of total additional risk of radiation		116000:1	

[a] Estimates based on data from Stewart and Kneale (1970) and Stewart et al. (1956, 1958).
[b] ABCC, Atomic Bomb Casualty Commission.
[c] Estimate based on data from Wright (1973), Rugh et al. (1966), and Brent and Bolden (1968).

child or adult. Until the mechanism is understood, there always will be doubt concerning the magnitude of the role of in utero radiation in leukemia induction. The increased incidence of cancer in children exposed in utero to diagnostic radiation has to be clarified in view of the fact that much higher doses of radiation to animal embryos and to the children exposed in utero at Hiroshima and Nagasaki have not resulted in a marked increase in the incidence of cancers, which one would expect if the embryo were as sensitive to the carcinogenic effects of radiation as Stewart and colleagues suggest (Tables 3.6, 3.7, 3.8). One cannot overemphasize either the importance of the multiplicity of factors involved or the difficulties in their identification and control. Even laboratory experiments concerned with tumor production are difficult to interpret. For example, Ross and Bras (1965) reported that the incidence of spontaneous tumors varied with the diet and weight of the animals. Heavier animals on high-protein diets had a higher incidence of tumors than did the lighter rats on low-protein diets. Hence there are many unanswered questions concerning the relationship between leukemia and malignancy and in utero radiation exposure.

Recently published results of the occurrence of cancer in adults who were irradiated in utero in Hiroshima and Nagasaki indicate that there is an increase in the incidence of cancer in the exposed population and that the cancers occur earlier than usual during adult life. Thus the long-term study in Hiroshima and Nagasaki does not support the increased incidence in childhood malignancies suggested by Stewart and colleagues, and the incidence of cancer in the adults does not support the markedly increased sensitivity of the fetus to radiation-induced cancer as suggested by those authors. There is little disagreement with the concept that low doses of radiation present a carcinogenic risk to the embryo and adult, and that there may be different risks per dose at different stages of development. The concept that is difficult to explain from a basic science viewpoint is why proliferating embryonic cells would be several orders of magnitude more sensitive to a carcinogenic effect of radiation than the child's or adult's cell.

Although a dose of less than 100 mGy to the implanted embryo is not associated with a significant increase in the incidence of congenital malformations, growth retardation, or fetal death, the possibility of the occurrence of these radiogenic effects in a particular case cannot be completely ruled out. Even if the radiogenic risks are real, let us examine how difficult it would be to use this information in counseling a patient who has received a dose of perhaps 20 mGy during her pregnancy. According to Stewart and colleagues, the risk of leukemia following this exposure in utero is 1:2000 versus 1:3000 in unexposed controls over a 10-year period (see Table 3.7). If one were inclined to recommend therapeutic abortion for this embryo because the probability of developing leukemia is 50% greater than controls, one would perform abortions in 1999 exposed nonleukemic subjects for every leukemic subject "saved." It is one thing to avoid radiation because of a potential hazard, but it is another matter to recommend therapeutic abortion on this basis. If a physician were inclined to accept this increased probability (1:2000) as a risk great enough to recommend therapeutic abortion, he or she would be placed in a serious dilemma because there are other epidemiological situations in which the risk of leukemia is greater. In fact, the hypothetical incremental risk for 20 mGy of in utero radiation is 1:6000 over a 10-year period. It is the combination of the control risk plus the incremental radiation risk that results in a 1:2000 risk for these patients. If one examines Table 3.7, it is obvious that the risk of leukemia is greater in "unirradiated" siblings of leukemics (1:720) than in patients subjected to diagnostic radiation (1:2000) according to the data of Stewart and colleagues.

Certainly, the position that all future pregnancies of parents with one leukemic child should be aborted would be untenable. One can carry this argument to its ridiculous extreme by advocating that all pregnancies should be aborted because the risk of malformation is approximately 40–60 per 1000 deliveries, and this does not include the probability of postnatal diseases occurring in these offspring. Some may interpret this as a facetious discussion, but the clinician and the patient must recognize that "spontaneous" risks of pregnancy are two orders of magnitude greater than the theoretical risks of diagnostic radiation (Tables 3.7, 3.8).

COUNSELING THE PREGNANT WOMAN EXPOSED TO RADIATION: ESTIMATING THE RISKS

It is appropriate to utilize the results of radiation studies in pregnant animals as part of the data on which to estimate the reproductive risks of embryonic radiation.

Radiation effects in the mammalian embryo are more similar than the effects of most other teratogens or embryopathologic agents because radiation has a direct effect on the developing embryo (Brent and Bolden 1967a, 1968, Jablon 1973, Ross and Bras 1965, Russell and Major 1957, Wood et al. 1967c). Variations in placental transport and maternal metabolism do not alter significantly the extrapolation of results of radiation experiments with pregnant mammals to humans as do those of experiments using other teratogens.

Table 3.9 lists the effects of low exposures to developing mammalian embryos. The data indicate that continuous exposures below 20 mGy/day to pregnant animals have no effect on numerous parameters such as fertility, growth, mortality, or the incidence of congenital malformations.

The biological effects of embryonic radiation are summarized for exposures of 100 mGy in Table 3.1 and 1 Gy in Table 3.2 for various species and different stages of gestation. Table 3.3 lists the estimated median lethal dose (LD_{50}), the minimal malforming dose, the minimal lethal dose, the minimal dose producing growth retardation, and other parameters for the human embryo, based on extrapolations from animal data.

The hazards of exposures in the range of diagnostic radiography (0.2 to 50 mGy) present an extremely low risk to the embryo, when compared with spontaneous mishaps that can befall humans (see Tables 3.6, 3.8). Approximately 30% to 50% of human embryos abort spontaneously. Many patients abort before pregnancy is diagnosed or recognized. Human infants have a 2.75% major malformation rate at term, which rises to approximately 6% to 10% once all malformations and genetic diseases become manifest. In spite of the fact that doses of 10 to 30 mGy can produce cellular effects and the fact that diagnostic exposure during pregnancy has been associated with malignancy in childhood, the maximum theoretical risk to the human embryos exposed to doses of 50 mGy or less is extremely small (see Tables 3.6, 3.8). In my experience, when the data and risks are explained to the patients, the family with a wanted pregnancy invariably continues with the pregnancy. The difficulty that frequently arises is that the radiation risk from diagnostic radiation is often evaluated without consideration of the significant normal risks of pregnancy, and the physician giving advice may be very concerned about radiation effects, especially if he or

TABLE 3.9. REPORTED EFFECTS OF LOW DOSES OF RADIATION ON THE EMBRYO WHEN ADMINISTERED THROUGHOUT PREGNANCY

Study	Organism	Source	Approximate Dose Rate Per Minute (mGy/min)	Approx. Dose Per Day (mGy/day)	Dose Per Pregnancy (mGy)	Comments	Effects
Russell et al. (1963)	Mouse	^{137}Cs	0.086	124	1700	Days 1–18	None except shortened breeding period in female
Ronnback (1965)	Mouse	^{137}Cs		84	1700	During gestation and in some instances 20 days postpartum	None
Vorisek (1965)	Rat	^{60}Co	0.017	25		Daily during pregnancy	None
Stadler and Gowen (1964)	Mouse		0.015	22		Continuous through 10 generations	None
Coppenger and Brown (1967)	Rat	^{60}Co	0.014	20		Continuous through 10 generations	None
Konerman (1969)	Mouse	^{60}Co	0.017	100	1800	Days 1–18	None
				200	3600	Days 1–18	Growth retardation
				200		Days 6–13 only	None
Wesley (1960)	Human	Background		0.003	1	Variation in background radiation	Increased malformations
Gentry, et al. (1959)	Human	Background		0.003	1	Variation in background radiation	Increased malformations
Grahn and Kratchman (1963)	Human	Background		0.003	1	Variation in background radiation	Background radiation level not a factor in incidence of congenital malformations

TABLE 3.9. *Continued*

Study	Organism	Source	Approximate Dose Rate Per Minute (mGy/min)	Approx. Dose Per Day (mGy/day)	Dose Per Pregnancy (mGy)	Comments	Effects
Segall et al. (1964)	Human	Background		0.003	1	Background radiation	Background radiation level not a factor in incidence of congenital malformations
Kriegal and Langendorff (1964)	Mouse	X-ray		25		Acute dose given daily	No influence
				50			No influence
				100			Malformations, resorptions, growth retardations
Piontkovskii (1961)	Rat	X-ray		200 10	200	Acute dose given daily	Functional changes in behavior and motor activity (questionable)

she has a poor background in radiation biology (Brent 1976). We have records in our files of patients advised to terminate pregnancy despite the fact that exposure was well below 50 mGy, and/or the fact that exposure took place prior to conception. Such problems may be avoided if physicians become more familiar with radiation biology, or if they consult with the appropriate specialist when questions arise.

The physician will never be able to guarantee the outcome of a pregnancy, but the basic logic of the practice of medicine would be undermined if some of its practitioners recommended therapeutic abortion for risks with probabilities of occurrence far below the expected spontaneous incidence in the embryo and infant. When counseling parents concerning the risks to an embryo of a particular exposure to radiation, the biological facts are only one facet to be considered. The risk of radiation damage is based on the estimated dose and the stage of gestation.

Risk perception by the parents depends upon the age of the parents, the number of children in the family, the religious and ethnic background of the family, and the options available to the family based on the state laws regarding therapeutic interruptions of pregnancy. There is also important information to be gathered, more difficult to acquire, on the emotional maturity of the family and whether the pregnancy was planned or unplanned. The following information should be obtained:

1. Stage of pregnancy of the exposure
2. Menstrual history
3. Previous pregnancy history
4. History of congenital malformations
5. Other potentially harmful environmental factors during the pregnancy
6. Age of the mother and father
7. Type of radiation study; dates and number of studies performed
8. Calculation of the embryonic exposure by a health physicist, medical physicist, or competent radiologist
9. Status of the pregnancy: wanted or unwanted

An evaluation should be made of this information, with both patient and counselor arriving at a decision. The physician should place in the medical record a summary of this information, that the patient has been informed, that every pregnancy has a significant risk of problems, and that the decision to continue the pregnancy does not mean that the counselor is guaranteeing the outcome of the pregnancy.

DIAGNOSTIC RADIATION EXPOSURES

Our experience has taught us that there are many variables involved in the evaluation of a radiation exposure to a pregnant or potentially pregnant woman. Therefore, it is impossible to provide the same routine response for every patient. The following examples are some typical cases of radiation exposure in which evaluation and counseling was provided.

Case Study A 19-year-old unmarried secretary went to her physician because of abdominal pain. She told him her menstrual periods were regular and the pain was periumbilical and persistent. An upper gastrointestinal series was performed but was unsatisfactory because the young girl had eaten breakfast before the study. When the second study was performed, she mentioned that she was 2 weeks past her scheduled menstrual periods and that she could be pregnant. The father of the child was married. The radiation therapist at the hospital recommended a therapeutic abortion because of radiation hazards to the embryo. The estimated dose to the embryo over a period of 3 days was 16 mGy.

Our laboratory was called and informed of the case history and recommendation. Our conclusion was that this dose of radiation would not result in embryonic malformation, growth retardation, or death, and that the nonexistent radiation hazard should not be used to obtain a therapeutic abortion for "social" reasons. The laws of that state, at that time, did not permit therapeutic abortions; so the girl flew to another state, where the therapeutic abortion was performed.

Case histories similar to this are transmitted to our laboratory frequently. In most instances, the dose to the embryo is less than 50 mGy and frequently less than 10 mGy. Unfortunately, variations in the approach of the physician in charge depend on many factors–(*a*) whether the pregnant woman is married, (*b*) whether the pregnancy was planned, (*c*) the mental status of the mother and/or father, and (*d*) the abortion laws of the state–rather than the knowledge of the physician regarding radiation hazards to the embryo.

We believe the specter of radiation hazards should not be invoked to circumvent a social or legal problem. If the abortion laws are archaic or unfair, physicians should work together to change them. They should not create an implied serious radiation hazard that is not justified by the facts to solve social, psychological, or legal problems.

Case Study A 29-year-old housewife (para 0, gravida 1, abortus 0) presented in her 6th month of pregnancy with the history that she had been to her dentist the week before and had had one radiograph of her left upper first molar. She was anxious and very worried about the effect of radiation on her baby. She was reassured but called back six times asking new questions. The estimated dose to the embryo was below 0.01 mGy. She continued to be anxious and hyperactive. She delivered a 3100-g baby boy with minimal syndactyly of the fourth and fifth fingers of the left hand, who was otherwise normal. A legal suit was almost instituted in this case.

A surprising number of cases are evaluated or brought to trial in which the only radiation exposure of the mother was to the head, the neck, or the chest, and yet there is both clinical and basic science information that indicates that embryopathological conditions are only a consequence of direct embryonic exposure. High doses of radiation to the mother, while shielding the embryo, result in offspring who have the same incidence of congenital malformation as do control embryos (Brent 1960, Brent and Bolden 1967b, Brent and McLaughlin 1960, Neifakh 1957).

There were many instances in the past in which a radiologist recommended a therapeutic abortion for a patient whose embryo received a fractionated dose of

0–50 mGy during the first trimester of pregnancy, because of the possibility that the embryo might be malformed. Frequently, this recommendation was made for unmarried women, primarily to assist them in eliminating unwanted pregnancies. It is this type of recommendation that will reap retribution when the same radiologist attempts to defend himself or herself against the contention that a series of diagnostic radiological examinations caused a particular malformation.

EXPOSURE FROM RADIATION THERAPY

There is ample evidence to indicate that radiation therapy in the region of the abdomen can result in doses of radiation that will affect the fetus deleteriously. In such a case, it is important to consult with either the radiotherapist or a qualified medical physicist to determine the total dose delivered to the embryo. If the fetus absorbs 500 mGy or more at any time during gestation, there is a significant possibility that the fetus may be damaged. Of course, there are instances in which human fetuses have survived greater doses and appeared normal, but at this dose the probability of CNS damage or other malformations is real, and the parents should be so informed (Murphy and Goldstein 1930, Ronderos 1961). If the dose absorbed by the embryo during the early organogenetic period amounts to several gray, there is reasonable possibility that the embryo may abort. As one proceeds into the second and third trimesters, the chance of abortion and malformations declines, but irreparable damage to the CNS can occur. Exposure to 1 Gy acute radiation from the 8th to 15th weeks of gestation presents a substantial risk for mental retardation to the surviving fetus (Dekaban 1968, Miller 1970, National Research Council 1972, Yamazaki et al. 1954).

When the radiation therapist fails to investigate the possibility of pregnancy or fails to record all the information regarding the pregnancy and the steps taken in arriving at a decision, the therapist places himself or herself in double jeopardy. He or she may be accused of poor medical practice in a suit over a malformed infant and thereby may be more likely to lose the case even when the radiation could not have been responsible for the abnormal embryonic development (Brent 1967). Let us examine some representative cases.

> ***Case Study*** A 31-year-old woman was in her 6th month of pregnancy when her physician noted an enlarged lymph node in the anterior cervical region. A diagnosis of Hodgkin's disease was made, and she received approximately 32 Gy to the anterior cervical region and adjacent areas over a period of 4 weeks. The fetus was viable at the time of the diagnosis, and the family wanted the baby. The dose of radiation to the fetus was approximately 250 mGy. The mother delivered a full-term baby who appeared to be normal.

In this case, the counselor told the family that he could not say that the fetus was not at increased risk. The mother's question was somewhat different; namely, was it very likely that the fetus would be severely affected? Of course the answer was that while the risks were increased, there was a greater chance that the embryo would not be seriously affected at this stage of gestation. Since the family was not certain of the mother's prognosis and, therefore, whether future children were a possibility, they elected to go through with the pregnancy.

> ***Case Study*** A 36-year-old woman (para 4, gravida 0, abortus 1) was 1 week past her menstrual period, which usually was regular. Examination revealed a moderate-sized lesion of the cervix. The pregnancy test was positive. A biopsy of the cervix revealed a carcinoma of the cervix that had invaded the muscle layer. Dilation and curettage was performed, and the patient was treated with the instillation of radium.

In this case, the options were presented to the family before the biopsy, and the above-mentioned course was taken.

THE USE OF DIAGNOSTIC OR THERAPEUTIC RADIATION IN WOMEN OF REPRODUCTIVE AGE WHEN THE OVARIES OR UTERUS WILL BE EXPOSED

In a woman of reproductive age, it is important for the patient and physician to be aware of the pregnancy status of the patient before any type of abdominal x-ray exposure is planned. If the embryonic exposure will be 50 mGy or less, the radiation risks to the embryo are minuscule when compared with the spontaneous risks. The patient will accept the information if it is offered as part of the preparation for the x-ray studies at a time when both the physician and patient are aware that a pregnancy exists.

The pregnancy status of the patient can be determined by several means: (*a*) The x-ray referral slip can request the date of the last menstrual period (LMP) and the previous menstrual period (PMP) and whether the patient possibly could be pregnant; (*b*) if the patient is uncertain about her pregnancy status, a pregnancy test should be performed; (*c*) pregnancy tests could be performed on all hospital admissions of women of reproductive age.

Because the risks of 50 mGy of fetal irradiation are so small, the immediate medical care of the mother should take priority over the risks of diagnostic radiation exposure of the embryo (Table 3.10). X-ray studies that are essential for optimal medical care of the mother and evaluation of medical problems that need to be diagnosed or treated should not be postponed. Elective procedures such as employment examinations or follow-up examinations, once a diagnosis has been made, need not be performed on a pregnant woman even though the risk to the embryo is very small. If other procedures (e.g., ultrasound) can provide adequate information without exposing the embryo to ionizing radiation, then of course they should be used. There is an initial period when a pregnancy test may be negative and the patient can still be pregnant. Furthermore, the menstrual history will be of little use in this circumstance, but the risks of 50 mGy or less of radiation are extremely small during this period of gestation. The patient will benefit from knowing that the diagnostic study was indicated and should be performed in spite of the fact that she may be pregnant.

In those instances in which elective x-ray studies need to be scheduled, it is difficult to know whether to schedule them during the first half of the menstrual cycle before ovulation or during the second half of the menstrual cycle, when

TABLE 3.10. ESTIMATED EMBRYONIC AND FETAL DOSE DURING DIAGNOSTIC RADIOLOGICAL EXAMS[a]

Examination	Dose (mGy)
Upper extremity	.01
Lower extremity	.01
Skull film	.04
Chest	
Radiography	.08
Fluoroscopy	.70
Cholecystography	2.0
Abdomen	2.9
Intravenous pyelography	4.0
Upper gastrointestinal	
Series	5.6
Barium enema	8.0

[a] Adapted from Brent RL. Irradiation in pregnancy. In: Sciarra JJ, ed. Davis' gynecology and obstetrics, vol. 2. New York: Harper & Row, 1972; 1–32.

[b] These are estimated means, but reported ranges vary substantially and may change with progressing gestation. Thus each should be assessed individually and consultation with a radiologist or radiation physicist is advisable when decisions have to be made.

most women will not be pregnant. The risk of diagnostic exposures to the oocyte or the preimplanted embryo is extremely small, and there are meager data with which to compare the relative risk of 50 mGy to the oocyte or the preimplanted embryo (Brent 1970, Brent and Bolden 1968, Brent and Gorson 1972). If the diagnostic study is performed in the 1st 14 days of the menstrual cycle, should the patient be advised to defer conception for several months, based on the assumption that the deleterious effect of radiation to the ovaries decreases with increasing time between radiation exposure and a subsequent ovulation? The physician is in a quandary because he or she is warning the patient about a very low-risk phenomenon. On the other hand, avoiding conception for several months is not an insurmountable hardship. This potential hazard is quite speculative for man, as indicated by the report of the National Council on Radiation Protection and Measurements (NCRP) dealing with preconception radiation:

> It is not known whether the interval between irradiation of the gonads and conception has a marked effect on the frequency of congenital changes in human offspring, as has been demonstrated in the female mouse. Nevertheless, it may be advisable for patients receiving high doses to the gonads (>250 mGy) to wait for several months after such exposures before conceiving additional offspring. (National Council on Radiation Protection and Measurements (NCRP) 1971)

During diagnostic radiological procedures patients absorb considerably less than 250 mGy; therefore, the recommendation made here may be unnecessary, but it involves no hardship to the patient or physician. Because both the NCRP and International Council for Radiation Protection have previously recommended that elective radiological examinations of the abdomen and pelvis be performed during the first part of the menstrual cycle to protect the zygote from possible but largely conjectural hazards, a recommendation to reduce another possible hazard by avoiding fertilization of recently irradiated ova perhaps should merit equal attention.

If a women is exposed to more than 50 mGy of radiation and then is found to be pregnant, a determination of the merits of continuing the pregnancy must be made by the physician, the patient, and an expert in radiation. A decision to terminate the pregnancy will depend on (*a*) the nonradiation-related hazard of the pregnancy to the expectant mother, (*b*) the extent and type of radiation hazard to embryo or fetus, (*c*) the ethnic and religious background of the family, (*d*) the laws of the state pertaining to legal abortion, and (*e*) any other relevant consideration.

If exposures of less than 50 mGy to the pelvis do not measurably affect the exposed embryos, why is there so much effort and energy spent in determining the pregnancy status of the patient? There are several reasons why the physician should have the burden of determining the pregnancy status before an x-ray or nuclear medicine procedure is performed. First, if the physician is forced to include the possibility of pregnancy in the differential diagnosis, a small percentage of diagnostic studies may no longer be considered necessary. Early symptoms of pregnancy may mimic certain types of gastrointestinal or genitourinary disease. Second, if the physician and patient are both aware that pregnancy is a possibility and the procedure is still performed, it is much less likely that the patient will be upset if she subsequently proves to be pregnant. Third, the careful evaluation of the reproductive status of women undergoing diagnostic procedures will prevent many unnecessary lawsuits. Many lawsuits are stimulated by the factor of surprise and won on the basis of the double jeopardy of the defendant (Brent 1977a). In some instances the jury is not concerned with cause and effect but with the fact that something was not done properly by the physician. In this day and age, failure to communicate adequately can be interpreted as less than adequate medical care. Both these factors are eliminated if the patient's pregnancy status has been evaluated properly and the situation discussed adequately with the patient. Physicians are going to have to learn that practicing good technical medicine may not be enough in a litigation-prone society. Even more important, the patient will have more confidence if the decision to continue the pregnancy is made before the medical x-ray procedure is performed, because the necessity of the procedure would have been determined with the knowledge that the patient may be pregnant.

EFFECTS OF INTERNALLY DEPOSITED RADIONUCLIDES

The effects of various radionuclides administered to the pregnant mammal have not been studied as thoroughly as the effects of external irradiation. It is difficult to generalize about the effects of radionuclides because (*a*) they may have specific target organs, (*b*) they may or may not cross the placenta, (*c*) the distribution of the radionuclide may be nonrandom, (*d*) the metabolism of radioactive elements or compounds may vary with individual biological differences or disease states, and (*e*) the radiation dose rate decreases exponentially with time.

The precise estimation of the hazards of a particular nuclide depends on accurate dosimetry, i.e., determination with greater or lesser precision of the total dose absorbed by the fetus or a particular fetal tissue, and the dose rate.

The use of particular radionuclides (and their frequency of utilization) has changed considerably. For example, the frequency of various nuclear medicine procedures at a 700-bed hospital has changed dramatically in the past 4 decades. A total of 13,136 studies were performed on 7682 patients during 1983–1984. In 1970 the same hospital performed 3134 studies on 2641 patients. Over the past 3 years the number of nuclear medicine procedures has remained constant. Over the past 15 years a number of procedures have been introduced, and others have been eliminated. Of interest to the obstetric patient is the elimination of nuclear medicine procedures to localize the placenta. In the earlier days isotopes other than technetium-99m (^{99m}Tc) were utilized, and they could deliver a measurable dose to the embryo. Although ^{99m}Tc placentography delivers a very low dose to the embryo (<0.1 mGy), ultrasonography has eliminated the need for placental localization with isotopes. Table 3.11 lists the various procedures in which radiopharmaceuticals are utilized.

Table 3.12 lists the commonly utilized radioisotope and embryonic exposure in mGy per millicurie of administered radionuclide. Table 3.13 lists the estimated dose to the fetal thyroid in mGy per millicurie of iodine-131 (^{131}I) administered at different stages of gestation.

It is obvious that most procedures will result in extremely low fetal exposures. In the case of iodine this may not be the case, especially if the patient is being treated for thyroid malignancy or hyperthyroidism. It is extremely important that a competent expert calculate the fetal dose once the question of fetal or embryonic exposure has been raised.

Radioactive Iodine

Radioactive iodine is frequently used in clinical studies and research, its usual forms being iodine-123 (^{131}I), iodine-125 (^{125}I), and ^{131}I. ^{131}I is used predominantly for uptake studies and radioactive scanning. The isotope may be administered as the inorganic ion or may be bound to protein through its affinity for the tyrosine molecule. Radioactive serum albumin (RISA) can be utilized for determining blood volume, protein leakage, placental localization, position of brain tumors, and other studies. ^{125}I has been used to label nanogram quantities of hormones for in vivo and in vitro assays. A reasonable estimate of the exposure of the fetal and maternal tissues for a given dose of radioactive iodine can be made (see Table 3.13).

The fetal thyroid absorbs and incorporates iodides readily by the 10th week of gestation and continues to do so thereafter. Uptake of iodine is greater in the fetal thyroid than in the maternal thyroid (Book and Goldman 1975, Jacobson and Brent 1959). Inorganic iodides readily cross the placenta; however, even in the case of bound iodides that do not cross the placenta, in time a substantial amount of iodide released from the binding molecule becomes available to the fetus. There probably is no form of radioactively labeled iodide compound that does not release some radioactivity that may reach the fetus.

Reported fetal effects from therapeutic (ablative) doses of ^{131}I administered to

TABLE 3.11. CLINICAL RADIOPHARMACEUTICAL PROCEDURES[a,b]

Radiopharmaceutical	Study	Adult Administered Activity
^{32}P-sodium phosphate	Therapy, polycythemia vera	2.3 mCi/m^2
^{51}Cr-albumin	GI protein loss	50 μCi IV
^{51}Cr-chromate	Red cell survival	160 μCi IV
^{51}Cr-chromate	Red cell mass	25 μCi IV
^{51}Cr-chromate	Red cell in vivo	160 μCi IV
^{51}Cr-chromate red blood cells	Spleen imaging	200 μCi IV
^{57}Co-vitamin B_{12}	Vitamin B_{12} absorption	0.5 μCi PO
^{60}Co-vitamin B_{12}	Vitamin B_{12} absorption	0.5 μCi PO
^{59}Fe-citrate	Iron absorption	5 μCi PO (700 mg ferrous ammonium sulfate, 300 mg ascorbic acid)
^{59}Fe-citrate	In vivo counting for effective hematopoiesis	20 μCi IV
^{59}Fe-citrate	Iron plasma clearance and turnover	20 μCi IV
^{59}Fe-citrate	Iron red blood cell uptake	20 μCi IV
^{67}Ga-citrate	Tumor imaging	3–4 mCi IV
^{67}Ga-citrate	Abscess imaging	3–4 mCi IV
^{75}Se-selenomethionine	Pancreas imaging	250 μCi IV or 4 μCi/kg, whichever is less
^{99m}Tc-diphosphonate	Myocardial imaging	15 mCi IV
^{99m}Tc-diphosphonate or pyrophosphate	Bone imaging	15 mCi IV
^{99m}Tc-DTPA	Brain imaging	15–20 mCi IV (no perchlorate)
^{99m}Tc-DTPA	Kidney imaging	15 mCi IV
^{99m}Tc-DTPA iron ascorbate	Kidney imaging	15 mCi IV
^{99m}Tc-human serum albumin	Pericardial imaging	10 mCi IV
^{99m}Tc-human albumin microspheres	Lung perfusion study	3 mCi IV
^{99m}Tc-human serum albumin	Placenta imaging	1–2 mCi IV
^{99m}Tc-macroaggregated albumin	Lung perfusion study	3 mCi IV
^{99m}Tc-macroaggregates	Venous imaging for thrombosis	6 mCi IV
^{99m}Tc-pertechnetate	Vascular flow	20 mCi IV
^{99m}Tc-pertechnetate	Thyroid uptake when uptake is low, organification blocked	1–3 mCi IV
^{99m}Tc-pertechnetate	Ectopic gastric tissue (e.g., Meckels diverticulum)	100 μCi/kg
^{99m}Tc-pertechnetate	Brain imaging	15–20 mCi IV (200 mg potassium perchlorate PO prior to exam)

TABLE 3.11. *Continued*

Radiopharmaceutical	Study	Adult Administered Activity
^{99m}Tc-pertechnetate	Carotid or cerebral hemisphere flow studies	20 mCi IV
^{99m}Tc-sulfur colloid	Bone marrow imaging	10 mCi IV
^{99m}Tc-sulfur colloid	Spleen imaging	3 mCi IV
^{99m}Tc-sulfur colloid	Liver imaging	3 mCi IV
^{111}In-DTPA	Cerebrospinal fluid rhinorrhea	0.5 mCi IT
^{111}In-DTPA	Normal pressure hydrocephalus	0.5 mCi IT
^{123}I-iodide	Thyroid uptake	100 μCi PO
^{123}I-iodide	TSH thyroid uptake study	100 μCi PO (10 units thyrotropin [Thytropar] IM × 3 days prior to test)
^{123}I-iodide	T_3 suppression thyroid uptake	100 μCi PO (25 mg liothyronine [Cytomel] TID × 7 days prior to test)
^{123}I-iodide	Thyroid imaging	100 μCi PO
^{123}I-iodohippurate	Kidney function	1–2 μCi IV
^{123}I-iodohippurate	Kidney imaging	1–2 μCi IV
^{125}I-human serum albumin	Plasma volume	4 μCi IV
^{131}I-fibrinogen	Venous imaging for thrombosis	100 μCi
^{131}I-iodide	Thyroid uptake	10 μCi PO
^{131}I-iodide	TSH uptake study	10 μCi PO (10 units thyrotropin IM × 3 days prior to test)
^{131}I-iodide	T_3 suppression	10 μCi PO (25 mg liothyronine sodium TID × 7 days prior to test)
^{131}I-iodide	Thyroid therapy	5–20 mCi for thyrotoxicosis 75–150 mCi for thyroid cancer

[a] Data from Kereiakes et al. 1976, Roedler et al. 1978, and Saenger 1976.

[b] DTPA diethylenetriamine pentacetic acid; TSH, thyroid-stimulating hormone; T_3, triiodothyronine.

TABLE 3.12. DOSE ESTIMATED TO EMBRYO FROM RADIOPHARMACEUTICAL ADMINISTRATION TO MOTHER[a]

Radiopharmaceutical	Embryo Dose (mGy/mCi Administered)
^{99m}Tc-human serum albumin	0.18
^{99m}Tc-lung aggregate	0.35
^{99m}Tc polyphosphate	0.36
^{99m}Tc-sodium pertechnetate	0.37
^{99m}Tc-stannous glucoheptonate	0.40
^{99m}Tc-sulfur coloid	0.32
^{123}I-sodium iodide (15% uptake)	0.32
^{123}I-sodium iodide (15% uptake)	0.10
^{123}I-rose bengal	1.3
^{131}I-rose bengal	6.8

[a] Reprinted with permission from Kereiakes, J. G. and M. Rosenstein, *Handbook of Radiation Doses in Nuclear Medicine and Diagnostic X-Ray*, Boca Raton, FL: CRC Press; 1980. Copyright CRC Press, Inc., Boca Raton, FL.

TABLE 3.13. THYROIDAL RADIOIODINE DOSE TO THE FETUS[a]

Gestation period	Fetal/Maternal ratio (thyroid gland)	Dose to thyroid (fetus) (mGy/mCi)[a,b]
10–12 weeks	—	0.01 (precursors)
12–13 weeks	1.2	7
2nd trimester	1.8	7
3rd trimester	7.5	—
Birth imminent	—	80

[a] Reprinted with permission from Kereiakes, J. G. and M. Rosenstein, *Handbook of Radiation Doses in Nuclear Medicine and Diagnostic X-Ray*, Boca Raton, FL: CRC Press; 1980. Copyright CRC Press, Inc., Boca Raton, FL.
[b] Dose of ^{131}I ingested by mother.

pregnant women include total fetal thyroid destruction. Although there are no reports of immediate deleterious fetal effects from tracer doses of radioactive iodine, a theoretical concern about the induction of thyroid cancer cannot be disregarded. The use of radioactive iodine should be avoided during pregnancy, unless its utilization is essential for the medical care of the mother and there is no substitute. Even if administered during the 1st 5–6 weeks of human gestation, when the fetal thyroid has not yet developed, the total dose to the embryo should be estimated. Ordinarily, the whole-body dose to the embryo from ^{131}I is insignificant when compared with the thyroid dose, because the whole-body dose primarily is due to gamma irradiation, while the thyroid dose is due to beta particles as well as gamma rays emitted from iodides concentrated in the thyroid tissue. But if large doses of ^{131}I are administered, significant embryonic or fetal exposure can occur because of the accumulation of ^{131}I in the bladder adjacent to the uterus.

OTHER RADIONUCLIDES

Inorganic radioactive potassium, sodium, phosphorus, cesium, thallium, selenium, chromium, iron, and strontium cross the placenta readily. Experiments in animals with radioactive phosphorus and strontium indicate that if the dose is large enough, embryonic abnormality and death can result (Sikov and Noonan 1958). These isotopes are used in less than 1% of procedures; only radioactive phosphorus or gold may be utilized therapeutically, e.g., in the treatment of polycythemia or management of malignancies involving peritoneal surfaces. Many of the newer radiopharmaceuticals are composed of complex macromolecules or microaggregates (e.g., ^{99m}Tc pertechnetate). The rate of placental transfer is very low for these newer agents and, consequently, they deliver extremely low doses to the embryo. ^{99m}Tc offers many advantages in diagnostic procedures. Total body dose is considerably lower with ^{99m}Tc bound to macromolecules than with ^{131}I-labeled protein and, with the absence of radioiodine, there is less concern about radiation exposure of the fetal thyroid.

For each procedure, the dose to the embryo must be calculated individually and is dependent on the form of the isotope, the site of administration, and the nature of the disease. Estimates of approximate fetal and maternal exposure for standard doses and procedures have been published (see Tables 3.12 and 3.13). Often the embryonic dose will be less than the maternal dose because the nature of the isotope preparation prevents its transfer across the placenta. Even in procedures in which the fetal thyroid dose is significant (e.g., >50 mGy), total fetal body exposure is quite low. Localization of the placenta with ^{99m}Tc results in approximately 1% of the radiation dose that would be absorbed from soft x-ray radiographic placentography. Of course neither procedure has been necessary since the advent of ultrasonography.

Rarely is a pregnant woman given a radioactive isotope after it is known that she is pregnant. It is much more common for a woman to be exposed to diagnostic radiographic procedures before the pregnancy becomes known. This is due to the fact that (*a*) diagnostic radiographic procedures are more frequently performed, (*b*) therapeutic doses of radioactive isotopes are primarily utilized in patients with malignancies, i.e., in an older population in whom pregnancy is unlikely, and (*c*) vague symptoms associated with pregnancy may simulate intraabdominal disease and warrant diagnostic radiographic procedures. In any event, the actual pregnancy is insignificant compared to the anxiety, chaos, and medicolegal suits resulting from exposure of pregnant women to diagnostic radiographic procedures (Brent 1967, 1977a, 1989). When radioisotopes are to be utilized in a woman of childbearing age, the following procedure is recommended if the procedure is necessary for the benefit of the patient and the risk is estimated to be very small.

1. Record the date of the LMP and determine whether the woman could be pregnant.
2. If pregnancy is a possibility, determine the stage of gestation and the estimated dose to the fetus or fetal target organs.
3. Communicate this information to the patient or a responsible member of the family, and record this information, the time and place of the communication, and an informed consent in the patient's record.

Let us examine some typical cases.

Case Study

Case 1 A 34-year old pregnant woman was admitted to the hospital on the 42nd day of gestation with fever of 38.4°C and pain and swelling of the right leg. There was extreme tenderness in the femoral region, and a diagnosis of thrombophlebitis was suggested. The obstetrician requested ^{131}I-labeled fibrinogen imaging to diagnose the problem and locate the extent of the thrombus. The radiologist canceled the procedure because the patient was pregnant.

The obstetrician called our laboratory to determine whether a significant risk was present, and if there was not a risk, whether we would convince the radiologist to perform the study. The radiologist was primarily concerned about fetal thyroid irradiation, not the whole-body radiation exposure. Both, of course, have to be considered with isotopes of iodine. In this particular patient the embryo's thyroid had not yet differentiated to the point of being able to concentrate iodine and therefore the only exposure that needed to be considered was the whole-body exposure to ^{131}I. Since the amount of ^{131}I-labeled fibrinogen utilized in this procedure was 100 mCi, the embryo would receive approximately 0.1 mGy. This was less than the 9-month dose of background radiation that the embryo would have received. Therefore, the radiologist was in error in refusing to do the procedure–at least on the basis of concern about a radiation risk.

Case 2 A 27-year-old woman was evaluated for a thyroid nodule. Thyroid imaging with 100 mCi of ^{131}I iodide revealed active uptake in the nodules and areas outside the thyroid. A biopsy revealed a thyroid carcinoma. Seventy-five mCi of ^{131}I iodide were administered orally. It was discovered 1 week later the patient had missed her menstrual period and was 36 days postconception when she received the radioactive iodine. Although the thyroid was not yet developed in the embryo at this stage of gestation, at least with regard to concentrating iodine, the embryo could have received a significant exposure, depending on the thyroid uptake of the mother. This could range between 25 and 75 mGy. Since 50 mGy is used frequently as the dose cutoff for advising continuation of pregnancy, at least with regard to risk, this was a difficult problem. Because the exposure was protracted, the patient was advised that if there was an increased risk, it was small compared with the spontaneous risk. She elected to continue the pregnancy, and she delivered a normal baby weighing 3300 g. The baby developed normally.

Case 3 A 16-year-old pregnant girl was administered 200 mCi of ^{131}I for the treatment of hyperthyroidism at a small community hospital. The administration of the radioiodine occurred with the knowledge that the girl was pregnant. She was referred to the nearest medical center for evaluation.

It was determined that the isotope was administered at approximately the 116th day of gestation. At this stage the fetal thyroid would receive a substantial dose of radiation. Using the data in Tables 3.12 and 3.13, the dose to the fetal thyroid can be estimated to be 1.4 Gy. This is a dose that could ablate the fetal thyroid and produce hypothyroidism in the newborn. The risk was presented to the prospective mother, who elected to continue the pregnancy. The obstetricians discussed whether thyroid hormone should be periodically instilled in the amniotic cavity, but this was rejected. After 40 weeks of gestation the mother delivered a normal baby who was euthyroid. The results were never resolved, but it is possible that the dose to the maternal thyroid was miscalculated since the patient's hyperthyroidism had to be medically managed by other means.

If a previously administered radiation dose is large enough to represent a significant hazard to the embryo, the physician and patient must decide on the relative merits of interrupting the pregnancy. This decision will depend on the perception of the risk to the mother, the extent of the radiation hazard to the fetus, the ethnic and religious background of the family, the laws of the state pertaining to abortion, and other individual considerations. In the vast majority of instances a careful analysis will reveal that the exposure is too low to present a significant risk to the embryo (Brent 1984).

SUMMARY

The term *radiation* evokes emotional responses both from laypersons and professionals. Many spokespersons are unfamiliar with radiation biology or the quantitative nature of the risks.

Embryonic and fetal exposures to internally deposited radionuclides present variable risks, depending on the isotope and the exposure, and have to be individually assessed. The calculations and risk assessment are frequently quite complicated.

In the field of ionizing radiation, we have a better comprehension of the biological effects and the quantitative maximum risks than for any other environmental hazard. Although animal and human data support the conclusion that there is a threshold dose for radiation-induced congenital malformations, intrauterine growth retardation, and abortion, this does not mean that there are no risks to the embryo exposed to doses less than 50 mGy. Whether there exists a linear or exponential dose-response relationship or threshold exposure for genetic, carcinogenic, cell-depleting, and life-shortening effects has not been determined. In establishing maximum permissible levels for the embryo at low exposures, refer to Tables 3.3, 3.4, 3.5, 3.6, and 3.8. It is obvious that the risks of a 10-mGy or 50-mGy dose are far below the spontaneous risks of the developing embryo because 15% of human embryos abort, 2.7% to 3.0% of human embryos have major malformations, 4% have intrauterine growth retardation, and 8% to 10% have early- or late-onset genetic disease. The maximal risk attributed to a 10-mGy dose, approximately 0.003%, is thousands of times smaller than the spontaneous risks of malformations, abortion, or genetic disease. Thus the present maximal permissible occupational dose of 5 mGy for pregnant women (see Table 3.9) is extremely conservative. Medically indicated diagnostic x-rays are appropriate for pregnant women, and there is no medical justification for terminating a pregnancy in women exposed to 50 mGy or less because of radiation exposure. On the other hand, diagnostic x-ray studies that can be replaced by ultrasound or other procedures should be avoided simply because the unnecessary use of x-ray procedures is not good medical practice. Counseling women of reproductive age should be based on sound information about the risks of radiation exposure.

REFERENCES

Ager, E., Schuman, L., Wallace, H., et al. An epidemiologic study of childhood leukemia. *J. Chron. Dis. 18:*113–132, 1965.

Blot, W. Growth and development following prenatal and childhood exposure to atomic radiation. *J. Radiat. Res. 16(suppl):*82–88, 1975.

Blot, W.J., and Miller, R.W. Mental retardation following in utero exposure to the atomic bombs of Hiroshima and Nagasaki. *Radiology 106:*617–619, 1973.

Book, S., and Goldman, M. Thyroidal radioiodine exposure of the fetus. *Health Phys. 29:*874–876, 1975.

Brent, R.L. The indirect effect of irradiation on embryonic development. II. Irradiation of the placenta. *Am. J. Dis. Child. 100:*103–108, 1960.

Brent, R.L. Medicolegal aspects of teratology. *J. Pediatr. 71:*288–289, 1967.

Brent, R.L. Effects of radiation on the foetus, newborn and child. In: *Late Effects of Radiation,* edited by R.J.M. Fry, D. Grahn, M.L. Griem, and J.H. Rust. London, Taylor & Francis, 1970, pp. 23–64.

Brent, R.L. The response of the 9½ day-old rat embryo to variations in dose rate of 150 R x-irradiation. *Radiat. Res. 45:*127–136, 1971.

Brent, R.L. Irradiation in pregnancy. In: *Davis' Gynecology and Obstetrics,* Vol. 2, edited by J.J. Sciarra. New York, Harper & Row, 1972, pp. 1–32.

Brent, R.L. Environmental factors: radiation. In: *Prevention of Embryonic, Fetal, and Perinatal Disease,* Vol. 3., edited by R.L. Brent and M.I. Harris. Bethesda, Fogarty International Center Series on Preventive Medicine, U.S. Department of Health, Education, and Welfare, DHEW Publication No. (NIH) 76–853, 1976M 00, pp. 179–197.

Brent, R.L. Litigation-produced pain, disease and suffering: an experience with congenital malformation lawsuit. *Teratology 16:*1–13, 1977a.

Brent, R.L. Radiations and other physical agents. In: *Handbook of Teratology,* Vol. 1, edited by J.G. Wilson and F.C. Fraser. New York, Plenum Press, 1977b, pp.153–223.

Brent, R.L. Radiation teratogenesis. *Teratology 21:*218–298, 1980.

Brent, R.L. The effects of ionizing radiation, microwaves and ultrasound in the developing embryo: clinical interpretations and applications of the data. In: *Current Problems in Pediatrics,* Vol. 14, No. 9, edited by J.C. Lockhart. Chicago, Year Book Medical Publishers, 1984, pp. 1–87.

Brent, R.L. The effect of embryonic and fetal exposure to x-ray, microwaves, and ultrasound: counseling the pregnant and nonpregnant patient about these risks. *Semin. Oncol. 16:*347–369, 1989.

Brent, R.L., and Bolden, B.T. The long-term effects of low-dosage embryonic irradiation. *Radiat. Res. 14:*453–454, 1961.

Brent, R.L., and Bolden, B.T. The indirect effect of irradiation on embryonic development. III. The contribution of ovarian irradiation, oviduct irradiation, and zygote irradiation to fetal mortality and growth retardation in the rat. *Radiat. Res. 30:*759–773, 1967a.

Brent, R.L., and Bolden, B.T. The indirect effect of irradiation on embryonic development. The lethal effects of maternal irradiation on the first day of gestation in the rat. *Proc. Soc. Exp. Biol. Med. 125:*709–712, 1967b.

Brent, R.L., and Bolden, B.T. Indirect effect on x-irradiation on embryonic development. V. Utilization of high doses of maternal irradiation on the first day of gestation. *Radiat. Res. 36:*563–570, 1968.

Brent, R.L., and Gorson, R.O. Radiation exposure in pregnancy. In: *Current Problems in Radiology,* Vol. 2, edited by R.D. Moseley, D.H. Baker, R.O. Gorson, et al. Chicago, Year Book Medical Publishers, 1972, pp. 1–48.

Brent, R.L., and McLaughlin, M.M. The indirect effect of irradiation on embryonic development. I. Irradiation of the mother while shielding the embryonic site. *Am. J. Dis. Child. 100:*94–102, 1960.

Brizzee, K.R., and Brannon, R.B. Cell recovery in foetal brain after ionizing radiation. *Int. J. Radiat. Biol. 21:*375–378, 1972.

Burrow, G., Hamilton, H. Hrubec, Z., et al. Study of adolescents exposed in utero to the atomic bomb. Nagasaki, Japan. I. General aspects: clinical laboratory data. *Yale J. Biol. Med. 36:*430–444, 1964.

Coppenger, C.J., and Brown, S.O. The gross manifestations of continuous gamma irradiation on the prenatal rat. *Radiat. Res. 31:*230–242, 1967.

Cowen, D., and Geller, L.M. Long-term pathological effects of prenatal x-irradiation on the central nervous system of the rat. *J. Neuropathol. Exp. Neurol. 19:*488–527, 1960.

Dekaban, A.S. Abnormalities in children exposed to x-irradiation during various stages of gestation: tentative timetable of radiation injury to the human fetus. *J. Nucl. Med. 9:*471–477, 1968.

Diamond, E.L., Schmerler, H., and Lilienfeld, A.M. The relationship of intrauterine radiation to subsequent mortality and development of leukemia in children. A prospective study. *Am. J. Epidemiol. 97:*283–313, 1973.

Fasal, E., Jackson, E.W., and Klauber, M.R. Birth characteristics and leukemia in children. *J. Natl. Cancer Inst. 47:*501–509, 1971.

Ford, D., and Patterson, T. Fetal exposure to diagnostic X-rays and leukemia and other malignant diseases in childhood. *J. Natl. Cancer Inst. 22:*1093–1104, 1959.

Furchgott, E. Behavioral effects of ionizing radiations, 1955–1961. *Psychol. Bull. 60:*157–200, 1963.

Gentry, J., Parkhurst, E., and Bulin, G. An epidemiological study of congenital malformations in New York State. *Am. J. Public Health 49:*497, 1959.

Goldstein. L., and Murphy, D.P.L. Etiology of ill health in children born after maternal pelvic irradiation. II. Defective children born after postconceptional maternal irradiation. *AJR 22:*322–331, 1929a.

Goldstein, L., and Murphy, D.P. Microcephalic idiocy following radium therapy for uterine cancer during pregnancy. *Am. J. Obstet. Gynecol. 18:*189–195, 281–283, 1929b.

Graham, S., Levin, M.I., Lilienfeld, A.M., et al. Preconception, intrauterine and postnatal irradiation as related to leukemia. *Natl. Cancer Inst. Mongr. 19:*347–371, 1966.

Grahn, D., and Kratchman, J. Variation in neonatal death rate and birth rate in United States and possible relations to environmental radiation, geology and attitude. *Am. J. Hum. Genet. 15:*329–352, 1963.

Hicks, S.P., and D'Amato, C.J. Effects of ionizing radiation on mammalian development. In: *Advances in Teratology,* Vol. 1, edited by D.H.M. Wollam. London, Logos Press, 1966, pp. 196–243.

Hoshino, T., Itoga, T., and Kato, H. Leukemia in the offspring of parents exposed to the atomic bomb at Hiroshima and Nagasaki. Paper read before the Japanese Association of Hematology, March 28–30, 1965.

Jablon, S. Comments. *Health Phys. 24:*257–258, 1973.

Jacobsen, L., and Mellemgaard, L. Anomalies of the eyes in descendants of women irradiated with small x-ray doses during age of fertility. *Acta Ophthalmol. (Copenh.) 46:*352–334, 1968.

Jacobson, A.G., and Brent, R.L. Radioiodine concentration by the fetal mouse thyroid. *Endocrinology 65:*408–416, 1959.

Kato, H. Mortality in children exposed to the A-bombs while in utero. *Am. J. Epidemiol. 93:*435–442, 1971.

Kereiakes, J.G., Feller, P.A., Ascoi, F.A., et al. Pediatric radiopharmaceutical dosimetry. In: *Radiopharmaceutical Dosimetry Symposium.* Rockville, MD, U.S. Department of Health, Education and Welfare, HEW Publication No. (FDA) 76–8044, Bureau of Radiological Health, Food and Drug Administration, l976.

Kinlen, L.J., and Acheson, F.D. Diagnostic irradiation, congenital malformations, and spontaneous abortion. *Br. J. Radiol. 41:*648–654, 1968.

Konerman, G. Die Keimesentwicklung der Maus nach Einwirkung wontinuierlicher Co^{60} Gammabestrahlung wahrend der Blastogenese, der Organogenese und fetalen period. *Strahlentherapie 137:*451–466, 1969.

Kriegal, H., and Langendorff, H. Wirkung einer fraktionierten Roentgenbestrahlung auf die Embryonalentwicklung der Maus. *Strahlentherapie 123:*429–437, 1964.

Laskey, J.W., Parrish, J.B., and Cahill, D.F. Some effects of lifetime parental exposure to low levels of tritium on the F_2 generation. *Radiat. Res. 56:*171–179, 1973.

Lewis, E.B. Leukemia and ionizing radiation. *Science 125:*865–972, 1957.

Lilienfeld, A.M. Epidemiological studies of the leukemogenic effects of radiation. *Yale J. Biol. Med. 39:*143–164, 1966.

Maroteaux, P., Spranger, J., Opitz, J.M., et al. Le syndrome campomelique. *Presse Med. 22:*1157–1162, 1971.

McMahon, B., and Hutchinson, G.B. Prenatal x-ray and childhood: a review. *Acta Union Int. Contra Cancrum 20:*1172–1174, 1964.

Miller, R.W. Delayed radiation effects in atomic bomb survivors. *Science 166:*569–574, 1969.

Miller, R.W. Epidemiological conclusions from radiation toxicity studies. In: *Late Effects of Radiation,* edited by R.J.M. Fry, D. Grahn, M.L. Griem, et al. London, Taylor & Francis, 1970, pp. 245–256.

Miller, R.W., and Mulvihill, J.J. Small head size after atomic irradiation. Teratogen update. *Teratology 14:*355–358, 1976.

Murphree, R., and Pace, H. The effects of prenatal radiation on postnatal development in the rat. *Radiat. Res 12:*495–504, 1960.

Murphy, D.P., and Goldstein, L. Micromelia in a child irradiated in utero. *Surg. Gynecol. Obstet. 40:*79–80, 1930.

National Council on Radiation Protection and Measurements. *Basic Radiation Criteria,* NCRP Report No. 39. Washington, DC, National Council on Radiation Protection and Measurements, 1971.

National Research Council. *The Effects of Populations of Exposure to Low Levels of Ionizing Radiation (BEIR I).* Washington, DC, National Academy Press, 1972.

National Research Council. *Health Effects of Exposure to Low Levels of Ionizing Radiation: BEIR V.* Washington, DC, National Academy Press, 1990.

Neifakh, A.A. Role of the maternal organism in the irradiation illness of fetal mice. *Dokl. Akad. Nauk. BSS. 116:*821–824, 1957.

Neutel, C.I., and Buck, C. Effects of smoking during pregnancy on the risk of cancer in children. *J. Natl. Cancer Inst. 47:*59–64, 1971.

Nokkentred, K. *Effect of Radiation Upon the Human Fetus.* Copenhagen, Munksgaard, 1968.

Okada, S., Hamilton, J.B., Egami, N., et al. A review of thirty-year study of Hiroshima and Nagasaki atomic bomb survivors. *J. Radiat. Res. (Tokyo) (suppl.):16:*89–96, 1975.

Otake, M., and Schull, W. In utero exposure to A-bomb radiation and mental retardation. A reassessment. *Br. J. Radiol. 57:*409–414; 1984.

Piontkovskii, I.A. Certain properties of the higher nervous activity in adult animals irradiated prenatally by ionizing radiations: the problem of the effect of ionizing irradiation on offspring. *Bull. Exp. Biol. Med. (Russian) 46:*77–80, 1958.

Piontkovskii, I.A. Some peculiarities of higher nervous activity in adult animals subjected to radiation in utero, pt. 3. *Bull. Exp. Biol. Med. (Russian) 51:*24–31, 1961.

Plummer, G. Anomalies occurring in children exposed in utero to the atomic bomb in Hiroshima. *Pediatrics 10:*687–692, 1952.

Polhemus, D., Koch, R. Leukemia and medical irradiation. *Pediatrics 23:*453–461, 1969.

Roedler, H.D., Kaul, A., and Hein, G.J. *Internal Radiation Dose in Diagnostic Nuclear Medicine.* Berlin, Verlag H. Hoffman, 1978.

Ronderos, A. Fetal tolerance to radiation. *Radiology 76:*454–456, 1961.

Ronnback, C. Effects of continuous irradiation during gestation and suckling period of mice. *Acta Radiol. Ther. 3:*169–176, 1965.

Ross, M.H., and Bras, G. Tumor incidence patterns and nutrition in the rat. *J. Nutr. 87:*245–260, 1965.

Rugh, R. Major radiobiological concepts and ionizing radiation on the embryo and fetus. In: *Response of the Nervous System to Ionizing Radiation,* Vol. 3, edited by T. Haley and R. Snider. New York, Academic Press, 1962.

Rugh, R. Effect of ionizing radiations, including radioisotopes, on the placenta and embryo. *Birth Defects 1:*64–73, 1965.

Rugh, R., Duhamel, L., and Skaredoff, L. Relation of embryonic and fetal x-irradiation to life-time average weights and tumor incidence in mice. *Proc. Soc. Exp. Biol. Med. 121:*714–718, 1966.

Rugh, R., and Wohlfromm, M. Can x-irradiation prior to sexual maturity affect the fertility of the male mammal (mouse)? *Atompraxis 10:*33–42, 1964a.

Rugh, R., and Wohlfromm, M. X-irradiation sterilization of the premature female mouse. *Atompraxis 10:*511–518, 1964b.

Russell, L.B. X-ray-induced developmental abnormalities in the mouse and their use in analysis of embryological patterns. I. External and gross visceral changes. *J. Exp. Zool. 114:*345–602, 1950.

Russell, L.B. X-ray-induced developmental abnormalities in the mouse and their use in analysis of embryological patterns. II. Abnormalities of the vertebral column and thorax. *J. Exp. Zool., 131:*329–395, 1956.

Russell, L.B., Badgett, S.K., and Saylors, C.L. Comparison of the effects of acute, continuous and fractionated irradiation during embryonic development. In: *Repair from Genetic Radiation,* edited by F.H. Sobels. New York, Pergamon Press, 1963, pp. 333–342.

Russell, L.B., and Major, M.H. Radiation-induced presumed somatic mutations in the house mouse. *Genetics 42:*161–175, 1957.

Russell, L.B., and Russell, W.L. The effects of radiation on the preimplantation stages of the mouse embryo. *Anat. Res. 108:*521, 1950.

Russell, L.B., and Russell, W.L. An analysis of the changing radiation response of the developing mouse embryo. *J. Cell Comp. Physiol. 43:*103–149, 1954.

Russell, L.B., and Saylors, C.L. The relative sensitivity of various germ-cell stages of the mouse to radiation-induced nondysfunction, chromosome losses and deficiencies. In: *Repair from Genetic Radiation,* edited by F.H. Sobels. New York, Pergamon Press, 1963, pp. 313–342.

Saenger, E.L. *Protocol Book.* Cincinnati, Radioisotope Laboratory, University of Cincinnati, Medical Center, 1976.

Segall, A., MacMahon, B., and Hannigan, M. Congenital malformations and background radiation in northern New England. *J. Chron. Dis. 17:*915–932, 1964.

Sikov, M.R., and Noonan, T.R. Anomalous development induced in embryonic rat by the maternal administration of radiophosphorus. *Am. J. Anat. 103:*137–162, 1958.

Stadler, J., and Gowen, J.W. Observations on the effects of continuous irradiation over 10 generations on reproductivities of different strains of mice. In: *Effects of Ionizing Radiation on Reproductive Systems,* edited by W.D. Carlson and F.X. Gassner. New York, Pergamon Press, 1964, pp. 111–122.

Stewart, A. Myeloid leukaemia and cot deaths. *Br. Med. J. 4:*423, 1972.

Stewart, A. The carcinogenic effects of low-level radiation: a reappraisal of epidemiologists' methods and observations. *Health Phys. 24:*223–240, 1973.

Stewart, A., and Kneale, G.W. Radiation dose effects in relations to obstetric X-rays and childhood cancers. *Lancet 1:*1185–1188, 1970.

Stewart, A., Webb, D., Giles, D., et al. Malignant disease in childhood and diagnostic irradiation in utero. *Lancet 2:*447, 1956.

Stewart, A., Webb, D., and Hewitt, D. A survey of childhood malignancies. *Br. Med. J. 1:*1495–1508, 1958.

Tabuchi, A. Fetal disorders due to ionizing radiation. *Hiroshima J. Med. Sci. 13:*125–173, 1964.

Tabuchi, A., Nakagawa, S., Hirai, T., et al. Fetal hazards due to x-ray diagnosis during pregnancy. *Hiroshima J. Med. Sci. 16:*49–66, 1967.

Vilumsen, A. *Environmental Factors in Congenital Malformations.* Copenhagen, FDAL's Forlag, 1970.

Vorisek, P. Einfluss der kontinuierlichen intrauterinen Bestrahlung auf die perinatale Mortalitat der Frucht. *Strahlentherapie 127:*112–120, 1965.

Wesley, J.P. Background radiation as a cause of congenital malformations. *Int. J. Radiat. Biol. 2:*97–118, 1960.

Wood, J.W., Johnson, K.G., and Omori, Y. In utero exposure to Hiroshima atomic bomb. An evaluation of head size and mental retardation-twenty years later. *Pediatrics 39:*385–392, 1967a.

Wood, J., Johnson, K., Omori, Y., et al. Mental retardation in children exposed in utero to the atomic bombs in Hiroshima and Nagasaki. *Am. J. Public Health 57:*1381–1390, 1967b.

Wood, J., Keehn, R., Kawamoto, S., et al. The growth and development of children exposed in utero to the atomic bombs in Hiroshima and Nagasaki. *Am. J. Public Health* 57:1374–1380, 1967c.

Wright, F.W. Diagnostic radiology and the fetus. *Br. Med. J. 3:*693–694, 1973.

Yamazaki, J., Wright, S., and Wright, P. Outcome of pregnancy in women exposed to the atomic bomb in Nagasaki. *Am. J. Dis. Child. 87:*448–463, 1954.

Zappert, J. Uber roentgenogene fetale microcephalie. *Monatsschr. Kinderheilkd. 34:*490–493, 1926.

Chapter 4

Radiation Exposure in Pregnancy: Health Physics Considerations

KENNETH L. MOSSMAN

ABSTRACT

Radiation exposure of the pregnant patient or worker can create significant challenges for the practicing health physicist. The purpose of this paper is to discuss some radiation protection concerns in the management of the exposed pregnant worker or patient. Dose assessment, risk assessment (including evaluation of competing risks), and legal issues are addressed.

INTRODUCTION

Radiation exposure of the pregnant patient or worker can be one of the most challenging problems in health physics practice. In many occupational and medical exposure situations, pregnancy status is unknown when exposure(s) occur. In some medical situations, conceptus irradiation occurs with the knowledge that the patient is pregnant. In such cases, special precautions regarding shielding and modification of planned procedures can be followed to minimize the dose to the fetus without compromising the value of the study. Exposure of the embryo or fetus is a serious problem not only because of the potential biological hazards but also because of the concerns and anxieties of the prospective parents. When a woman learns that radiation exposure has occurred or will occur during pregnancy, serious concerns about the health of the baby arise. Resolution of a particular case may be complex because of the interplay of medical, legal, and personal factors.

The purpose of this paper is to discuss some practical health physics problems in the management of the exposed pregnant worker or patient. Specifically, dose assessment, risk assessment (including evaluation of competing risks), and legal issues are addressed in the context of medical and occupational exposures. Typically, health physicists are involved in dose reconstruction and other aspects of radiation risk assessment. Medical physicists are also frequently involved in dose assessment, particularly for patients. The health physicist's or medical physicist's participation in a particular pregnancy case may also include assistance in patient/worker counseling, and in the review and analysis of employment and other institutional policies, including interpretation of applicable laws and regulations.

This paper is concerned only with x-rays and other ionizing radiations. However, many pregnant patients have ultrasound and/or magnetic resonance imaging procedures involving nonionizing radiation, and many pregnant women are exposed to nonionizing radiation in the workplace environment. The risks of nonionizing radiation exposure are not discussed in this paper.

DOSE ASSESSMENT

Radiation dose and gestational age at exposure are the principal determinants of radiogenic effects in the embryo and fetus. The time during gestation when the exposure occurred determines the type(s) of radiation effects to be expected; the embryo/fetal dose determines the probability of occurrence of effects (Brent 1980, Mossman and Hill 1982). Radiation exposure of the embryo or fetus occurs primarily as a result of medical diagnostic or therapeutic procedures or in occupational situations in which a worker is exposed to radioactive materials or to radiation from radiation-producing equipment in the work environment.

Estimation of radiation dose may carry substantial uncertainties because of the need to do retrospective calculations. This problem is especially relevant in medical radiodiagnostic situations where accurate records of exposure variables may not be kept, which necessitates use of generic technique charts. Other factors such as differing x-ray output from machines of various manufacturers and use of published values of exposure and dose may also contribute to errors in dose estimation (Mossman and Hill 1982). In many situations the radiographic procedure may have been repeated for technical reasons. Often records of these repeated procedures are not kept. Many, if not most, radiographic procedures are performed with automatic exposure termination of the radiation, and exposure time may not be accessible on some equipment. Fluoroscopy poses an additional problem in dose assessment because of the lack of a record (e.g., permanent images) of the anatomical sites exposed and the irradiation times.

Assumptions should be conservative in the dose calculation so that the embryo or fetal dose is not underestimated. The reasonableness of these assumptions strongly influences the accuracy of the dose calculations (Mossman and Hill 1982). Calculations of dose should include an error estimate to characterize the precision of the dose assessment. Methods for estimating fetal dose following diagnostic x-ray procedures or nuclear medicine procedures have been previously published (Cloutier et al., 1984, Husak and Wiedermann 1980, Jacobson and Conley 1976, Kereiakes and Rosenstein 1980, National Council on Radiation Protection and Measurements [NCRP] 54 1977, Wagner et al., 1985).

Typical values of radiation doses to the embryo/fetus for various diagnostic x-ray and nuclear medicine procedures are shown in Tables 4.1 and 4.2. Doses for actual radiodiagnostic procedures may vary greatly from values in Table 4.1 because of differences in the x-ray equipment and the exposure factors used. Table 4.1 refers to radiographic procedures only and not fluoroscopy. Depending on the procedure, fluoroscopy may add substantially to the overall dose. For example, the total dose from an upper gastrointestinal tract or barium enema examination is higher than indicated in Table 4.1 because fluoroscopy is a usual part of these studies. For most radiodiagnostic procedures, fetal doses (excluding

TABLE 4.1. EMBRYO DOSES FROM DIAGNOSTIC RADIOGRAPHY (EXCLUDING FLUOROSCOPY)[a]

Selected Examinations[b]	Images[c]	Embryo Dose (mGy)
Cholecystogram	3.2	0.1
Lumbar spine	2.9	4.1
Upper gastrointestinal tract	4.3	0.5
Kidney, ureters, bladder	1.7	2.6
Barium enema	4.0	8.2
Lumbosacral spine	3.4	6.4
Intravenous pyelogram	5.5	8.2
Pelvis	1.3	2.0
Hip	2.0	1.3

[a] From Hall EJ. Radiobiology for the radiologist, 3rd ed. Philadelphia: JB Lippincott, 1988.
[b] Certain examinations, including chest, skull, cervical spine, ribs, shoulder, thoracic spine, breast, and oral cavity, provide negligible dose to the embryo/fetus.
[c] Average number of films per examination.

TABLE 4.2. DOSE TO EMBRYO FROM VARIOUS RADIOPHARMACEUTICALS[a]

Radiopharmaceutical	Activity (MBq)	Embryo Dose (mGy)
Tc-99m pertechnetate	370	3
Tc-99m polyphosphate	370	3
Tc-99m DTPA	370	4
Tc-99m sulfur colloid	74	<1
Ga-67 citrate	74	5
Se-75 methionine	9	10

[a] From Husak V, Wiedermann M. Radiation absorbed dose estimates to the embryo from some nuclear medicine procedures. Eur J Nucl Med 5:205–207, 1980.

procedures with long fluoroscopy times) are less than 10 mGy. For nuclear medicine procedures (Table 4.2), the major factors determining the fetal dose include the radionuclide used, its chemical form, total activity, and placental transfer and other maternal metabolic characteristics. Furthermore, a patient's physiological condition may have a marked effect on the uptake of particular radiopharmaceuticals in specific tissues (including the fetal compartment), thus modifying the embryo dose shown in Table 4.2.

In occupational settings, exposure simulations may be necessary to supplement dose estimates from personal dosimetry and bioassay data. Furthermore, estimation of health effects may be difficult when a number of bioeffects must be considered because fetal exposure was protracted over a significant portion of the gestation period.

RISK ASSESSMENT

Risk assessment in the evaluation of radiation exposure in pregnancy is not limited to the numerical determination of radiation dose and the corresponding radiation risk. A complete risk assessment includes consideration of maternal exposure to other teratogenic and carcinogenic agents and an analysis of the probability of naturally occurring adverse pregnancy outcomes. In the medical setting, the patient's medical condition may need to be considered as a potential pregnancy risk. Furthermore, any discussion of the radiation contribution to the total risk should include an evaluation of the uncertainties in radiation dose and risk coefficients.

The principal health effects in humans following in utero radiation exposure are postnatal neoplasia and mental retardation (National Research Council 1990, United Nations Scientific Committee on the Effects of Atomic Radiation [UNSCEAR] 1986). The probability of occurrence of these conditions depends primarily on the embryo/fetal dose and the time during gestation at which the exposure(s) occurred. It is unclear whether these effects are stochastic in nature or are associated with a small threshold dose (National Research Council 1990, UNSCEAR 1986). Other radiogenic effects, such as growth retardation and congenital malformations, are nonstochastic in nature and are believed to be associated with a threshold dose of at least 50 mGy (Brent 1980). Risk of radiation-induced mental retardation or postnatal neoplasia is not an all-or-none proposition. No matter what the dose level, the occurrence, or lack of occurrence, of these effects cannot be predicted with certainty.

Radiogenic effects, such as postnatal neoplasia, may occur in the absence of radiation. Furthermore, they are indistinguishable from effects that occur naturally. Consequently, it is impossible to conclude, in an individual case, that a particular effect was caused by radiation. Because developmental effects may have multiple etiologies, risk assessment must include an evaluation of other possible risk factors in addition to the radiation exposure in question. For example, as shown in Table 4.3, intrauterine growth retardation (IUGR), malformations, mental retardation, and postnatal neoplasia can occur naturally or

TABLE 4.3. RISKS IN PREGNANCY[a]

Effect	Risk (Cases/1000 Live Births)	References
Natural Incidence		Mossman and Hill 1982, UNSCEAR 1986, Young et al. 1986
IUGR[a]	20–30	
Malformations	60	
Mental retardation	10	
Cancer in childhood	0.13	
Radiation		UNSCEAR 1986, National Research Council 1990
Mental retardation[b]	0.4/mGy at 8–15 wk 0.1/mGy at 15–25 wk	
Cancer in childhood	0.02/mGy	
Other risks[c]		Mossman and Hill 1982
IUGR–smoking	40–60	
IUGR–alcohol	300–500	

[a] IUGR is defined as birth weight less than 2500 g in full-term infants.

[b] The risk of mental retardation is four times higher if exposure occurs at 8–15 weeks of gestation rather than at 15–25 weeks of gestation. Exposure at other times during in utero development is not associated with mental retardation.

[c] Heavy indulgence in alcohol and smoking during pregnancy.

as a result of radiation exposure or heavy indulgence in social activities such as smoking and drinking. A numerical comparison of the risks in Table 4.3 indicate that radiation risks usually encountered occupationally or in medical tests are small compared with other risks of pregnancy. A complete risk assessment should involve a review of the individual's medical records, previous pregnancy history, and social activities such as cigarette and alcohol consumption (Mossman and Hill 1982). Although it may be impossible to quantify some or all of these risks, knowledge of their presence is valuable in understanding the total risk profile and in putting the radiation risk into perspective.

In occupational exposure settings, risk assessment should include an evaluation of potential exposures to all physical, chemical, and biological agents that may enhance the possibility of an untoward pregnancy outcome. A careful inspection of the workplace environment is needed to identify possible exposure to agents (e.g., chemicals) that may have teratogenic activities in humans. Table 4.4 lists some agents that cause, or are suspected of causing (as indicated by the question marks), malformations in humans that may be found in the general environment or in the workplace. Agents associated with other kinds of developmental effects, such as intrauterine growth retardation and functional abnormalities, are not included in this list.

Estimates of dose and risk coefficients have substantial uncertainties associated with them. It must be recognized that the uncertainty associated with the product of these factors (the total radiation risk) is even larger than the uncertainty associated with either term alone because of propagation of errors. For example, if the dose estimate and risk coefficient each have fractional errors of 100% (i.e., the standard deviation is equal to the mean), the fractional error associated with the total radiation risk (the product of the dose estimate and risk coefficient) will be 141%. In evaluating the radiation risk in the context of other risks, careful consideration must be given to the uncertainties.

LEGAL AND POLICY CONSIDERATIONS

The sensitivity of the embryo and fetus to ionizing radiation injury necessitates consideration of fetal protection policies in the workplace and in medical applications. Accordingly, various policies, recommendations, and laws have been promulgated to reduce exposure of pregnant workers and patients. It is important

TABLE 4.4. SOME TERATOGENIC AGENTS IN HUMANS[a]

Biological Agents	*Drugs*	*Environmental Agents*
Rubella virus	Busulfan	Chlorobiphenyls
Cytomegalovirus	Cyclophosphamide	Mercury, Organic lead?
Toxoplasmosis	Diethylstilbestrol	Diphenylhydantoin
	Lithium	
Social Factors	Penicillamine	
Alcoholism	Tetracycline	
Cigarette smoking?	Thalidomide	
Cocaine?	Streptomycin?	

[a] From Shepard TH. Catalog of teratogenic agents, 6th ed. Baltimore: The Johns Hopkins University Press, 1989.

to distinguish between fetal irradiation from medical procedures and exposures as a consequence of occupational activities. Regulations exist that limit occupational exposures; patient exposures from medical procedures are not regulated.

Medical Exposures

In radiographic examinations of the mother, the embryo/fetus may be incidentally exposed; doses from such examinations are less than 10 mGy (Tables 4.1 and 4.2) and pose little, if any, risk to the developing embryo or fetus. In fluoroscopy studies, doses to the conceptus may exceed 10 mGy. However, rarely does the total dose reach 50 mGy, believed to be the threshold dose for nonstochastic effects such as growth retardation and malformations (Brent 1980). In therapeutic radiological procedures (e.g., for the treatment of cancer), the embryo/fetus may receive large doses even though the uterine contents may not be in the primary radiation beam. Because of the substantial probability of adverse pregnancy outcomes under such circumstances, radiation oncologists may postpone treatment until the end of pregnancy, if possible, or recommend termination of the pregnancy.

In the medical arena, the clinical use of radiation in diagnosis and therapy is based primarily on professional standards of medical care and judgment in individual cases. However, federal regulations exist that govern various aspects of the use of radiation-producing equipment in medicine. By authority of the Radiation Control for Health and Safety Act of 1968, the Food and Drug Administration regulates quality control of radiation-producing equipment, including x-ray machines and accelerators (Food and Drug Administration 21 CFR 1000, 1990). The Nuclear Regulatory Commission regulates reactor but not accelerator produced byproduct materials used in diagnostic imaging and therapy procedures (U.S. Nuclear Regulatory Commission 10 CFR 35, 1990).

Radiology and nuclear medicine examination request forms frequently address pregnancy status to eliminate or reduce the possibility of ionizing radiation exposure of the pregnant patient. The decision to perform a particular radiological or nuclear medicine procedure should be made on a case-by-case basis, with careful consideration of all risks (including the possibility of pregnancy) of the procedure.

In human subjects research, pregnant females are a protected class. Federal law requires that pregnant women be excluded from participation in clinical trials or human research protocols (including studies using diagnostic or therapeutic radiological procedures) except when the purpose of the activity is to meet the health needs of the mother or the risk to the fetus is minimal (Department of Health and Human Services 45 CFR 46, 1990).

Occupational Exposures

The question of permissible occupational exposures to pregnant workers has been addressed in a number of recommendations, guidelines, and laws. NCRP Report 53 provides a history (to 1977) of recommendations by the NCRP and the International Commission on Radiological Protection (ICRP) regarding dose limits to the embryo/fetus (NCRP 53 1977b). Because of the enhanced radiosensitivity of the embryo and fetus, the NCRP recommended that the dose equivalent to the embryo/fetus should not exceed 5 mSv for the entire pregnancy period. In this same report, the NCRP recommended that fertile women should be employed in occupational settings in which the annual dose equivalent would not be likely to exceed 20–30 mSv (NCRP 39 1971). The 1987 Presidential Guidance specifies that "the dose to an unborn as a result of occupational exposure of a woman who has declared that she is pregnant should be maintained as low as reasonably achievable, and in any case should not exceed 5 mSv during the entire gestation period." The guidance goes on further to say that efforts should be made to avoid substantial variations in the monthly exposure rate (U.S. Environmental Protection Agency 52 FR 2822, 1987). NCRP Report #91, which replaced NCRP Report 39, also recommends the 5-mSv limit with uniform exposure rate (NCRP 91 1987).

The recently revised 10 CFR 20 adopted the 1987 Presidential Guidance (U.S. Nuclear Regulatory Commission 10 CFR 20 1991). Prior to this revision, the Code of Federal Regulations did not have occupational dose limits for pregnant workers. Part 20.1208, "Dose to an embryo/fetus," requires that (*a*) the dose to the embryo/fetus during the entire pregnancy, due to occupational exposure of a declared pregnant woman, not exceed 5 mSv, (*b*) substantial variation above a uniform monthly exposure rate be avoided, and (*c*) the dose to the embryo/fetus be the sum of the deep dose equivalent to the declared pregnant woman and the dose to the embryo/fetus from radionuclides in the embryo/fetus and radionuclides in the maternal compartment (U.S. Nuclear Regulatory Commission 10 CFR 20 1991).

In the occupational setting, protecting the embryo/fetus from unnecessary radiation exposure may necessitate the institution of employer policies that conflict with the statutory prohibition against special treatment of pregnant women. Title VII of the Civil Rights Act of 1964, which forbids discrimination in employment on the basis of sex (and other characteristics of protected classes), was amended in 1978, through the Pregnancy Discrimination Act, to include explicit discrimination on the basis of pregnancy, childbirth, and related medical conditions (Equal Employment Opportunity Commission 29 CFR 1604 1990). Establishing different occupational exposure limits for pregnant workers may satisfy fetal protection goals but may discriminate against the pregnant worker in the workplace. Should fetal protection policies be directed solely toward females? Developmental anomalies, functional abnormalities (e.g., mental retardation), cancer, and other effects in pregnancy are assumed to be primarily the result of toxic agent exposure during in utero development. However, recent studies by Gardner and colleagues suggest that radiation exposure of male workers may be linked to an increased incidence of leukemia in their offspring (Gardner et al. 1990). Although a link between paternal exposure and childhood cancer has not been firmly established, the possibility of a paternal-related fetal injury begs the legal question of establishing a gender-neutral fetal protection policy.

In the face of Title VII provisions, enforcement of a lower occupational exposure limit for pregnant workers presents a potentially serious legal conflict. Nuclear industry employers have a legitimate concern. Available radiobiology data suggest that the unborn child is more radiosensitive than the adult and, therefore, that additional protection should be afforded the fetus. Under the federal Occupational Safety and Health Act, employers have an obligation to provide a safe working environment. However, imposing employment restrictions to meet fetal protection goals may result in violation of employee rights as stipulated in Title VII. Thus policies to protect the health of unborn children would seem to require policies that restrict employment opportunities for women. The Presidential Guidance of 1987 directly addressed the pregnancy discrimination question. The guidance noted that "the limiting value for the unborn does not create a basis for discrimination, and should be achieved in conformance with the provisions of Title VII of the Civil Rights Act of 1964, as amended, regarding discrimination in employment practices." (U.S. Environmental Protection Agency 52 FR 2822 1987).

The key provision in the Presidential Guidance, and in the newly revised Part 20, is that the pregnancy be declared (U.S. Environmental Protection Agency 52 FR 2822 1987, U.S. Nuclear Regulatory Commission 10 CFR 20 1991). Only when the pregnancy is declared shall the employer ensure that the dose equivalent to the embryo/fetus does not exceed 5 mSv. Accordingly, the law does not appear to require the employer to inquire as to pregnancy status (and thus infringe on the employee's right to privacy) and leaves to the judgment of the pregnant worker the decision to declare the pregnancy and thus be subject to the administrative limit. Implementation of the declaration provision raises a number of interesting questions, including: (*a*) What if the pregnancy is never declared? Can regulatory requirements be enforced (or claims made) in the absence of a voluntary, written declaration of pregnancy? (*b*) Does the voluntary disclosure of pregnancy preclude the employer's asking about pregnancy status, that is, from affirmatively seeking the declaration? (*c*) Can an employer be liable for fetal injury because of failing to know of an undeclared pregnancy that may be deemed to be reasonably apparent or knowable?

Of particular interest to the nuclear industry in establishing policies governing pregnant worker rights is the recently decided Johnson Controls case (*Automobile Workers v. Johnson Controls, Inc.* 1991). This case has been widely regarded as one of the most significant tests of women's rights in the workplace since the 1964 Civil Rights Act. Johnson Controls, Inc., is a battery-manufacturing firm. A primary ingredient in the manufacturing process is lead, occupational exposure to which entails health risks, including the risk of harm to any fetus carried by a female employee. Respondent established a policy barring all women except those whose infertility was medically documented from jobs involving actual or potential lead exposure exceeding the Occupational Safety and Health Administration (OSHA) standard. Petitioners, a group including employees affected by respondent's fetal-protection policy, filed a class action suit claiming the policy constituted sex discrimination in violation of Title VII of the Civil Rights Act of 1964, as amended. The U.S. Supreme Court found in favor of the petitioners, holding that Title VII, as amended by the Pregnancy Discrimination Act, forbids sex-specific fetal-protection policies. The key test in the decision was the so-called BFOQ (bona fide occupational qualification) provision of Title VII, which allows an employer to discriminate on the basis of sex "in those certain instances where . . . sex . . . is a [BFOQ] reasonably necessary to the normal operation of the particular business." The Pregnancy Discrimination Act states that unless pregnant employees differ from others "in ability or inability to work," they must be "treated the same" as other employees "for all employment related purposes." The Supreme Court found that the respondent could not establish a BFOQ. Furthermore, the Court suggested that an employer's tort liability for potential fetal injuries ought not be affected by this decision. If the employer fully informs the woman of the risk, and the employer has not acted negligently, the basis for holding the employer liable would seem remote (*Automobile Workers v. Johnson Controls, Inc.* 1991). The upshot of this case is that women may not be forced to choose between having a child and having a job and that the decisions about the health and welfare of children must be left to the parents. Furthermore, female employees should not be subjected to different terms and conditions of employment based on gender.

Employers must be careful in formulating policies regarding pregnant workers in order to avoid conflict with laws governing privacy and discrimination. A recent labor arbitration case involving a worker who was denied overtime work illustrates the point (Kelly 1990). In this case, a worker who had the potential for radiation exposure in excess of 1.5 mSv was required, as a matter of policy, to certify that she was not pregnant in order to work overtime. She refused to sign the certification and was denied overtime work. The employee filed a grievance charging that she was refused the opportunity to work overtime because of her refusal to sign the certification. She further claimed that the policy was an unnecessary invasion of privacy and violated the union contract and the firm's affirmative action policies. Management, on the other hand, claimed it had a responsibility to protect the unborn and that the company had a right to protect itself against future civil liabilities that could arise from untoward pregnancy outcomes to occupationally exposed female employees. The arbitration board decided in favor of the employee, claiming that the policy provided no additional assurance of fetal protection and that it interfered with the union contract clause providing for equalization of overtime. As a consequence, the employer revised its policy by eliminating the requirement for pregnancy status certification (Kelly 1990).

The company's concern, in this case, for tort liability apparently is not compelling as reflected in the Supreme Court opinion in the subsequent Johnson Controls decision. The Supreme Court noted that if the employer fully informs the woman of the risks and does not act in a negligent fashion, there is little basis, if any, for holding the employer liable (*Automobile Workers v. Johnsons Controls 1991*). The Presidential Guidance of 1987 addresses the question of informing workers of radiation risks. Individuals who are occupationally exposed to radiation and managers of activities involving radiation should be instructed on health risks from ionizing radiation exposure, including in utero effects (U.S. Environmental Protection Agency 52 FR 2822 1987).

SUMMARY

Several radiation protection concerns in the management of the exposed worker or patient have been reviewed. The major points developed in the discussion are as follows:

1. Dose assessment and risk assessment may carry substantial uncertainties. When calculating dose, assumptions should be reasonable so as not to seriously underestimate or overestimate dose. When assessing risk, uncertainties in factors such as dose and risk coefficients must be carefully considered.
2. A complete risk assessment should include consideration of maternal exposures to other teratogenic and carcinogenic agents and an analysis of the probability of naturally occurring adverse pregnancy outcomes.
3. Fetal protection policies must be carefully formulated and followed to avoid conflict with laws governing privacy and discrimination.

REFERENCES

*Automobile Workers v. Johnson Controls, Inc. Law Week 59:*4209–4219, 1991.

Brent, R. Radiation teratogenesis. *Teratology 21:*281–298, 1980.

Cloutier, R.J., Watson, E.E., and Coffey, J.L. Radiopharmaceutical dose calculations. In: *Textbook of Nuclear Medicine*, Vol. 1: *Basic Science*, edited by J.W. Harbert and A.F.G. DaRocha. Philadelphia, Lea & Febiger, 1984, pp. 267–282.

Department of Health and Human Services. *Protection of Human Subjects.* Washington, DC, U.S. Government Printing Office, Title 45, Code of Federal Regulations, Part 46 (45 CFR 46), revised as of 1 October 1990.

Equal Employment Opportunity Commission. *Guidelines on Discrimination Because of Sex.* Washington, DC, U.S. Government Printing Office, Title 29, Code of Federal Regulations, Part 1604 (29 CFR 1604), revised as of 1 July 1990.

Food and Drug Administration. *Radiological Health.* Washington, DC, U.S. Government Printing Office, Title 21, Code of Federal Regulations, Part 1000 (21 CFR 1000), revised as of 1 April 1990.

Gardner, M.J., Snee, M.P., Hall, A.J., Powell, C.A., Downes, S., and Terrell, J.D. Results of case-control study of leukaemia and lymphoma among young people near Sellafield nuclear plant in West Cumbria. *Br. Med. J. 300:*423–429, 1990.

Hall, E.J. *Radiobiology for the Radiologist,* 3rd Ed. Philadelphia, JB Lippincott, 1988.

Husak, V., and Wiedermann, M. Radiation absorbed dose estimates to the embryo from some nuclear medicine procedures. *Eur. J. Nucl. Med.* 5:205–207, 1980.

Jacobson, A., and Conley, J. Estimation of fetal dose to patients undergoing diagnostic x-ray procedures. *Radiology 120:*683–685, 1976.

Kelly, J.J. Prenatal radiation exposure policy: a labor arbitration. *Health Physics 59:*103–107, 1990.

Kereiakes, J.G., and Rosenstein, M. *Handbook of Radiation Doses in Nuclear Medicine and Diagnostic X-Ray.* Boca Raton, FLA, CRC Press, 1980.

Mossman, K.L., and Hill, L.T. Radiation risks in pregnancy. *Obstet. Gynecol. 60:*237–242, 1982.

National Council on Radiation Protection and Measurements. *Basic Radiation Protection Criteria,* NCRP Report No. 39. Bethesda, MD, NCRP, 1971.

National Council on Radiation Protection and Measurements. *Medical Radiation Exposure of Pregnant and Potentially Pregnant Women,* NCRP Report No. 54. Bethesda, MD, NCRP, 1977a.

National Council on Radiation Protection and Measurements. *Review of NCRP Radiation Dose Limit for Embryo and Fetus in Occupationally-Exposed Women,* NCRP Report No. 53. Bethesda, MD, NCRP, 1977b.

National Council on Radiation Protection and Measurements. *Recommendations on Limits for Exposure to Ionizing Radiation,* NCRP Report No. 91. Bethesda, MD, NCRP, 1987.

National Research Council. *Health Effects of Exposure to Low Levels of Ionizing Radiation: BEIR V.* Washington, DC, National Academy Press, 1990.

United Nations Scientific Committee on the Effects of Atomic Radiation. *Genetic and Somatic Effects of Ionizing Radiation,* 1986 Report to the General Assembly, with Annexes. New York, United Nations, 1986.

U.S. Environmental Protection Agency. Radiation protection guidance to federal agencies for occupational exposure. *Fed. Register 52:*2822–2834, 1987 (52 FR 2822).

U.S. Nuclear Regulatory Commission. *Medical Use of Byproduct Material.* Washington, DC, U.S. Government Printing Office, Title 10, Code of Federal Regulations, Part 35 (10 CFR 35), revised as of 1 January 1990.

U.S. Nuclear Regulatory Commission. *Standards for Protection against Radiation.* Washington, DC, U.S. Government Printing Office, Title 10, Code of Federal Regulations, Part 20 (10 CFR 20), revised as of 21 May 1991.

Wagner, L.K., Lester, R.K., and Saldana, L.R. *Exposure of the Pregnant Patient to Diagnostic Radiations.* Philadelphia, JB Lippincott, 1985.

Young, J.L., Ries, L.G., Silverberg, E., Horm, J.W., and Miller, J.W. Cancer incidence, survival, and mortality for children younger than age 15 years. *Cancer 58:*598–602, 1986.

Section II

Radiation and Cancer

Chapter 5

Cancer and Carcinogenesis

ANATOLY DRITSCHILO

ABSTRACT

Cancer is the term used to generically describe malignant neoplasia. Cancers are made up of cells that have lost their responsiveness to regulatory factors that permit normal functioning in a complex, multicellular organism. Cancers grow by appropriating a blood supply and invading surrounding structures, frequently destroying the useful function of an organ. Cancers can also metastasize by traveling through lymphatics and blood vessels to develop secondary tumors at remote sites from the primary. To the radiation oncologist, cancer is a disease that can be treated successfully with ionizing radiation, frequently achieving a cure when the disease is localized. On the other hand, ionizing radiation is a known and well-studied physical agent that can cause cancer and should not be used indiscriminately for diagnostic or therapeutic purposes. In this review, the current radiobiological principles as well as the molecularly based current understanding of carcinogenesis and radiation therapy of cancers will be examined. Using cancers of head and neck origin as a model clinical system, radiobiological principles will be extended to the analysis of the treatment of such patients, and by extrapolation to other disease sites.

INTRODUCTION: SCOPE OF THE CLINICAL PROBLEM

Each year, approximately 1,000,000 cases of cancer are diagnosed in the United States (American Cancer Society 1989). Patients are treated with surgery, radiation therapy, and chemotherapy. It has been estimated that the majority of cancer cures (80% to 85%) are achieved as the result of surgical treatment.

More than half of all patients will be treated with ionizing radiation at some point in their disease, half of whom (27%) are treated with curative intent, resulting in an overall cure (defined for these purposes as survival free of disease for 5 years or more) of about 10% to 15% (Brady et al. 1984, Cumberlin et al. 1989). The remainder of cures are achieved with combination chemotherapy, where the current major research thrust is to identify new effective drugs for clinical use.

The malignant process can affect almost any tissue or organ in the body. The process, however, is not static within the population and, as a result of environmental factors, changes in the incidence of various types of cancers have been observed. Specifically, decreases in gastric and cervical carcinomas have been observed in the United States. However, cancers of the lung, breast, and colon have been observed to increase in incidence in recent years, and this is felt to reflect the increased use of tobacco and other environmental exposures.

A patient with a solid tumor is first evaluated surgically. This may include a biopsy or a resection for cure, depending on the specific disease site and extent. Patients who are not amenable to surgical resection or who would not benefit from resection are referred for radiation therapy management. This may be the principal treatment for cure or an adjuvant treatment to decrease the risk of local-regional recurrence. Chemotherapy is offered for certain tumors not amenable to surgery or radiation therapy and generally involves the management of disseminated disease. The use of adjuvant chemotherapy to prevent or delay the appearance of disseminated disease has been shown to be of particular value in the treatment of breast and colorectal cancers.

HEAD AND NECK CANCERS

Cancers of head and neck origin are generally squamous cell carcinomas. A variety of etiological factors have been implicated including viruses and chemical carcinogens; however, tobacco and alcohol are the most prominent. The clinical presentation is usually a mass or a nonhealing ulceration, generally visible on physical examination. For this reason, initial evaluation and follow-up management for recurrence are easily achieved, and tumors in these sites are ideal for investigation of responses to therapeutic intervention. These tumors are an excellent solid tumor clinical model that permits insight into initial evaluation and staging, therapeutic strategies, and follow-up management and late effects.

Therapeutic approaches for these cancers are based on the anatomical site of presentation and stage of disease. Staging consists of determination of the tumor primary extent (T), lymph node involvement (N), and presence or absence of metastases (M) (American Joint Committee on Cancer 1988). For example, a cancer of the tongue with a primary tumor size of 2 cm and a single palpable small ipsilateral lymph node less than 3 cm in greatest dimension and no evidence of metastases will be staged $T_1N_1M_0$ or stage III. Treatment may then consist of surgery, radiation therapy, or both modalities together. Generally, radiation therapy is offered for early lesions (stages I and II) in an effort to cure the cancer and to preserve function and cosmesis, with surgery reserved for salvage of recurrences. More advanced lesions (T_3 and T_4) are frequently treated with a planned combined modality approach, i.e., surgery followed by postoperative irradiation. Chemotherapy may also play a role in the initial management of advanced tumors, but this role is incompletely defined. Chemotherapy may be effectively used in the palliative management of recurrent disease after surgical and radiation therapy failure.

Radiation therapy is usually delivered in 1.8–2.0 Gy daily fractions to cumulative doses of 60–70 Gy. A recent trend, promoted at several prominent university medical centers, involves the use of smaller (1.2–1.6 Gy) fractions delivered twice daily (Parsons et al. 1988, Wang 1988). This fractionation appears to offer an improved local control rate for advanced tumors (Wang 1990). Tumor regression is observed during the course of radiation therapy, but occasionally tumors will progress in the face of radiation therapy, or shortly after completion of treatment. These tumors are said to be radiation resistant and have been the subject of much investigation. Cell lines established from biopsies of head and neck carcinomas have been used in various clonogenic survival analyses in efforts to develop clinically useful predictive assays (Brock et al. 1990, Fertil and Malaise 1981, Weichselbaum et al. 1986).

RADIATION SENSITIVITY, RESISTANCE, AND RESPONSIVENESS

Strictly speaking, radiation sensitivity is defined by the terminal slope of the survival curve (single hit, multitarget model) following exposure of cells to graded doses of ionizing radiation (Fig. 5.1). The steeper the slope, the smaller the value of D_0, and the more sensitive the cell. Conversely, a shallow slope results in a larger D_0 value and reflects a more resistant cell. Over the last decade, as human tumor cell lines have been established from different histological types, an apparent correlation has been observed between clinical tumor responsiveness and the survival fraction at 2 Gy (Goffman et al. 1990). Of necessity, some relationship between radiation sensitivity and responsiveness may exist; however, the exact nature of this relationship may be confounded by the state of tumor oxygenation, cell cycle distribution, nutritional state, the redox status of the cells, and other factors.

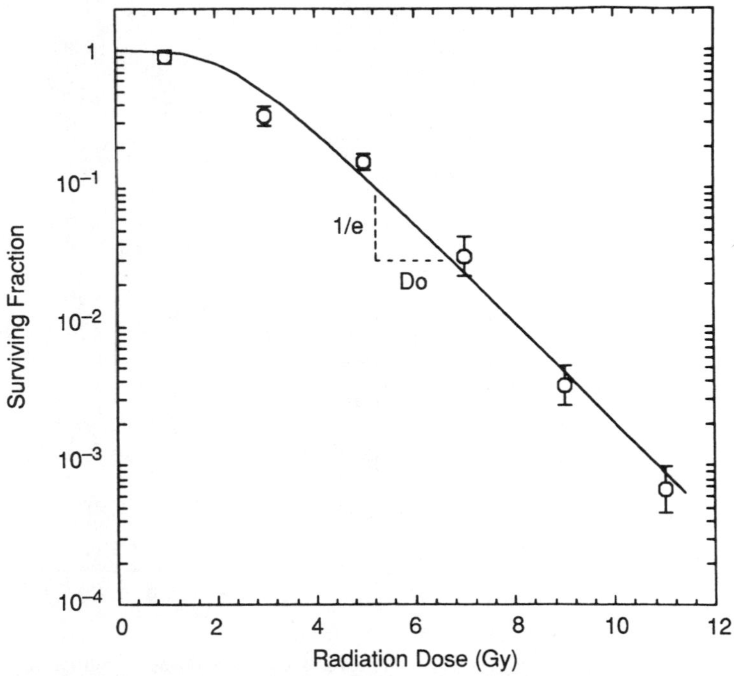

FIG. 5.1. The radiation survival for adrenal carcinoma cell line KS-6, curve fit using the single hit, multitarget model. D_0, obtained from the terminal slope, is 1.72 Gy, and n, obtained from extrapolation of the straight line portion of the curve to the ordinate, is 4.8.

Fertil et al. have used the mean inactivation dose ($\bar{D}$) to characterize cellular responses to ionizing radiation (Fertil et al. 1984). This involves the use of the linear quadratic cell survival model on linear coordinates and integrating the area under the curve. A greater weight is given to cell survival in the initial (shoulder) portion of the survival curve, and arguably, the $\bar{D}$ may be more representative of the cellular intrinsic ability to survive ionizing radiation. It is interesting that when the values of $\bar{D}$ are calculated for a number of squamous carcinoma cell lines, the resultant frequency histogram (Fig. 5.2) shows a distribution with a resistant tail.

The genetic basis for the radiation response of cells is supported by several lines of evidence. The most compelling is the observation that patients with the genetic disease ataxia telangiectasia (AT) are exquisitely sensitive to ionizing radiation, and standard doses of radiation are fatal for these patients (Gotoff et al. 1967). Fibroblast cell lines grown from biopsies from patients with AT show a marked sensitivity, as compared to squamous cell carcinomas, as illustrated in Figure 5.3. Furthermore, the survival responses for two different squamous cell carcinomas also show a large difference in radiation sensitivities (1.3 versus 2.4). Cell line SCC-9 was obtained from a "radiation-sensitive" tumor, amenable to "cure" by radiation therapy, whereas cell line SQ-20B was obtained from a "radiation-resistant" tumor that progressed in the field of radiation shortly after treatment was completed. Furthermore, four clonally derived cell lines established

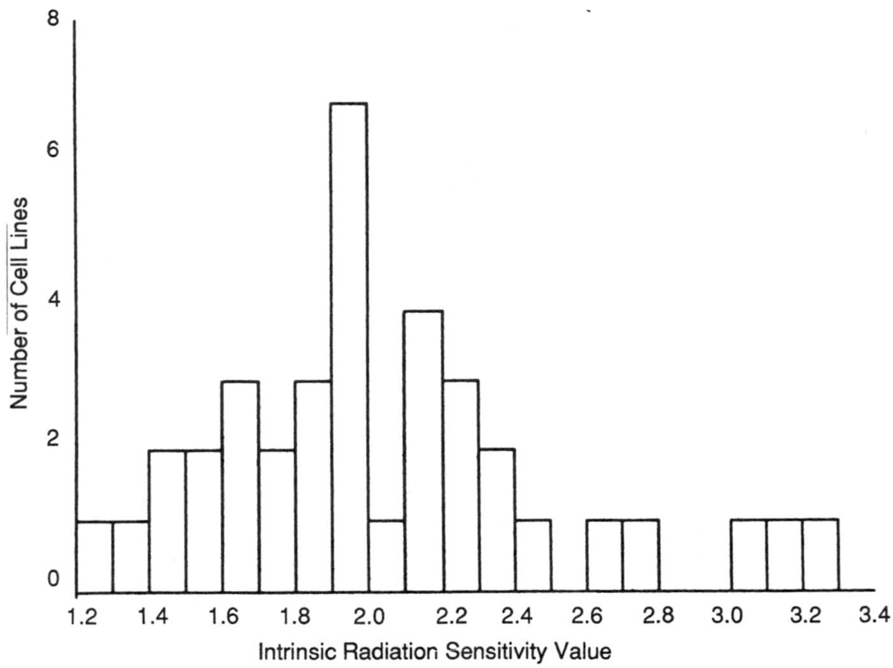

FIG. 5.2. Frequency distribution of calculated mean inactivation doses for human tumor cell lines. (Adapted from Weichselbaum RR, Beckett MA, Schwartz JL, Dritschilo A. Radioresistant tumor cells are present in head and neck carcinomas that recur after radiotherapy. Int J Radiat Oncol Biol Phys 1988;14:575–579.)

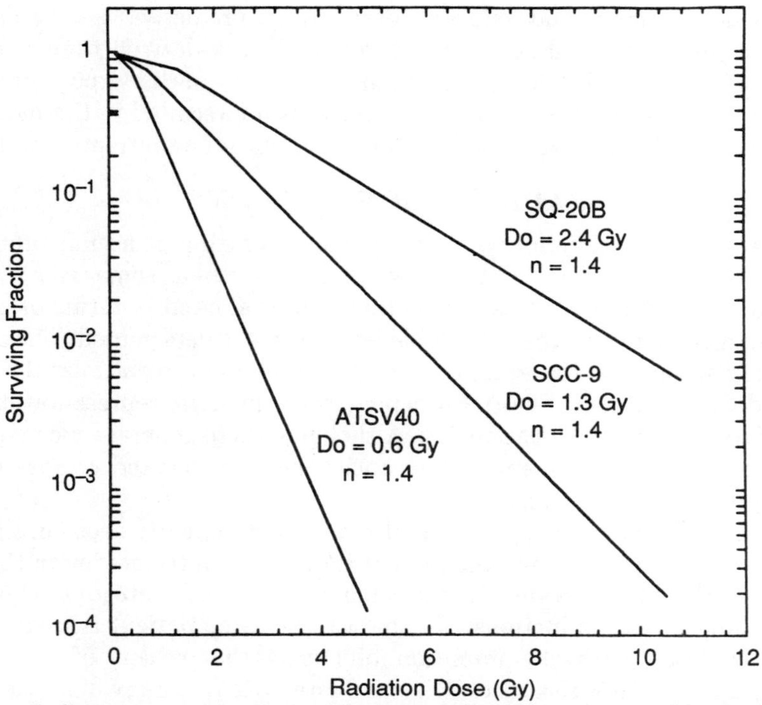

FIG. 5.3. Radiation response curves for two squamous carcinoma cell lines (SQ-20B and SCC-9) and for an SV40 virus immortalized ataxia telangiectasia cell line (AT SV40).

from the same parent epidermoid carcinoma cell line showed a substantial heterogeneity, with D_0 values ranging from 1.31–2.66 Gy (Gotoff et al. 1967).

There is an axiom in radiation biology that the less differentiated a tumor cell, the more radiation sensitive is that tumor. Furthermore, if transformation of a normal cell results in a less differentiated phenotype, it may be predicted that as cells go through the steps required for malignant transformation, some increase in radiation sensitivity may be expected. However, the transformation state of a cell does not, per se, predict for radiation sensitivity or resistance. Kasid et al. used the epidermal keratinocyte model system to determine the radiation response of normal cells, immortalized cells, and malignant transformed cells (Kasid et al. 1987a). The radiation responses of these cells showed little variation with enhanced malignancy. Furthermore, although there appears to be an association of the radiation-resistant phenotype with activity of some oncogenes (ras, raf), the presence of an activated oncogene is not always sufficient to result in cellular radiation resistance. Such studies of the genetic factors associated with radiation response have become a fruitful research area, explaining cellular observations at the level of the genome.

That radiation-resistant cells may be implicated in clinical treatment failure has also been suggested by radiobiological analysis of head and neck carcinomas (Weichselbaum et al. 1989). It has been proposed that since daily radiation fractions are usually 1.8–2.0 Gy, determination of the normal fraction at 2 Gy may provide a simple and relevant basis for a predictive assay. Table 5.1 summarizes calculated values for the head- and neck-derived tumor cell lines from a group that failed radiation treatment and from another group prior to any treatment. A substantial increment in survival is appreciated in the mean value of the radiation therapy failure group as compared to the pretreatment group.

RADICAL OR PALLIATIVE

The treatment of patients with cancer using ionizing radiation offers some benefits (cure of the tumor or relief of symptoms) as well as some risks (radiation injury and late effects). The patient is therefore evaluated in terms of intent of treatment, and therefore, the acceptable level of risk is determined. The clinician weighs the risks against the benefits of treatment in order to optimize the therapy for an individual patient. Figure 5.4 depicts the schematic representation of the nature of the risk-benefit relationship. Although data to generate such curves are only available for the treatment of skin cancer, they are accepted as representative of many other clinical settings.

When the radiation oncologist evaluates a patient, an early decision that must be made is the classification of the patient into a radical (treated with the intent to cure) or a palliative (treated for the relief of symptoms) category. This results in several important distinctions. The radically treated patient is treated to full therapeutic doses, with the intention of curing the patient of disease, and accepting a 5% to 10% risk of toxicity and late effects. The palliatively treated patient, however, is treated for relief of a specific local symptom, and moderate radiation doses are used in an effort to avoid treatment-related toxicity. Such

TABLE 5.1. SUMMARY OF CALCULATED SURVIVING FRACTIONS AFTER 2 GY RADIATION EXPOSURE FOR 14 CELL LINES OBTAINED FROM PATIENTS WHO FAILED RADIATION THERAPY, COMPARED TO 8 CELL LINES FROM PATIENTS PRIOR TO ANY TREATMENT[a]

RT[b] Failure Cell Line	Calculated 2 Gy Surviving Fraction	No RT[b] Cell Line	Calculated 2 Gy Surviving Fraction
SCC-4	0.485	SCC-9	0.306
SCC-25	0.490	SCC-61	0.409
SCC-35	0.571	SCC-73	0.146
SCC-13	0.594	SCC-71	0.445
SCC-49	0.609	SCC-66	0.585
SCC-50	0.450	SQ-29	0.645
SCC-76	0.684	SQ-38	0.648
SQ-9G	0.384	SQ-39	0.377
SQ-20B	0.715		
SQ-31	0.601		
SQ-43	0.704		
HN-SCC-29	1.02		
HN-SCC-68	0.936		
JSQ-3	0.777		
(Mean ± SEM)	0.643 ± 0.048		0.4451 ± 0.0620

[a] Adopted from Weichselbaum RR, Beckett MA, Schwartz JL, Dritschilo A. Radioresistant tumor cells are present in head and neck carcinomas that recur after radiotherapy. Int J Radiat Oncol Biol Phys 1988;14:575–579.

[b] RT, radiation therapy.

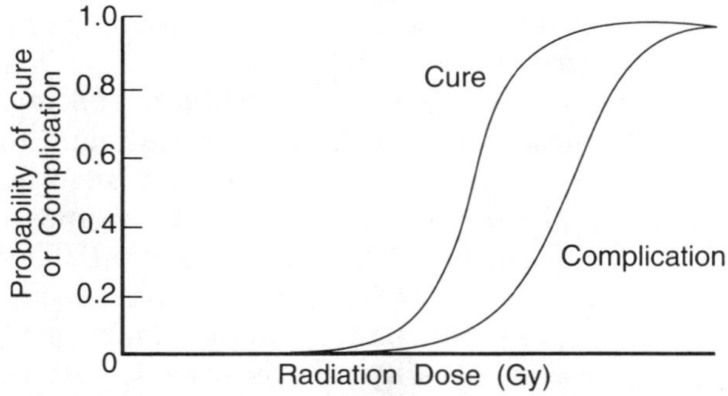

FIG. 5.4. The sigmoid dose-response curves for tumor cure and for radiation complications lie in close proximity for most tumors of epidermal origin. Generally for a selected dose with a cure probability of 80% to 90%, the risk of serious complications is 5% to 10%.

patients generally have a shorter life expectancy, and it is felt to be inappropriate to subject patients to significant complications in the process of relieving a symptom. Radiation carcinogenesis becomes an issue for patients treated radically, but generally not for patients treated palliatively.

ONCOGENES AND TUMOR SUPPRESSOR GENES

The origin of cancer is hypothesized to be the result of mutational event(s) to critical genetic loci that lead to the development of cancer. Currently, this "somatic mutational hypothesis" includes several relevant implications: (*a*) Spontaneous mutations produce an irreducible "background" level of cancer; (*b*) agents that can change the host genome by mutation or by addition or deletion of genetic material should increase this background incidence; (*c*) abnormalities in a genome that favor mutations should result in an increased risk of cancer; and (*d*) inheritance of an initiating mutation should strongly predispose to cancer (Knudson 1985).

To understand the molecular genetic events that underlie the carcinogenesis process, reproducible induction of the malignant phenotype in a test system was essential. To this end, rodent cell systems have been used. Many of the known oncogenes have been identified by the DNA-mediated gene transfer assay (Wigler et al. 1979).

DNA-mediated gene transfer involves the use of high molecular weight DNA from donor cells, which is precipitated in calcium phosphate and incubated with recipient cells (usually mouse fibroblasts, NIH 3T3 cells). The calcium phosphate-DNA matrix is phagocytized by the recipient cells, transferring the DNA to the nucleus and stably incorporating it into the genome (Krontiris 1983). The transferred DNA may then cause the recipient cells to assume a transformed morphology, to grow into colonies in agar, and to cause tumors when injected into nude mice. Such cells can be identified when they grow to form large, overgrown colonies. Subsequent isolation of these cells permits gene cloning and identification.

Similarly, cells may be transformed by ionizing radiation, presumably effecting a genomic alteration that results in a dominant transforming phenotype. Efforts to isolate such genes have pointed to non-ras transforming genes that have not been otherwise characterized (Borek et al. 1987).

Oncogenes were first isolated as parts of the genomes of acutely transforming viruses (Bishop 1987). Subsequently, homologies were found to cellular genes (proto-oncogenes) that could become "activated" and identified using DNA-mediated gene transfer experiments in rodent cells. Oncogenes are activated by DNA amplification, rearrangement, deletion, or mutation. These genes play a significant role in cellular growth, development, and carcinogenesis. These findings offered a much sought for molecular explanation for tumor formation.

Tumor suppressor genes, or recessive anti-oncogenes, were first predicted to exist by Knudson in his seminal work with the hereditary form of childhood retinoblastoma (Knudson 1978). The genes are postulated to work in a stabilizing fashion on the genome, although the exact mechanisms have not been elucidated. Complementation studies, fusing malignant cells with cells containing the chromosome containing the putative suppressor gene, result in loss of some aspects of the malignant phenotype of hybrid cells (Stanbridge et al. 1982). The retinoblastoma gene and the p53 gene are known examples of tumor suppressor genes, or recessive anti-oncogenes.

HOW DO ONCOGENES WORK IN CELLS?

Oncogenes exert their effects on the genetic programs of cells by regulating signal transduction pathways, resulting in multifactorial genetic responses. These genes can be classified functionally into four different groups. *Growth factor genes* may act as oncogenes following constitutive activation. This is exemplified by the sis oncogene and the autocrine model of activation. *Growth factor receptors* may be truncated and fixed in a state of continuous faulty signaling, as is the case with the erb-B, a receptor for epidermal growth factor. *Genes coding for cytoplasmic or membrane proteins* are involved in signal transduction. Examples include the src, ras, raf and others and can be activated by point mutations in certain domains. *DNA-binding proteins* are presumably involved in DNA replication and transcriptional activation. Examples of this category include the myc family, fos, jun, and others.

MODULATION OF THE RADIATION RESPONSE BY ONCOGENES

Through their effects on signal transduction pathways oncogene products are capable of communicating and effecting multiple gene products (Weichselbaum et al. 1991) and may include those associated with the ability of cells to survive exposure to ionizing radiation. We do not imply that the oncogene protein(s) are directly involved in DNA repair, but they may play a role as upstream regulatory factors for gene products that are more directly involved in the DNA repair cascade. This may also involve genes that may enhance cellular tolerance for radiation-induced toxic free radicals (Burt et al. 1988).

Based on antibody-blocking experiments, Rapp and coworkers have proposed that ras, fes, or src oncogene products depend on serine/threonine kinases (raf and mos) for transmission of their growth- and proliferation-related signals

(Morrison et al. 1988). These and other reports provide evidence that suggests c-raf protein is an important signal transducer for a variety of growth factor receptors and membrane-bound oncoproteins. Although the existing information is insufficient to define gamma radiation-repair molecules that might respond to signals from raf or ras oncogene products, it seems reasonable to propose a possible modulation of repair gene(s) function as one of the consequences of the activation of these oncogenes. Table 5.2 summarizes the positive reports of oncogene expression modulating the radiation responses of cells.

MODEL HUMAN CELL SYSTEMS–FUTURE DIRECTIONS IN CARCINOGENESIS STUDIES

While the majority of studies of transformation have relied on the use of rodent cells in culture, experimental models using human cells have been only recently developed. Unlike rodent cells, normal human cells in culture do not undergo spontaneous transformation and usually prove resistant to neoplastic transformation by carcinogens, including ionizing radiation. Through the concerted use of viruses as the immortalizing agents, several epithelial cell lines have been established that are suitable for the analysis of the promotional steps of the carcinogenesis process. SV40 large T antigen has been used to transform human liver cells and bronchial epithelial cells (Zeitlin et al. 1991). Human epidermal keratinocytes have been immortalized by the AD12-SV40 hybrid virus (Rhim et al. 1985). Human papilloma virus has been used to immortalize and transform normal human cervical epithelial cells (Woodworth et al. 1988).

Using immortalized human epidermal keratinocytes, Rhim et al. (1985) have developed a primary epithelial cell line that has acquired an indefinite life span in culture but does not undergo malignant conversion. Subsequent addition of the Kirsten murine sarcoma virus, or chemical carcinogen (N-methyl-N′-nitro-N nitrosoguanidine or 4-nitroquinodine-1-oxide) results in striking morphological changes and the acquisition of neoplastic properties. We have employed this cell line in studies of x-ray transformation (Thraves et al. 1990). Several observations are noted from these experiments.

TABLE 5.2. ONCOGENES AND RADIATION RESISTANCE (RR)

Raf
Kasid et al. 1987b, radiation resistance associated with a squamous cell carcinoma of the larynx
Chang et al. 1987 radiation resistance associated with fibroblasts from Li-Fraumeni syndrome
Kasid et al. 1989, antisense modulation of tumorigenicity and radiosensitivity of SQ-20B
Pirollo et al. 1989, modulation of NIH/3T3 cell radiation sensitivity
Ras
Fitzgerald et al. 1985, ras transformed NIH/3T3 fibroblasts show a larger D_0
Sklar 1988, resistance associated with mutated ras in NIH/3T3 cells
Ling and Endlich 1989, ras, myc or both lead to resistance in rat embryo cells
McKenna et al. 1990, H-ras, or v-myc and H-ras lead to resistance in rat embryo cells
Myc
Carney et al. 1983, variant small cell lung carcinoma (SCLC) more resistant than SCLC
Ling and Endlich 1989, McKenna et al. 1990, combined effect with ras lead to radiation resistance.

Transformation of cells was not observed after exposure of primary human epidermal keratinocytes to graded doses of ionizing radiation. RHEK-1 cells (immortalized with the AD12-SV40 virus), however, demonstrated morphological change after being irradiated twice with 2- and 4-Gy radiation doses. These cells formed colonies in soft agar and grew tumors in nude mice. Chromosomal analysis showed an aneuploid male karyotype with five marker chromosomes. Analysis for mutation or overexpression of known genes (ras, myc) was negative. The gene was found to be weakly transforming in the rodent tumor system, permitting efforts to clone and characterize the gene to proceed.

RADIATION CARCINOGENESIS

Radiation transformation of cells into the malignant phenotype and tumor induction are viewed as stochastic effects. Much of the data for predicting the risk of radiation-induced cancers is obtained from patient populations exposed to high doses and extrapolated to low doses and low dose rates. This requires the use of mathematical models and assumptions as to the shape of the dose-response curve in the low-dose region. Several such dose-response models have been developed and have been applied to predict carcinogenic risk. The three models most often used are the linear, quadratic, and linear-quadratic. In practice, the linear and the quadratic models are frequently used to predict the upper and lower boundaries of risk estimate, and the linear-quadratic appears to provide the best fit to animal data and perhaps to humans. Since other factors complicate the analysis of carcinogenic risk, most cancers have an age-related spontaneous risk of development. For this reason, a relative risk, or the multiple of the natural age-specific cancer risk for a particular population, is generally used in clinical risk estimates. This is a relatively useful concept for discussing risk-benefit consideration with a patient who may be considering therapeutic options, since other risks, such as surgical complications, medication complications and everyday life activities can be similarly quantitated.

Radiation therapy of patients exposes patients to large radiation doses to small volumes of tissue, and to small doses to large volumes of tissue. The carcinogenic risk for patients is small but not insignificant. In an analysis of the overall carcinogenic risk to patients, Cumberlin et al. estimate a lifetime leukemia evidence of 0.7% and a solid tumor risk of 0.3% (Cumberlin et al. 1989). These data are used in weighing the risks and benefits of alternative treatments for cancer.

REFERENCES

American Cancer Society. *Facts and Figures–1989.* New York, American Cancer Society, 1989.

American Joint Committee on Cancer. *Manual for Staging of Cancer*, 3rd Ed. Philadelphia, JB Lippincott, 1988.

Bishop, J.M. The molecular genetics of cancer. *Science 235:*305–310, 1987.

Borek, C., Ong, A., and Mason, H. Distinctive transforming genes in x-ray transformed mammalian cells. *Proc. Natl. Acad. Sci. (USA) 84:*797–798, 1987.

Brady, L.W., Sheline, G.E., Suntharalingam, N., and Sutherland, R.M. Introduction and overview to the interdisciplinary program for radiation oncology research. *Cancer Treat. Symp. 1:*1–11, 1984.

Brock, W.A., Baker, F.L., Wike, J.L., Sivon, S.L., and Peters, L.J. Cellular radiosensitivity of primary head and neck squamous cell carcinomas and local tumor control. *Int. J. Radiat. Oncol. Biol. Phys. 18:*1283–1286, 1990.

Burt, R.K., Garfield, S., Johnson, K., and Thorgeirsson, S.S. Transformation of rat lines epithelial cells with v-H-ras or v-raf causes expression of MDR-1, glutathione-S-transferases and increased resistance to cytotoxic chemicals. *Carcinogenesis 9:*2329–2332, 1988.

Carney, D.N., Mitchell, J.B., and Kinsella, T.J. *In vitro* radiation and chemotherapy sensitivity of established cell lines: large cell morphological variants. *Cancer Res. 43:*2806–2811, 1983.

Chang, E.M., Pirollo, K.F., Zon, Z.Q., Cheung, H.Y., Lawler, E.L., Garner, R., White, E., Bernstein, W.B., Fraumeni, J.W., Jr., and Blattner, W.A. Oncogenes in radioresistant noncancerous skin fibroblasts from a cancer-prone family. *Science 237:*1036–1039, 1987.

Cumberlin, R.L., Dritschilo, A., and Mossman, K.L. Carcinogenic effects of scattered dose associated with radiation therapy. *Int. J. Radiat. Oncol. Biol. Phys. 17:*623–629, 1989.

Fertil, B., Dertinger, H., Courdi, A., and Malaise, E.P. Mean inactivation dose: a useful concept for intercomparison of human cell survival curves. *Radiat. Res. 99:*73–84, 1984.

Fertil, F., and Malaise, E.P. Inherent radiosensitivity as a basic concept for human tumor radiotherapy. *Int. J. Radiat. Oncol. Biol. Phys. 7:*621–629, 1981.

Fitzgerald, T.J., Daugherty, C., Kase, K., Rothstein, L., and Greenberger, J.S. Activated human n-ras oncogene enhances x-irradiation repair of mammalian cells in vitro less effectively at low dose rate. *Am. J. Clin. Oncol. 8:*517–522, 1985.

Goffman, T.E., Raubitschek, A., Mitchell, J.B., and Glatstein, E. The emerging biology of modern radiation oncology. *Cancer Res. 50:*7735–7744, 1990.

Gotoff, S.P., Arimokri, E., and Liebner, E.J. Ataxia telangiectasia, neoplasia, untoward response to x-irradiation, and tuberous sclerosis. *Am. J. Dis. Child 114:*617–625, 1967.

Kasid, U., Dritschilo, A., and Rhim, J.S. Human epidermal keratinocytes retain radiation resistance following in vitro immortalization and malignant transformation. *Radiat. Res. 111:*565–571, 1987a.

Kasid, U., Pfeifer, A., Brennan, T., Beckett, M., Weichselbaum, R.R., Dritschilo, A., and Mark, G.E. Effect of antisense c-raf-1 on tumorigenicity and radiation sensitivity of a human squamous carcinoma. *Science 243:*1354–1356, 1989.

Kasid, U.N., Pfeifer, A., Weichselbaum, R.R., Dritschilo, A., and Mark, G. The raf oncogene is associated with a radiation-resistant human laryngeal cancer. *Science 237:*1039–1041, 1987b.

Knudson, A.G. Retinoblastoma: a prototype hereditary neoplasm. *Sem. Oncol. 5:*57–60, 1978.

Knudson, A.G. Hereditary cancer, oncogenes, and anti-oncogenes. *Cancer Res. 45:*1437–1443, 1985.

Krontiris, T.G. The emerging genetics of human cancer. *N. Engl. J. Med. 309:*404–409, 1983.

Ling, C.C., and Endlich, B. Radioresistance induced by oncogenic transformation. *Radiat. Res. 120:*267–279, 1989.

McKenna, W.G., Weiss, M.C., Endlich, B., Ling C.C., Bakananskas, V.J., Kelsten, M.L., and Muschel, R.J. Synergistic effect of the V-myc oncogene with H-ras on radioresistance. *Cancer Res. 50:*97–102, 1990.

Morrison, D.K., Kaplan, D.R., Rapp, U., and Roberts, T.M. Signal transduction from membrane to cytoplasm: growth factors and membrane-bound oncogene products increase raf-1 phosphorylation and associated protein kinase activity. *Proc. Natl. Acad. Sci. (USA) 85:*8855–8859, 1988.

Parsons, J.T., Million, R.R., Cassisi, N.J., et al. Hyperfractionation for head and neck cancer. *Int. J. Radiat. Oncol. Biol. Phys. 14:*649–658, 1988.

Pirollo, K.F., Garner, R., Yuan, S.Y., Li, L., Blattner, W.A., and Chang, E.H. raf involvement in the simultaneous genetic transfer of the radioresistant and transforming phenotypes. *Int. J. Radiat. Oncol. Biol. Phys. 55:*783–796, 1989.

Rhim, J.S., Jay, G., Arnstein, P., Price, F.M., Sanford, K., and Aaronson, S.A. Neoplastic transformation of human epidermal keratinocytes by AD12-SV40 and Kirsten sarcoma viruses. *Science 227:*1250–1252, 1985.

Sklar, M.D. The ras oncogenes increase the intrinsic resistance of NIH/3T3 cells to ionizing radiation. *Science 239:*645–647, 1988.

Stanbridge, E.J., Der, C.J., Doersen, C.J., Nishimi, R.P., Peehl, D.M., Weinstein, B.E., and Wilkinson,

J.E. Human cell hybrids: analysis of transformation and tumorigenicity. *Science 215:*2259–2262, 1982.

Thraves, P., Salehi, Z., Dritschilo, A., and Rhim, J.S. Neoplastic transformation of immortalized human epidermal keratinocytes by ionizing radiation. *Proc. Natl. Acad. Sci. (USA) 87:*1174–1177, 1990.

Wang, C.C. Twice-daily radiation therapy for head and neck carcinomas. *Front. Radiat. Ther. Oncol. 22:*93–98, 1988.

Wang, C.C. *Radiation Therapy for Head and Neck Neoplasms: Indications, Techniques, and Results.* Chicago, Year Book Medical Publishers, 1990, pp. 66–74.

Weichselbaum, R.R., Beckett, M.A., Vijayakumar, S., Ahmed-Swan, S., Dritschilo, A., et al. Radioresistant tumor cell lines derived from head and neck radiation failures. *Head-Neck 11:*343–348, 1989.

Weichselbaum, R.R., Dahlberg, W., Beckett, M., Karrison, T., Miller, D., Clark, J., Ervin, T.: Radiation-resistant and repair-proficient human tumor cells may be associated with radiotherapy failure in head-and-neck cancer patients. *Proc. Natl. Acad. Sci. (USA)* 83:2684–2688, 1986.

Weichselbaum, R.R., Hallahan, D.E., Sukhatme, V., Dritschilo, A., Sherman, M.L., and Kufe, D.W. Biological consequences of gene regulation after ionizing radiation exposure. *J. Natl. Cancer Inst. 83:*480–484, 1991.

Wigler, M., Pellicer, A., Silverstein, S., Axel, R., Urlaub, G., and Chasin, L. DNA-mediated transfer of the adenine phosphoribosyltransferase locus into mammalian cells. *Proc. Natl. Acad. Sci. (USA) 76:*1373–1376, 1979.

Woodworth, C.D., Bowden, P.E., Doniger, J., Pirisi, L., Barnes, W., Lancaster, W.D., and DiPaolo, J.A. Characterization of normal human exocervical epithelial cells immortalized in vitro by papillomavirus types 16 and 18 DNA. *Cancer Res. 48:*4620–4628, 1988.

Zeitlin, P.L., Lu, L., Rhim, J., Cuting, G., Stetten, G., Kieffer, K.A., Craig, R., and Guggino, W.B. A cystic fibrosis bronchial epithelial cell line: immortalization by adeno-12-SV40 infection. *Am. J. Respir. Cell Mol. Biol. 4:*313–319, 1991.

Chapter 6

Radiation Carcinogenesis–Human Epidemiology

JOHN D. BOICE, JR.

ABSTRACT

Epidemiological studies of irradiated populations have become the basis for estimating cancer risks in man. Leukemia is the most prominent radiogenic cancer, occurring within 2 years after exposure and showing a wave-like pattern of excess risk over time. In contrast, other cancers do not arise until about 10 years have past and appear to remain increased for 30 or more years. The bone marrow, thyroid, and breast are tissues that seem particularly susceptible to the carcinogenic action of radiation. Radiation has not been reported to increase the risk of chronic lymphocytic leukemia, Hodgkin's disease, or cancer of the pancreas, prostate, testis, or cervix. Cancers of the rectum, uterus, bone, and connective tissue appear elevated only after extremely high radiation exposures. Liver cancer has been excessive only among patients given Thorotrast, a colloid of alpha-particle emitting thorium. Bone cancer has occurred following very high exposures to internally ingested radium. Iodine-131 has not been convincingly linked to either leukemia or thyroid cancer, suggesting the possible protective effect of protracted exposures. Prenatal exposures likely result in excess childhood cancer, but it is unclear whether the risk is substantially greater than that seen following infant exposure. Diagnostic x-rays do not appear to be a major contributor to the cancer burden. Living near a nuclear facility or in an area of high background radiation has not been confirmed as a cancer risk factor. The link between preconception irradiation and childhood leukemia is a provocative but unconfirmed hypothesis. Few consistent patterns of cancer risk have been seen in occupational studies of workers at nuclear facilities. Radon in underground mines clearly causes lung cancer, especially in combination with cigarette smoking, but the level of risk associated with residential exposures is not known and may be small. Given conservative assumptions on cancer risk, perhaps less than 5% of all cancers might be attributable to all sources of exposure to ionizing radiation.

INTRODUCTION

Epidemiological studies of populations irradiated for medical, occupational, or military reasons have conclusively linked ionizing radiations of all types with increased rates of cancer (National Academy of Sciences [NAS] 1988, 1990; UNSCEAR 1986, 1988). The attention given to the health effects of ionizing radiation, however, has intensified for several reasons: the Chernobyl reactor accident, indoor radon gas (NAS 1988), the revised dosimetry for the atomic bomb survivors in Japan (Fry and Sinclair 1987), radioepidemiological tables on the probability of causation (NIH 1985), living near nuclear installations (Forman et al. 1987), preconception radiation and childhood leukemia (Gardner et al. 1990), nuclear weapons tests (Rallison et al. 1990, Stevens et al. 1990), and new studies of medically irradiated populations (Boice 1988; Boice et al. 1991a) and of radiation workers (Gilbert et al. 1989a; Wing et al. 1991). An NAS committee recently concluded that estimates of lifetime cancer risk following relatively low doses may be as much as four times larger than previously thought (NAS 1990). Continuous lifetime exposure of 100,000 persons to 1 mSv of ionizing radiation per year was estimated to induce about 65 leukemias and 495 fatal cancers of other sites. On this basis, 4%–5% of all cancers might be attributable to all sources of radiation exposure (1.6 mSv/yr), excluding indoor radon, which has been suggested as an important cause of lung cancer (NAS 1988). Previous estimates of attributable risk were lower, but all such estimates should be interpreted with caution given the substantial uncertainties in applying risks from high-dose studies to low-dose and low dose rate situations. Epidemiological studies of human populations exposed to ionizing radiation will be discussed below, with emphasis on the most recent investigations that have increased our understanding of radiation carcinogenesis.

MEDICAL EXPOSURES–RADIOTHERAPY FOR CANCER (Table 6.1)

Cervical Cancer

An international study of cancer incidence following radiotherapy for cervical cancer was conducted in more than 200,000 women from 15 countries (Boice et al. 1985, 1987). A small but significant risk of leukemia was detected, together with the characteristic wave-like pattern of risk over time. Risk was modeled to account for the inhomogeneous distribution of dose to active bone marrow throughout the body, and for the possibility of cell killing at high doses. Risk increased with increasing dose up to about 4 Gy and then decreased at the highest doses. The relative risk (RR) at 1 Gy was estimated to be 1.7.

Table 6.1. Studies of Cancer Patients Treated with Radiation, with Quantitative Estimates of Risk

Population	Excess Cancer
Cervical cancer	Bladder, rectum, uterus, ovary, stomach, sarcomas, kidney; small leukemia risk
Childhood cancer	Sarcomas, thyroid, brain, breast; no leukemia excess
Hodgkin's disease	Possibly breast, stomach, thyroid, lung; no leukemia excess
Non-Hodgkin's lymphoma	Leukemia
Breast	Leukemia, uterus; small risk of second breast cancer

After a minimum latent interval of about 10 years, risk of second tumors following radiotherapy increased with time and reached nearly twofold among long-term survivors. A comprehensive dosimetry program provided organ dose estimates for individual patients based on their actual radiotherapy records (Stovall et al. 1989). New information was provided on dose response for 18 solid tumors (Boice et al. 1988). Among women with intact ovaries, radiotherapy was linked to a significant 35% reduction in breast cancer risk, attributable in all likelihood to the cessation of ovarian function (Boice et al. 1989). Among women without ovaries, there was a slight increase in breast cancer risk and a suggestion of a dose response. If the low doses of scatter radiation increased the risk of breast cancer among these women over age 40, the increase per unit dose was much lower than seen in studies of younger women exposed to higher doses. Risk estimates for this study of incident cancers and leukemia were generally lower than those from the mortality studies of atomic bomb survivors and spondylitics (UNSCEAR 1988). Possibly (*a*) the protracted nature of the treatment for cervical cancer allowed more time for the repair of radiation damage than is possible from acute exposures, (*b*) the high doses might have resulted in substantial cell killing, or (*c*) factors associated with cervical cancer might have confounded the observations.

Childhood Cancer

Only a few series have estimated radiation doses to specific organs or evaluated dose-response relationships following irradiation in childhood for cancer (de Vathaire et al. 1988, 1989; Tucker et al. 1987a, 1987b, 1991). Radiotherapy was not found to increase the risk of leukemia, possibly because of the predominance of cell-killing effects over oncogenic transformation at such high levels (Tucker et al. 1987b). Radiogenic thyroid cancer, however, has been reported at 30 Gy (Tucker et al. 1991). A dose response over the range of 10–60 Gy has been seen for bone cancer (Tucker et al. 1987a). Interestingly, the RR for radiogenic bone cancer in children with bilateral retinoblastoma (who possess an underlying predisposition to develop osteosarcoma) was similar to that among children irradiated for other cancers; their cumulative and absolute risks, however, were much greater. Children with familial retinoblastoma have a deletion in chromosome 13 that predisposes to osteosarcoma, and radiation is thought to cause a second mutation in an osteoblast that leads to a high rate of osteosarcoma development (Brachman et al. 1991).

Other Cancers

Whole-body irradiation for non-Hodgkin's lymphoma (NHL) was seen to increase the risk of leukemia (Greene et al. 1983) . A radiation dose response was not evident among breast cancer patients who subsequently developed leukemia (Curtis et al. 1989), although an increased risk was reported in a clinical trial series (Fisher et al. 1985) . Breast cancer was excessive following radiation treatment for Hodgkin's disease (Curtis and Boice, 1988). A small risk of contralateral breast cancer was recently identified among long-term survivors of breast cancer who were treated with radiation under age 45 (Boice et al., in press).

MEDICAL EXPOSURES–RADIOTHERAPY FOR BENIGN DISEASE (Table 6.2)

ANKYLOSING SPONDYLITIS

The mortality experience of 14,558 persons treated between 1935 and 1954 in 87 British radiotherapy clinics for ankylosing spondylitis, a rheumatoid condition of the spine, has been carefully evaluated through 1983 (Darby et al. 1987). Radiation doses, estimated for each leukemia fatality and from a 7% sample of the population, averaged about 3.8 Gy for the active bone marrow and several gray for organs in the treatment fields (Lewis et al. 1988). Leukemia risk peaked 3–5 years after exposure and gradually declined but not to baseline levels; chronic lymphocytic leukemia (CLL) was not increased. The dose response for leukemia was irregular and essentially flat, possibly reflecting reduced leukemogenesis in the most heavily irradiated portions of the marrow due to cell killing or related to the fractionated nature of the exposures (Smith and Doll 1982). The RR was estimated as 4.5 at 1 Gy and the absolute excess risk as 2.02 per 10^4 person-years per gray (PY-Gy) (UNSCEAR 1988). For nonleukemia cancers, the estimated RR was 1.14 at 1 Gy, and the absolute excess risk was 4.67 per 10^4 PY-Gy. The RR for nonleukemia cancer was as great during the first 5 years after exposure as it was later and then declined to near normal levels after 25 years. Some of the early excess may reflect conditions associated with the underlying disease, other therapies such as cytotoxic medications, or possibly preexisting metastatic lesions causing pain that was misdiagnosed as spondylitis. The temporal pattern was dominated by lung cancer, for which variations in smoking habits, as well as the influence of cigarette smoke on relatively immobile lungs, may have played a role.

MASTITIS

Breast cancer was excessive among 601 women treated with radiation for acute postpartum mastitis compared with 1239 women not irradiated (Shore et al. 1986). A treatment involved 1–11 exposures (mean, 3.4), and 3.8 Gy to the irradiated breast. Excess risk was not apparent until 10 years after treatment and was approximately linear in dose up to about 3.5 Gy, when it declined. The absolute risk was estimated as 3.5 cases/10^4 PY-Gy, and the RR at 1 Gy as 1.4.

TABLE 6.2. STUDIES OF PATIENTS IRRADIATED FOR BENIGN CONDITIONS WITH QUANTITATIVE ESTIMATES OF RISK

Population	Excess Cancer
Spondylitis	Leukemia, lymphoma, lung, esophagus, stomach, CNS, bone
Tinea capitis	Thyroid, brain, skin, leukemia
Thymus enlargement	Thyroid, breast
Benign gynecological disease	Leukemia, colon, uterus, bladder
Benign breast disease	Breast

The age-specific risk estimates obtained for mastitis patients, tuberculosis patients fluoroscopically examined, and atomic bomb survivors are remarkably similar (Boice et al. 1979).

Benign Breast Disease

Risk of breast cancer among women given radiation therapy for benign breast disorders in Sweden was inversely related to age at exposure (Baral et al. 1977). Without a suitable comparison group, however, there was no way to determine whether the underlying breast disease predisposed toward cancer.

Thymus

In the 1930s and 1940s, it was common for newborn children to receive radiation therapy to shrink enlarged thymus glands. The fifth mail survey of 2856 irradiated persons identified 30 thyroid cancers versus 1 in 5055 untreated siblings (Shore et al. 1985). Females were at two to three times greater absolute risk than males, and the risk among Jews seemed especially high. The data were consistent with a linear dose response (mean, 1.2 Gy), risk remained high even after 40 years, and fractionation did not appear to reduce risk. Risk estimates were 3.5 cases/10^4 PY-Gy and RR = 1.58 at 1 Gy. Benign thyroid adenomas also occurred more frequently among exposed persons than among their siblings, 59 versus 8, respectively (5.7/10^4 PY-Gy). The incidence of thyroid neoplasms rose abruptly during adolescence, suggesting the influence of thyroid-stimulating hormone as a promoting or secondary factor. Childhood irradiation may also be particularly damaging if rapidly proliferating cells injured by radiation are more likely to develop abnormally than cells irradiated in later life with limited growth potential (Hempelmann 1968). Breast cancer was also excessive following thymic irradiation (Hildreth et al. 1989).

Ringworm of the Scalp

Among 10,834 children in Israel who received x-ray therapy to the scalp for tinea capitis, 43 thyroid cancers were observed versus 11 expected based on two comparison groups (Ron et al. 1989). The dose to the thyroid was particularly low, 9 cGy on average, although alignment errors might have greatly increased the exposure for some children. The dose response was consistent with linearity up to 50 cGy. The absolute excess risk was 13 cancers/10^4 PY-Gy, and the RR at 1 Gy was 31. Comparable numbers for benign thyroid tumors were 15 tumors/10^4 PY-Gy and RR = 11 at 1 Gy. Risk was greatest among persons under age 5 at irradiation. A dose-response relationship for brain cancer and other neural tumors was also reported (Ron et al. 1988b). The brain dose was estimated as 1.5 Gy. For all neural tumors, the absolute excess risk was 1.14 tumors/10^4 PY-Gy, and the RR at 1 Gy was 5.9. Significant excesses of leukemia and cancers of bone and connective tissue also occurred (Ron et al. 1988a).

Among 2215 tinea capitis patients treated in New York, no excess thyroid cancer was found, but thyroid adenomas, leukemia, and brain cancers were elevated (Shore et al. 1976). Radiotherapy likely contributed to excess skin cancers, especially for anatomical areas exposed to ultraviolet radiation from the sun (Shore et al. 1984). Basal cell carcinomas of the face were significantly increased among white but not black patients.

Tonsils

Excess thyroid cancer has occurred among 5379 persons irradiated in Chicago during childhood for mostly tonsil and nasopharyngeal conditions (Favus et al. 1976; Schneider et al. 1985). Intensive clinical screening of 1922 persons included thyroid scans. About 37.5% (1108) of the 3610 persons for whom vital status was known had nodular thyroid disease, and in 297 of these (or 8% overall) it was malignant. A large number of small tumors were apparently detected only at screening; about 80% were less than 1.5 cm. RR appeared to decrease among those followed for more than 30 years. Estimation of radiation risks is difficult because individual doses were not known; a nonirradiated comparison group, comparably screened, was not available; the significance of small, clinically silent cancers is uncertain; and the follow-up was incomplete.

Based on clinical examinations, an excess of thyroid nodularity was recently reported among 1590 individuals treated with radiation for lymphoid hyperplasia in Boston and 1499 persons treated with surgery only (Pottern et al. 1990). There was a strong dose-response gradient (mean, 24 cGy), similar RR at 1 Gy for males and females (8.0 and 7.0, respectively), and an inverse relationship with age at exposure. Much higher risks were suggested, however, by self-reported conditions from a mailed questionnaire, apparently due to underreporting among the nonexposed.

Uterine Bleeding

Among 2068 women who received x-ray treatment of 5–10 Gy to their ovaries for metropathia hemorrhagica in Scotland, 7 leukemia deaths occurred (2.6 expected) following an estimated bone marrow dose of 1.3 Gy (Smith and Doll 1976). Cancers of heavily irradiated sites, taken together, were also increased. Breast cancer occurred significantly below expectation, possibly related to the cessation of ovarian function after radiation castration. In a new study of 4483 women treated mainly with radium implants for bleeding disorders in New England, a linear dose response for leukemia was reported (Inskip et al. 1990b). The RR at 1 Gy was estimated to be 2.9 based on 34 leukemia deaths (17.3 expected) and a median dose of 53 cGy. The lower doses used to treat benign menstrual disorders appear to have been more leukemogenic than the higher doses used to treat cervical cancer, presumably because of less cell inactivation. Cancers of heavily irradiated sites were also significantly elevated, including the colon, uterus, and bladder, but not cancers of the rectum or cervix, lymphomas, or myeloma (Inskip et al. 1990a).

Other Benign Diseases

In a survey of 2049 patients with peptic ulcer given radiotherapy (16 Gy), a significant increase of stomach cancer was reported compared to 763 nonexposed patients (Griem et al. 1984). Patients treated with radiation for skin hemangioma were reported to be at increased risk for thyroid cancer (Fürst et al. 1988).

MEDICAL EXPOSURES–DIAGNOSTIC RADIATION (Table 6.3)

Tuberculosis

X-ray fluoroscopies were frequently used to monitor lung-collapse treatments for tuberculosis during 1935–1954. The most recent study included 2573 women in Massachusetts who received an average of 88 fluoroscopies, and a comparison group of 2367 nonexposed tuberculosis patients (Boice et al. 1991b). Radiation-absorbed dose to the breast per fluoroscopy was estimated as 0.9 cGy and the average total dose as 79 cGy (Boice et al. 1978). Excess breast cancers were related to dose in a linear manner. The excess did not appear until 10 to 15 years after exposure and remained high throughout 50 years of observation. Exposures during the adolescent and teenage years carried the greatest risk, and exposures after age 40 the least. The excess absolute risk was estimated as 10.7 cases/10^4 PY-Gy, and the RR as 1.61 at 1 Gy. A large Canadian series was generally consistent with these findings, although based on mortality rather than incidence data (Miller et al. 1989). The breast appears to be one of the most sensitive tissues to the carcinogenic action of radiation, and, in comparison with other studies, fractionated exposures seem similar to single exposures of the same total dose in their ability to induce breast cancer.

No excesses of leukemia, lymphoma, or lung cancer have been reported among tuberculosis patients fluoroscopically monitored (Davis et al. 1989). The dose to the lung was approximately 100 cGy. In animal experiments, splitting dose over time also appears to reduce the risk of radiogenic lung cancer to a much greater extent than the risk of radiogenic breast cancer (Ullrich et al. 1987, UNSCEAR 1986). Uncertainties about the dose estimates limit precise quantification of radiation risks from these tuberculosis series.

Scoliosis

Women with scoliosis exposed to multiple diagnostic x-rays (mean, 12 cGy) during adolescence were at increased risk of breast cancer (Hoffman et al. 1989). The numbers were small, however, and factors associated with scoliosis, such as nulliparity, could have influenced risk.

Prenatal Irradiation

Most studies seem consistent with a 40% increased risk of childhood leukemia associated with low-dose intrauterine exposure to diagnostic radiation of between 1 and 10 cGy just before birth (UNSCEAR 1986). A New England study was

Table 6.3. Studies of Patients Exposed to Diagnostic Radiation with Quantitative Estimates of Risk

Population (Exposure)	Excess Cancer
Tuberculosis (fluoroscopy)	Breast, no lung or leukemia
Scoliosis (spinal x-rays)	Breast
Children (prenatal)	Leukemia, childhood cancer
Adults (general diagnostic procedures)	Possibly multiple myeloma, possibly CMI, but not other types of leukemia, or lymphoma

extended to include 1342 childhood cancers among 1,429,499 births (Monson and MacMahon 1984). The RR associated with prenatal x-ray was 1.52 for leukemia and 1.27 for other cancers, and there was no evidence that the associations were due to confounding. A reanalysis of the large Oxford Survey of childhood cancer in England concluded that x-raying 1000 fetuses with 1 cGy would yield about 2–3 extra cases of childhood cancer in the first 15 years of life (Bithell and Stiller 1988). It remains to be established whether adult cancer is increased by fetal irradiation, although recent evidence suggests this possibility (Yoshimoto et al. 1988).

It has been postulated, however, that selection factors, related to the medical reasons why women receive prenatal x-rays, were responsible for at least some of the increased leukemia risk. The absence of any childhood leukemia (and only one childhood cancer) in atomic bomb survivors exposed in utero (Jablon and Kato 1970) supported the selection hypothesis. Animal experiments do not suggest an enhanced sensitivity to leukemia induction following irradiation during fetal stages (UNSCEAR 1986). Evidence against the selection hypothesis comes from the demonstration of a dose-response relationship for childhood leukemia based on number of x-ray films taken (Stewart and Kneale 1970), and from the observation that the excess risk was as great among twins, for whom x-ray pelvimetry was far more frequent (55%), as among singletons (15%), simply because of a greater likelihood to determine fetal positioning before delivery (Mole 1974). This observation was confirmed in a case-control study of twins born in Connecticut (Harvey et al. 1985). While it is established that prenatal x-ray is associated with an increased risk of childhood cancer, the causal nature of this association and the level of risk remain uncertain (MacMahon 1989, UNSCEAR 1986).

General Diagnostic Radiation

Studies linking diagnostic radiation with adult leukemia are inconsistent (Boice and Land 1979). A recent report from California found an association between diagnostic radiography, particularly low-back x-rays, and chronic myelogenous leukemia (CML) based on personal interviews of 136 cases and 136 neighborhood controls (Preston-Martin et al. 1989b). Based on x-ray records on over 1000 patients with leukemia and lymphoma treated at two prepaid health plans, however, little evidence was found that diagnostic x-ray procedures were causally associated with leukemia or NHL, although a possible increased risk of multiple myeloma was suggested among persons who received rather extensive x-ray exposures (Boice et al. 1991a). Interestingly, persons with leukemia and NHL were x-rayed frequently just prior to diagnosis for conditions related to the development or natural history of their disease. Diagnostic x-rays were not linked to multiple myeloma in an interview study of 399 cases and 399 controls in the United Kingdom (Cuzick and DeStavola 1988).

Past exposures to dental or medical radiography of the head and neck have been linked to elevated risks of meningiomas, gliomas, and salivary gland tumors in several case-control interview studies from California (Preston-Martin et al. 1980, 1988, 1989a). Radiation doses were not known but might have been substantial. Limitations of many of these studies on diagnostic x-irradiation include the potential for response bias in interview surveys; incomplete verification of the actual numbers of x-rays; limited dosimetry; and study sizes too small to detect risks on the order of currently accepted estimates. The possible contribution of diagnostic radiology to the cancer burden appears small in comparison with other causes (Evans et al. 1986).

EXPOSURES FROM ATOMIC AND THERMONUCLEAR WEAPONS (Table 6.4)

Japanese Atomic Bomb Survivors

The atomic bomb survivor studies provide an important human experience from which estimates of radiation risk can be derived. The population is large and not selected because of disease or occupation. Estimates of doses are based upon a comprehensive program that included interviews with virtually all proximal survivors. The Life Span Study (LSS) includes about 93,000 atomic bomb survivors and 27,000 nonexposed comparison subjects. Recent analyses of cancer mortality cover 1950–1985 (Pierce and Vaeth 1991, Shimizu et al. 1990), while reports on cancer incidence are based on data from Hiroshima and Nagasaki tumor registries from 1959–1978 (Wakabayashi et al. 1983). Site-specific incidence studies include cancers of the breast (Tokunaga et al. 1987), lung (Yamamoto et al. 1987), thyroid (Prentice et al. 1983), and leukemia (Ichimaru et al. 1986). In the late 1980s, a new dosimetry, called Dosimetry System 1986 or DS86, was developed, and revised estimates were computed for 76,000 of the 93,000 exposed survivors in the LSS. Risk estimates based upon the new dosimetry are 1.5 to 2 times greater than those under the old dosimetry (Preston and Pierce 1988, Shimizu et al. 1989). Neutron exposures now appear much smaller than previously estimated, and the Hiroshima and Nagasaki dose-response curves are no longer significantly different.

Leukemia

Compared to other tumors, leukemia has one of the highest relative risk coefficients. It also has the shortest time to onset, with the minimum latency being about 2 years. At 1 Gy, whole body exposure, a 6.21 RR is estimated, whereas it is 1.41 for all other cancers combined. More than half of the 200 leukemias (112 of 202) occurring among atomic bomb survivors exposed to more than 1 cGy were attributed to radiation (Shimizu et al. 1990). No diagnoses of chronic lymphocytic leukemia have been made. This risk of CML was high, and the wave-like response quite evident, especially in Hiroshima. CML has been

Table 6.4. Studies of Patients Exposed to Atomic and Thermonuclear Weapons with Quantitative Estimates of Risk

Population	Excess Cancer
Japanese atomic bomb survivors	Leukemia, lung, breast, thyroid, esophagus, stomach, colon, ovary, bladder, myeloma
Marshall Islanders (fallout)	Thyroid

thought to be the most characteristic leukemia of the atomic bomb survivors. Acute myelogenous leukemia (AML), with over 100 cases, is the most common leukemia, with excesses occurring at all ages. The risk for acute lymphocytic leukemia was somewhat higher than for AML and decreased more rapidly, with the excesses predominantly among the younger survivors. No association with radiation was found for adult T-cell leukemia (Ichimaru et al. 1986, NAS 1990).

Breast

The dose response for breast cancer incidence was linear in both cities, and age at exposure strongly influenced risk (Tokunaga et al. 1987). The risk was highest among women irradiated before age 20. Increased risks were also seen for exposures prior to age 10. A lesser risk was observed among survivors aged 20–40 in 1945, whereas no risk was apparent in women exposed after age 40. Excess risk did not appear until 10 years after exposure and not before about age 30. Estimates of risk at 1 Gy are roughly 2.68 for the RR, and 4.1 cancers/10^4 PY-Gy.

Lung

All major types of lung cancer have occurred in excess, including adenocarcinoma, epidermoid carcinoma, and small-cell carcinoma (Yamamoto et al. 1987). Age at exposure had little influence on the level of risk, although those under age 10 at exposure have not shown a risk. RR diminished nonsignificantly with time after exposure, most noticeably during the latest follow-up periods (NAS 1990, Shimizu et al. 1990). Females had higher RR than males; however, this difference was substantially reduced when adjusted for smoking history (Prentice et al. 1983). Lung cancer case-control data appeared consistent with an additive relation between smoking and radiation (Blot et al. 1984), although most analyses could not clearly distinguish between additive and multiplicative models (Prentice et al. 1983). The RR at 1 Gy was 1.63, and the absolute excess was 1.68/10^4 PY-Gy.

Thyroid

Significant excesses of papillary and follicular carcinoma of the thyroid, but not medullary or anaplastic cancer, have been reported (Prentice et al. 1982, Wakabayashi et al. 1983). The dose response was consistent with linearity; RRs were similar for males and females and highest among those under age 20 at exposure. Because women are more likely to develop thyroid cancer than men, equality of the RRs implies higher absolute risks among women. This is one of the few studies suggesting an excess of thyroid cancer following exposures in adult life.

Other Cancers

Increased deaths were reported for cancers of the esophagus, stomach, colon, lung, breast, ovary, and urinary bladder (Shimizu et al. 1990). Multiple myeloma, but not malignant lymphoma, was excessive. There was a suggested increase for tumors of the central nervous system, excluding the brain. No excesses were seen for cancers of the rectum, gallbladder, pancreas, liver, uterus, bone, oral cavity and pharynx, nasal passages, larynx, or skin. The minimal latent period for radiogenic cancers other than leukemia was at least 10 years, and subsequent excess risks were roughly proportional to the background rates (Pierce et al. 1991). Relative and absolute risks differed significantly by age at exposure, with younger survivors having higher risks. For nonleukemia deaths, the dose-response gradient appeared to be linear. The RR at 1 Gy was 1.41, and the absolute excess was $10.1/10^4$ PY-Gy. Among the 41,791 survivors exposed to more than 1 cGy and followed through 1985, 8.1% of the nonleukemia deaths have been attributed to the atomic radiation in 1945 (Shimizu et al. 1990).

General Comments

Despite its strengths, the study of atomic bomb survivors is not large enough to provide direct evidence of the effects of low doses (less than about 0.2 Gy) and thus can be used at best to limit the range of plausible dose-response models. At high doses, there were only about 6000 survivors who received over 1 Gy. Also, since the exposures were acute and the dose rate was high, the data provide no direct information on the effects of protracted low dose rate exposures. Additional factors that might affect interpretations include the relative absence of healthy men of military age in the cities in 1945, restricting the study population to 5-year survivors, possible effects of thermal or mechanical injury on those close to the hypocenter, possible effects of poor nutrition or health problems on subsequent cancer risks, and inaccuracies in dose assessment and death certificate diagnoses of cancer. Current efforts to study cancer incidence with registry data should provide new information on radiation risks.

Radioactive Fallout

Marshall Islands

Over 200 native residents of four inhabited atolls east of Bikini Island were accidentally exposed to nuclear fallout from the BRAVO weapons test in 1954 (Adams et al. 1989, Conard 1984). Thyroid dose, from gamma radiation plus radioiodines, was estimated as 7–14 Gy to children, depending upon age, and 4 Gy to adults. During 30 years of observation, 48 (or 17%) of 253 subjects developed thyroid neoplasia. The earliest thyroid lesions appeared 9 years after exposure. Nodularity was highest for children under age 10. Possibly, the rapid growth of the thyroid gland, from 1–2 g at birth to 18 g at maturity, influenced risk. Dosimetry for the radioiodines is complex. Most beta particle energy from iodine-131 is supposedly deposited in the colloid of the large follicles without reaching the critical follicular cells, and the low dose rate would tend to minimize risk. In contrast, the shorter-lived and more energetic isotopes (iodine-132, iodine-133, and iodine-135) contributed two to three times the dose of ^{131}I, and more uniformly exposed the thyroid at a higher dose rate. A new study of 7000 Marshall Islanders from 14 atolls, including several not previously studied, revealed a linear relationship between thyroid nodules and proximity to Bikini (Hamilton et al. 1987). An estimate of 11 nodules/10^4 PY-Gy was based on clinical examinations.

UTAH

Thyroid. Clinical examination of 2945 children in the 6th–12th grades from 1965–1971 were conducted in two counties in Utah and Nevada that received fallout from nuclear weapons tests at the Nevada Test Site (NTS) in the 1950s, and on 2271 children in Arizona exposed to negligible fallout. There was little suggestion of increased thyroid neoplasm among children exposed to fallout (Rallison et al. 1975). Assuming that ^{131}I was concentrated in milk and subsequently drunk by children, the thyroid dose was estimated to be between 0.3 and 2.4 Gy (mean, 1.2 Gy). An excess of 73–100 thyroid nodules at these doses would have been expected based on radiation risk estimates from studies of x-irradiation in childhood.

Nearly 75% of the original cohort of children were re-examined in 1985–1986. Again, there was no evidence to suggest that living near the NTS was associated with increased rates of thyroid neoplasms (Rallison et al. 1990). The absence of excess thyroid cancer or nodules could be partially explained if ^{131}I is not as biologically effective in inducing thyroid tumors as external x-rays, possibly because the low dose rate permits biological repair or because much of the beta particle energy is deposited in the colloid and not the follicular cells; or the actual thyroid dose was much lower than estimated.

Leukemia. A geographical study of childhood cancer mortality within Utah suggested an increase of leukemia in "high-fallout" counties near the NTS (Lyon et al. 1979). However, there was a significantly low risk of other childhood cancers in the high-exposure counties, and no overall association with total childhood cancers (Land 1979, National Council on Radiation Protection and Measurements [NCRP] 1980). Attempts to duplicate the leukemia findings using county mortality statistics from the National Center for Health Statistics (NCHS) also failed (Land et al. 1984).

A recent case-control study of over 1000 individuals who died of leukemia in southwestern Utah reported a weak association between estimated bone marrow dose and all leukemia, though the trend was not significant (Stevens et al. 1990). Significant risks, however, were observed for acute leukemia among those under age 20 when exposed to fallout, consistent with that expected based on other studies of exposed populations. The increasing trends seen for CLL, a tumor not known to be elevated after irradiation, and the difficulty in estimating doses retrospectively add caution to the causal nature of the associations reported.

NUCLEAR WEAPONS TEST PARTICIPANTS

NEVADA

No excess in total cancer mortality (112 versus 117.5) was found among 3017 of 3217 participants in military maneuvers during the 1957 nuclear test SMOKY (Caldwell et al. 1983). Leukemia, however, was significantly elevated; 10 cases were observed, including the index case that prompted the investigation and 1 case that developed after radiation therapy for lymphoma, versus 4.0 expected. Lower cancer frequencies were generally noted among the military units with the highest exposures based on film badge doses (mean, 0.46 cGy). A survey of 46,186 military participants in two test series conducted at the NTS and three in the Pacific also reported no excess of nonleukemia deaths (990 versus 1187) (Robinette et al. 1985). Excluding SMOKY, 46 leukemia deaths occurred versus 52.4 expected, suggesting that the leukemia excess among SMOKY participants was either due to chance or to circumstances peculiar to that shot (or its participants).

United Kingdom

Cancer mortality and incidence among 22,347 participants in the United Kingdom's atmospheric nuclear weapons tests in Australia and the Pacific Ocean between 1952 and 1967 and in 22,326 matched controls were recently evaluated (Darby et al. 1988). Mortality from all causes (SMR = 80 and 79) and from all cancers (SMR = 80 and 83) was comparable. Leukemia and myeloma risks were significantly higher among participants than controls, whereas controls had significantly higher levels of prostate and kidney cancer. The findings are difficult to interpret because the possible increased risks of leukemia and multiple myeloma were related more to significant deficits among the controls than to excesses among the exposed.

EXPOSURE TO RADIONUCLIDES (Table 6.5)

Radium-224

Injection with radium-224 to treat bone tuberculosis and ankylosing spondylitis in Germany resulted in 56 osteosarcomas versus 0.3 expected (Chmelevsky et al. 1988, Mays and Spiess 1984, Spiess et al. 1989). Similar to the pattern seen for radiogenic leukemia, bone cancer risk peaked at 6–8 years and decreased to normal level after about 33 years. Children and adults were at similar risk, as were males and females. A linear-quadratic-exponential equation best described the data. Average dose to the bone volume was estimated as 4.2 Gy, the excess risk as 0.8 cancers/10^4 PY-Gy, and the lifetime risk as 0.8% per person per Gy. Protracted exposure to radium alpha particles appeared more carcinogenic than acute exposures. This "protraction-enhancement" from alpha-emitting radiation might be related to less killing of premalignant cells, exposing more cells, increasing the stimulus for cell division, and/or preventing repair of local damage. For x- and gamma rays, a decrease in the carcinogenic effectiveness of a given dose generally occurs when dose is delivered over longer periods of time (NCRP 1980).

Table 6.5. Studies of Patient Populations Exposed to Internal Radionuclides

Population (Isotope)	Excess Cancer
German patients with bone disease (^{224}Ra)	Bone, possibly liver and breast
Thyroid cancer (^{131}I)	Leukemia, bladder
Polycythemia vera (^{32}P)	Leukemia
Thorotrast patients (^{232}Th)	Liver, leukemia, lung, possibly bone and others

Risk per unit dose was 10 times greater than that seen in radium dial painters who ingested radium-226 and radium-228 (Rowland et al. 1978), reflecting, perhaps, different dose distributions in bone. ^{224}Ra has a short half-life (3.62 days) and releases its energy on bone surfaces where the "critical" cells for osteosarcoma induction, the endosteal cells, are located. In contrast, ^{226}Ra is a bone-volume seeker with a very long half-life (1600 years), and distributes its energy more uniformly throughout the bone, the dose to the bone matrix being essentially irrelevant to risk. For the same average bone dose, the risk from ^{224}Ra will always be greater than that from ^{226}Ra because more radiation will reach the endosteal cells.

Therapeutic Iodine-131

No radiation-induced leukemia risk was seen in a study of 30,000 patients treated for hyperthyroidism (Saenger et al. 1968). Among 18,400 patients treated with ^{131}I, 17 leukemias developed, contrasted with 16 in 10,700 patients who received surgery only. The value of a nonexposed comparison group was underscored in that an excess of leukemia was suggested when general population rates were used for comparison. The dose to the whole body was low (7–13 cGy). Other hyperthyroid series have also failed to link ^{131}I with leukemia (Hoffman 1984, Holm et al. 1991).

Thyroid cancer has not been correlated with ^{131}I therapy (Dobyns et al. 1974; Holm et al. 1991), possibly because of the cellular destruction and loss of thyroid function that follows a dose of 10–100 Gy to the thyroid. Excess cancer of organs that concentrate iodine (Hoffman 1984) and of the breast (Goldman et al. 1988) has been suggested but not confirmed in a large incidence series of 10,552 patients followed for up to 30 years (Holm et al. 1991). In a study of 258 persons treated with high-dose ^{131}I for inoperable thyroid cancer, 4 leukemias were observed versus 0.08 expected based on general population rates (Edmonds and Smith 1986). No excess leukemia (3 versus 1.6) was reported among 834 patients with thyroid cancer in Sweden (Hall et al. 1991). The doses to the bone marrow and other organs in these series were large and likely between 50 and 100 cGy.

Diagnostic Iodine-131

A recent study of 35,074 Swedish patients failed to link the incidence of any cancer with diagnostic doses of ^{131}I (Holm et al. 1988, 1989). The dose to the thyroid was 50 cGy, so that a substantial excess of thyroid cancer was expected. The absence of an effect supports the notion that internal ^{131}I beta particles are much less carcinogenic than external x-rays or gamma rays, perhaps due to the protracted nature of the exposure or to the distribution of dose within the gland from ^{131}I (half-life = 8 days).

Phosphorus-32

Leukemia has been linked to phosphorus-32 treatment for polycythemia vera (PV), a blood disease characterized by overproduction of red cells (Modan and Lilienfeld 1964). The level of risk associated with ^{32}P exposure cannot be estimated with certainty because the bone marrow of patients with PV may be unusually sensitive to radiation; the underlying myeloproliferative disease may predispose to leukemia; and PV patients are exposed to other medications, including powerful cytotoxic drugs, that could increase leukemia risk. However, a randomized clinical trial found that 9 of 156 (6%) patients treated with ^{32}P developed leukemia in contrast to 1 of 134 (1%) treated by phlebotomy (Berk et al. 1981). Patients treated with chlorambucil were at highest risk (11%).

Thorotrast

A colloidal solution of thorium dioxide (Thorotrast) was used between 1928 and 1955 as a contrast agent during radiographic procedures. The thorium, however, remained in body tissue and resulted in continuous alpha particle exposure throughout life at a low dose rate. The annual dose from a typical injection of 25 ml of Thorotrast was 25 cGy to liver and 16 cGy to bone marrow. Surveys in Denmark, Germany, Japan, and Portugal show substantial excesses of liver cancer, including angiosarcoma and cholangiocarcinoma, and AML (NAS 1988; Taylor et al. 1989). Among 2326 exposed persons in the German Thorotrast study, 396 (17%) have died from liver cancer in contrast to only 2 (0.1%) among 1890 controls (NAS 1988, van Kaick et al. 1989). Hemangioendothelioma of the liver appears uniquely related to Thorotrast. Excess lung cancer has been reported in some series, suggesting a possible effect of exhaled thoron (radon-220) (Olsen et al. 1989). Bone cancer has also been occasionally reported to be increased, possibly due to translocating ^{224}Ra from Thorotrast deposits (NAS 1988).

The nonuniform deposition of thorium in the liver likely resulted in very high local doses, which may be the important determinant of cancer risk. If so, the convention of averaging dose over the entire organ would be misleading. Risk estimation is also hindered because the chemical nature of thorium, a heavy metal, may be related to risk. The average dose to the liver was about 5 Gy, and a portion of this radiant energy, expended in necrotic tissue, was probably not essential for carcinogenesis. The completeness of patient follow-up was generally poor. The combination of necrosis and liver regeneration might influence risk. Nonetheless, Thorotrast appears to be one of the most carcinogenic exposures known to man, with cumulative lifetime incidences of cancer estimated to be as high as 60%.

OCCUPATIONAL EXPOSURE (Table 6.6)

Luminous Dial Painters

Among 1474 women employed in the U.S. radium dial industry before 1930, 61 bone sarcomas and 21 head carcinomas have occurred (Rowland et al. 1978,

Table 6.6. Studies of Populations Occupationally Exposed to Radiation

Population	Excess Cancer
Radiologists	Leukemia, skin; possibly breast, thyroid, myeloma; no lung
Underground miners	Lung
Luminous dial painters (^{226}Ra)	Bone, possibly myeloma

Rundo et al. 1986, Stebbings et al. 1984). The habit of licking paint brushes to make fine tips resulted in the ingestion of large quantities (mean, 17 Gy) of bone-seeking ^{226}Ra, and some quantities of ^{228}Ra. Cancers in mastoid air cells or paranasal sinuses (head carcinomas) likely were caused by radon gas emitted as a decay product of radium. The "latency period" for osteosarcoma was not related to dose (Polednak 1978). Age at first exposure did not influence risk. Risk was estimated as 0.1 bone cancers/10^4 PY-Gy. An equation of the form $I = (c + \beta D^2)e^{-\alpha D}$ fits the osteosarcoma data, whereas a linear form, $I = c + \beta D$, fits the head carcinoma data. The actual dose-incidence curve determined for radium dial painters must be considered tentative for the following reasons: The estimation of dose was made many years after the ingestion of radium; the nonuniform distribution of radium in bone likely resulted in "hot spots" that caused extensive cell killing; the dose responsible for tumor induction cannot be distinguished from the "irrelevant" dose received after initiation; and the fraction of the total dose to the endosteal cells cannot be specified precisely.

No excess of leukemia was observed (10 versus 9.24 expected) among U.S. dial painters (Spiers et al. 1983) or among 1100 English radium luminisers (Baverstock and Papworth 1989). Early reports linking breast cancer with radium or external gamma ray exposures were not confirmed (Baverstock and Papworth 1989, Stebbings et al. 1984). Multiple myeloma was increased in the U.S. study but correlated with duration of employment (a surrogate for gamma ray exposure) rather than radium intake. Liver cancer was not increased. The British study reported only 1 osteosarcoma, but the systemic intake of radium was much lower than for the United States.

Radiologists

The first cancer attributed to radiation occurred on the hand of a radiologist in 1902, and leukemia was first associated with chronic exposure in studies of radiologists (March, 1944). Leukemia, aplastic anemia, and skin cancer were excessive among radiologists who practiced during the early part of this century before radiation protection guidelines were in use, but these risks appear to have disappeared among more recent radiologists (Matanoski et al. 1975, Smith and Doll 1981, Wang et al. 1990a). Multiple myeloma was increased among U.S. radiologists practicing in later years (Lewis 1963, Matanoski et al. 1975), but not among English or Chinese radiologists. Cancers of the pancreas and lung were increased among the pioneering radiologists in the United Kingdom, but not in the United States or China. Suggested increases of breast, thyroid, and bone cancers were correlated with radiation work in China only (Wang et al. 1990a). Neither leukemia nor cancer was reported to be in excess among U.S. Army x-ray technologists, who likely received much lower total doses (Jablon and Miller 1978).

The absence of radiation dose estimates is a serious limitation of these studies. Cumulative doses were likely between 1 and 8 Gy during the early part of this century, and it is possible that radiologists who developed cancer were those who scorned safety measures and received even greater doses. It was not uncommon for x-ray workers to be given time off from work because of severe depression of white blood cell counts. Radiologists also receive more personal (non occupational) exposures to diagnostic and therapeutic radiation than other specialists (Jessup and Silverman 1981). Nonetheless, these studies indicate that leukemia and skin cancer can result from repeated, presumably small, radiation exposures received over a period of many years.

Workers at Nuclear Installations

The mortality experience of about 31,500 male and 12,600 female workers employed between 1944 and 1978 at the Hanford nuclear installation in Richland, Washington, has been reported by several investigators. An early proportional mortality analysis on 3520 certified deaths (Mancuso et al. 1977) was widely criticized (Hutchison et al. 1979). The most recent analyses revealed a strong healthy worker effect, a significant deficit of cancer mortality, including leukemia, and no evidence for increasing risk with increasing film badge exposure for any cancer, except multiple myeloma (Gilbert 1989a, 1989b).

Studies of workers at nuclear installations are generally inconsistent. Leukemia was not in excess among Sellafield workers in the United Kingdom, although an increase was suggested when exposure was lagged 15 years (Smith and Douglas 1986). An inverse association between leukemia and film badge dose was seen for all workers at the Oak Ridge National Laboratory (Checkoway et al. 1985, Gilbert et al. 1989b). Multiple myeloma was elevated in two studies, although based on only two cases receiving over 50 cGy at Sellafield (Smith and Douglas 1986), and only three cases over 5 cGy at Hanford (Gilbert et al. 1989b). Prostate cancer was increased at the U.K. Atomic Energy Authority and the U.K. Atomic Weapons Establishment (Beral et al. 1985, 1988), and bladder cancer at Sellafield (Smith and Douglas 1986). The pancreatic cancer excess, reported in earlier Hanford studies has now disappeared (Gilbert et al. 1989a). Significant deficits of lung cancer are often found (Gilbert et al. 1989b), and an elevation has been suggested in only one series (Beral et al. 1988).

Recently, radiation risks were asserted to be an order of magnitude higher than currently accepted (NAS 1990) in a report of workers at the Oak Ridge National Laboratory (Wing et al. 1991). Considerable caution, however, must be exercised before accepting this interpretation (Howe 1991). An excess of leukemia, as reported previously, was highlighted and radiation risk coefficients computed, but the risk was seen to decrease with increasing levels of exposure. Further, CLL was grouped with other cell types in most analyses, although this malignant condition has never been found to be increased following radiation (NAS 1990). Analyses also apparently failed to take proper account of the confounding influence of socioeconomic factors, which, if adjusted for, likely would have negated the significance of the findings for lung cancer, which dominated the analyses. Compared to the general population, the Oak Ridge workers were at a 21% significantly reduced risk of dying from cancer.

A proportional mortality study of naval shipyard workers suggested an increased risk of cancer and leukemia among nuclear workers in Portsmouth, New Hampshire (Najarian and Colton 1978), which was not borne out in a subsequent cohort study (Rinsky et al. 1981). It was found that radiation exposure histories, ascertained from next of kin by newspaper reporters, did not correlate at all with employment records. Further, relatives were more likely to be located and interviewed for workers who died of cancer, which, in combination with a lower all-cause mortality among nuclear workers, contributed to the spurious result (Greenberg et al. 1985). Case-control studies of leukemia and lung cancer also have failed to find an association with radiation work (Rinsky et al. 1988, Stern et al. 1986).

Combining the various studies of nuclear facility workers might improve interpretation. The first comprehensive effort pooled data from four U.S. studies (Gilbert et al. 1989b). Risk estimates were negative for all cancer and for leukemia, and only multiple myeloma, excessive at only the Hanford site, was increased. The authors concluded that data from higher dose studies are unlikely to overestimate risks at lower doses. Studies of nuclear utility workers may eventually provide useful information on radiation risks because of relatively higher exposures and larger numbers (Goldsmith et al. 1989). Occupational studies of radiation workers, however, must be interpreted with caution because film badge exposures are imperfect measures of organ doses; the dose from natural background radiation (about 7 cGy in 70 years) is often greater than the occupational dose; other occupational and nonoccupational carcinogens are usually not considered; and ascertainment bias is possible if the working population receives slightly better medical care and more accurate cancer diagnoses recorded on death certificates than the general population.

Workers Exposed to Plutonium

At one time, plutonium was thought to be one of the most toxic elements known to man. Animal experiments clearly indicate that excessive exposure to plutonium can cause cancers of the lung, bone, and liver; however, the evidence in man is sketchy (NAS 1988). Twenty-six heavily exposed individuals (1.4 to 86 nCi) who worked on the Manhattan project in Los Alamos and who received extensive clinical examinations over the last 40 years, have not died at a higher rate than expected (Voelz et al. 1991). However, three lung cancers occurred in long-term smokers, and the one osteosarcoma is noteworthy. In 224 males with plutonium body burdens of 10 nCi or more, a lower than expected cancer mortality was reported, and lung cancer and osteosarcoma were not excessive (Voelz et al. 1983). Studies of large populations of workers exposed to lower levels of plutonium also find no correlation between plutonium intake and cancer mortality (Gilbert et al. 1989a, Wilkinson et al. 1987).

Underground Miners

Among 3366 white and 780 nonwhite U.S. uranium miners who worked at least 1 month underground, a substantial excess of respiratory cancer was observed: 155 deaths versus 30.4 expected based on general population rates (Hornung and Meinhardt 1987). Lung irradiation was from inhaled radon gas, and the mean dose was about 4.7 Gy (NAS 1972). Most lung cancers developed in cigarette smokers, but an excess risk was also seen among 516 nonsmoking miners exposed to very high radon levels (Roscoe et al. 1989). Radiation and smoking appear to interact in a way that enhances risk, though somewhat less than multiplicatively (NAS 1988). Dose-response data for lung cancer in the U.S. miners is difficult to interpret because of uncertainties about doses to individual miners, which had to be estimated based on infrequent measurements in over 2500 mines (Lundin et al. 1971); dose to respiratory tissue, which could vary depending on the thickness of the epithelial and overlapping mucous layers, and the clearance rate of absorbed radioactive particles; and the uncertain contribution to risk of cigarette smoking and of pollutants, like diesel exhaust, in mine atmospheres.

Many of the above concerns were addressed in miner studies in Sweden (Radford et al. 1984), Canada (Howe et al. 1986, 1987), Czechoslovakia (Sevc et al. 1988), and other countries (NAS 1988) for which radon exposures were lower and better characterized than in Colorado mines. The most comprehensive analysis to date from four major studies concluded that the dose-response relationship between cumulative radon exposure and lung cancer was linear, that the increase in RR per working level month (WLM) was approximately 2.5%, that age at first exposure had little influence on risk, and that relative (but not absolute) risk declined with both time since exposure stopped and attained age (NAS 1988). Further, several studies suggest that exposures at a low dose rate may be more harmful than at a high dose rate for the same total dose (Hornung and Meinhardt 1987, Howe et al. 1987, Lubin et al. 1990a).

ENVIRONMENTAL EXPOSURES

Indoor Radon

Although it has long been recognized that radon causes lung cancer among underground miners (NAS 1988, Samet 1989), the possible hazard to homeowners exposed to much lower levels was not appreciated until recently. Indoor radon accounts for over half of all radiation exposures received by the general population, and, based on extrapolations from high-dose studies, may cause between 6600 to 24,000 lung cancer deaths per year in the United States (Lubin and Boice 1989).

The evidence that radon causes lung cancer comes mainly from miners exposed to well in excess of 100 WLM, and no risk below about 50 WLM has been convincingly reported (NAS 1988). In contrast, average levels of indoor radon are estimated as 0.2 WLM/yr (NCRP 1984). Several case-control studies in Sweden showed positive associations between lung cancer and living in structures where radon levels were supposedly high (Samet 1989). In New Jersey, lung cancer risk was nonsignificantly increased more than twofold among women living in homes with 4+ versus <1 pCi/liter radon levels, but the 4+ exposure group consisted of only six cases and two controls (Schoenberg et al. 1990).

The largest and most comprehensive case-control study to date consisted of 308 women recently diagnosed with lung cancer in China. Little evidence was found for an increase in risk of lung cancer with increasing radon exposure (Blot et al. 1990). Year-long measurements of radon were made in current residences, and 20% of the readings exceeded 4 pCi/liter, the level above which remedial action is suggested in the United States. It was concluded that projections from surveys of miners exposed to high radon levels may overestimate the risks of lung cancer associated with levels typically seen in homes. Studies of indoor radon, however, have to be interpreted with caution because of the inherent difficulties in accurately estimating past exposures and in adequately adjusting for the effect of smoking. Exposure reconstruction is complicated further because of migration, home modifications, and uncertain estimates of occupancy (Lubin et al. 1990b).

Natural Background Radiation

Correlation studies attempting to link cancer mortality with natural background radiation have generally been negative but are fraught with uncertainties about dose levels, migration patterns, selection factors for place of residence, and geographic variations in the accuracy of cancer diagnoses (Boice and Land 1982). An investigation in China failed to detect an excess of cancer, including lung cancer, in a stable population of 73,000 persons who received three times the amount of background radiation as 77,000 inhabitants in a comparison region (Hofmann et al. 1986, Wei 1980). Although the accuracy of cancer diagnoses in these areas may be in doubt, a study of women that included physical examinations also found no difference in thyroid nodularity, a sensitive indicator of low-dose radiation effects, among female residents of the high (14 cGy) and low (5 cGy) background areas (Wang et al. 1990b). A dose of 9 cGy accumulated gradually over a lifetime thus produced many fewer tumors than seen following a similar dose of x-rays received briefly in childhood (Ron et al. 1989). Recently, childhood cancer has been correlated with maternal irradiation from background sources (Knox et al. 1988) and leukemia with radon (Henshaw et al. 1990) in England, but the interpretations remain clouded by the serious limitations of ecological correlation analyses (NAS 1990).

Cancer Clusters around Nuclear Installations

Reports of small clusters of childhood leukemia around nuclear installations in the United Kingdom began appearing in 1984, most notably in the village of Seascale, near the Sellafield (Windscale) nuclear fuel reprocessing plant in West Cumbria. Interestingly, the excess of leukemia occurred entirely among individuals born in Seascale (5 versus 0.53) and not among children born elsewhere (0 versus 0.54) (Gardner et al. 1987). A systematic survey in 1987 also found lymphoid leukemia among persons under age 25 to be generally increased in populations living near nuclear fuel reprocessing or weapons production facilities (Cook-Mozaffari et al. 1987, Forman et al. 1987). Mortality from Hodgkin's disease at ages 0–24 was also increased, whereas mortality from lymphoid leukemia at ages 25–64 was significantly reduced. Overall, there was no general increase in cancer deaths in the vicinity of nuclear installations.

In the largest systematic survey to date, over 900,000 cancer deaths in 113 counties in the United States containing or adjacent to 62 nuclear facilities were compared to 1,800,000 cancer deaths in control counties with similar population and socioeconomic characteristics (Jablon et al. 1990, 1991). Overall, and for specific groups of nuclear installations, there was no evidence that mortality for any cancer, including childhood leukemia, was higher in the counties with nuclear reactors than in the control counties. For childhood leukemia, the RR in the study counties versus their controls after plant start-up was 1.03, while before start-up it was 1.08. For all leukemia, the RRs were 0.98 after start-up and 1.02 before. Leukemia was also not excessive around French nuclear sites (Hill and Laplanche 1990).

Annual doses received by persons living near nuclear installations are estimated to be much below 0.1 mSv/yr, or just a fraction of natural background radiation (Darby and Doll 1987, NCRP 1987, UNSCEAR 1988). Such levels are much lower than those expected to cause a detectable increase in childhood leukemia. Interestingly, a recent study from Britain evaluated residents of areas where construction of nuclear power stations had only been considered, or just recently built. Excesses of childhood leukemia and Hodgkin's disease, and deficits of adult leukemia, were reported that were similar to those previously identified in areas with operating nuclear facilities (Cook-Mozaffari et al. 1989). The authors concluded that the unexpected increases in some childhood cancers around nuclear installations are unlikely to be due to environmental radiation pollution but rather to other risk factors as yet unidentified. An infective agent associated with large migrations of people into these areas, for example, has been proposed as one possible explanation (Kinlen 1988).

The ecological correlation analyses are not without problems since radiation dose to the population is unknown, although likely much below natural background; mortality is not the best indicator of cancer hazard due to inaccuracies of death certificates, while cancer registration may also be incomplete and variable between areas; other important risk factors often cannot be identified; the county or district may be too large an administrative unit to detect localized increases in cancer rates; and the many comparisons made with regard to individual cancers, ages, and time frames increase the likelihood that chance plays a role in highlighting areas with seemingly high, or low, cancer rates (Boice 1991).

Preconception

Occupational exposure at Sellafield of fathers prior to conception was recently linked to exceptionally high (six- to eightfold) risks of leukemia among offspring under age 25 (Gardner et al. 1990). The study is difficult to interpret, however, since the number of cases was small, only 52; cases occurred over a 35-year period; over 150 comparisons were made; significantly high risks were based on only four cases with paternal exposures prior to conception; these same four cases lived within 4 km of a nuclear reprocessing facility; information on potential confounders such as prenatal x-ray was poor; information was obtained more completely on cases than on controls; and the data seem at odds with the large series of atomic bomb survivors in which no excess cancers (Yoshimoto et al. 1990), chromosome aberrations, or genetic mutations in blood proteins were found in over 5000 children born to proximally exposed survivors compared to over 5000 children born to distally exposed survivors (Neel et al. 1990, UNSCEAR 1988). Experimental studies also indicate that chronic exposures, such as experienced among Sellafield workers, should produce fewer mutations than equivalent acute doses. Further, there are virtually no human data to suggest that leukemia is an inherited disease. If radiation were acting to cause heritable mutations, then an epidemic of known heritable diseases and congenital malformations might be expected in the Sellafield area whereas none was noted (Evans 1990). Congenital malformations have also not been linked to occupational or environmental radiation exposures at the Hanford nuclear installation in Washington state (Sever 1988a, 1988b).

On the other hand, paternal occupation was a prior hypothesis, although the scientific justification for this hypothesis seems unclear. One experimental study suggests that an acute radiation exposure (or chemical exposure) might increase the risk of cancer, primarily of the lung, in subsequent progeny, but the experiment lacked concurrent controls and has not been replicated (Nomura 1982). It was unlikely that there would be bias for occupational variables obtained through birth certificates and record linkage. The dose-response relationship adds to the possibility that an association of etiological significance was found. An attempt to replicate these provocative findings, however, was then made at the Dounreay nuclear facility, but the results were negative (Urquhart et al. 1991) .

Another recent report purports to corroborate the preconception effect found at Sellafield (McKinney et al. 1991). Unfortunately, this new study must be interpreted with caution because radiation exposures were not verified, there was no information on the dose of radiation received, and, perhaps more importantly, the study included many of the same fathers previously evaluated in the Gardner et al. investigation that raised the hypothesis (Smith 1991). Excluding these overlapping individuals, as would be appropriate for an independent assessment, meaningfully reduces the evidence for an association.

The initial reports of clusters of childhood leukemia around nuclear facilities in the United Kingdom produced a flurry of research activity but, to date, no clear insights into the possible causes for the apparent clusterings. The excess leukemias do not appear related to the radiation doses received by children as a result of radioactive emissions or discharges. An explanation in terms of preconception radiation of the fathers remains a provocative hypothesis, and additional studies have been recommended, such as among pioneering radiologists or cancer patients treated with radiation who later became parents. No excess childhood cancer, however, has been reported among offspring of long-term survivors of cancer (Mulvihill et al. 1987). Another hypothesis being pursued is the possibility that childhood leukemia may occur as a rare response to an unidentified infection whose transmission is facilitated when large numbers of people come together, such as might occur when large industrial complexes are built in rural areas (Kinlen et al. 1990).

Nuclear Reactor Accidents

The nuclear reactor accident at Three Mile Island released little radioactivity into the environment, and much less than the annual population exposure to natural background. Any presumed increase in cancer cases at these levels would be negligible and undetectable. An ecological survey did not link increased cancer rates with possible patterns of radiation releases (Hatch et al. 1990), nor were any peculiar mortality patterns noted (Jablon et al. 1990). In contrast, the accident at Chernobyl resulted in a massive release of radioactivity, and thus far 237 cases of acute radiation sickness, including 31 deaths, have been reported. The extent to which persons who resided near the reactor site are being studied is unclear, but the purported doses of between 10 and 50 cGy are of scientific interest (Ginzburg and Reis 1991). Further, the many thousands of workers who were involved in the environmental cleanup should be studied for possible late effects. Doses received by many of these workers have been estimated to be well over 20 cGy.

CAUTIONS IN INTERPRETING EPIDEMIOLOGICAL STUDIES

Epidemiological studies have conclusively linked high-dose radiation to increased cancer risks. Estimates of risk for low-level exposures, however, are based on uncertain extrapolations and require assumptions about the shape of the dose-effect relationship and about the mechanisms of radiation carcinogenesis (Upton 1987, 1990). These assumptions are often guided by experimental investigations and radiobiology theory and then applied to epidemiological data. Studies of radiogenic cancers, however, are also susceptible to the problems of any epidemiological investigation: inappropriate comparison groups; biases of selection, observation, and response; multiple comparisons; inadequate control of potential confounding; inadequate measures of dose and cancer outcome; inappropriate cluster analyses; and low statistical power due to small numbers at low doses (Beebe 1981, Land 1980, MacMahon 1989, NAS 1990, UNSCEAR 1988).

An appropriate comparison group for irradiated populations is important. Comparisons with general population rates could be misleading when the irradiated population is ill, since the disease itself could be related to the occurrence of cancer, and other competing risks or treatments could influence subsequent cancer incidence. Many employed populations are also self-selected, with different expectations of disease than the general population. Interpretation should be made cautiously when an elevated "risk" is related more to a decreased risk in the comparison group than to an increased risk among the exposed.

Searching for cancers more diligently among an irradiated population than among the general population used to compute expected numbers could bias study results in a positive direction (ascertainment bias). Special screening of exposed but not nonexposed persons is an extreme but not uncommon example of observational bias. It could, for example, account for many of the excess benign tumors seen in persons with histories of head and neck irradiation during childhood. In case-control studies of cancer and diagnostic x-rays, a respondent might be a more reliable witness if he or she recently developed a cancer (response bias).

Radiation dose should be accurately estimated, and dose-response gradients add meaningfully to causal interpretations. If doses are not known for individuals, an excess in a presumably low-dose group could be concentrated in those few individuals who received high doses. When the number of excess cancers is low, chance events are more likely to play a role, and the influence of confounding factors is more worrisome. Care should be exercised in performing statistical tests on investigations generated because of an apparent cluster of disease in the study population that prompted the investigation. The statistical problems of cluster analysis are complex, especially in light of possible gerrymandered observations. When multiple comparisons are made, investigators should be careful not to overly interpret unexpected findings.

The confounding influence of other carcinogenic exposures should be evaluated. Estimates of lung cancer risks obtained without adjusting for smoking or estimates of occupational risks obtained without controlling for toxic chemical exposures may be misleading. Other radiation exposures, such as excessive medical x-rays or environmental background radiation, are not often considered in low-dose studies, despite the possibility that their contribution to the total radiation burden may be greater than the exposure under study (Spiers 1979). It is important that exposed populations be followed for long periods of time to detect excesses of solid tumors. If the atomic bomb survivor studies had been terminated after about 15 years, the leukemia effect would have been observed but not the solid tumor excess. Evidence that the exposure preceded the disease is obviously important. In the case of diagnostic radiation and adult leukemia, however, it was likely that many x-rays occurred during the early stages of cancer when symptoms were being evaluated but before any diagnosis was actually made or onset recognized. Consistency with other studies is important. Excesses in sites not frequently seen after radiation, such as prostate cancer or Hodgkin's disease, in the absence of elevations in leukemia or other sensitive sites, should be interpreted cautiously.

There are substantial difficulties in studying directly the carcinogenic effects of very small radiation doses, such as impracticably large study size requirements and the inability to control for other factors that influence cancer incidence. An unfortunate consequence of small study sizes (and in studies of cancer and low radiation dose, one million persons may be a small study size) is that when a positive result is obtained (and reported), it is likely to be mainly the result of random variation and, if so, must yield an overestimate of risk (Land 1980). Epidemiological methods are capable of directly detecting RRs perhaps as low as 1.4 (i.e., a 40% relative excess). However, the RRs of interest following low doses of radiation (1–10 cGy) are on the order of 1.01–1.05. Thus not much should be anticipated from direct observations at 1 cGy, and indirect approaches must be used to estimate low-dose effects. Such approaches include studies of populations exposed to a wide range of doses, both low and high, and the development of models to interpolate the effects at high and moderate doses to those at low doses.

Because of the multifactorial nature of carcinogenesis, studies should be encouraged that integrate biochemical and molecular measures of response (Brachman et al. 1991, Elkind et al. 1991, Harris 1991, Langlois et al. 1987, Weichselbaum et al. 1989). Such studies are needed to provide insights into the mechanisms of carcinogenesis, and to help in the setting of guidelines for occupational, medical, and environmental exposures to radiation.

REFERENCES

Adams, W. H., Heotis, P. M., Scott, W. A., et al. *Medical Status of Marshallese Accidentally Exposed to 1954 BRAVO Fallout Radiation: January 1985 through December 1987,* BNL 52192. Upton, NY, Brookhaven National Laboratory Report, 1989 (available from NTIS).

Baral, E., Larsson, L.-E., and Mattsson, B. Breast cancer following irradiation of the breast. *Cancer 40:*2905–2910, 1977.

Baverstock, K. F., and Papworth, D. G. The UK radium luminiser survey. In: *Risks from Radium and Thorotrast,* edited by D.M. Taylor, C.W. Mays, G.B. Gerber, and R. G. Thomas, BIR Report 21. London, British Institute of Radiology, 1989, pp. 72–76.

Beebe, G. W. The atomic bomb survivors and the problem of low-dose radiation effects. *Am. J. Epidemiol. 114:*761–783, 1981.

Beral, V., Fraser, P., Carpenter, L., et al. Mortality of employees of the atomic weapons establishment. *Br. Med. J. 297:*757–770, 1988.

Beral, V., Inskip, H., Fraser, P., et al. Mortality of employees of the United Kingdom Atomic Energy Authority, 1946–1979. *Br. Med. J. 291:*440–447, 1985.

Berk, P. D., Goldberg, J. D., Silverstein, M. N., et al. Increased incidence of acute leukemia in polycythemia vera associated with chlorambucil therapy. *N. Engl. J. Med. 304:*441–447, 1981.

Bithell, J. F., and Stiller, C. A. A new calculation of the carcinogenic risk of obstetric X-raying. *Stat. Med. 7:*857–864, 1988.

Blot, W. J., Akiba, S., and Kato, H. Ionizing radiation and lung cancer: A review including preliminary results from a case-control study among A-bomb survivors. In: *Atomic Bomb Survivor Data: Utilization and Analysis,* R. L. Prentice and D. J Thompson. Philadelphia, SIAM, 1984, pp. 235–248.

Blot, W. J., Xu, Z.-Y., Boice, J. D., Jr., et al. Indoor radon and lung cancer in China. *J. Natl. Cancer Inst. 82:*1025–330, 1990.

Boice, J. D., Jr. Carcinogenesis–a synopsis of human experience with external exposure in medicine. *Health Phys. 55:*621–630, 1988.

Boice, J. D., Jr. Epidemiologic studies of radioactively contaminated environments and cancer clusters. *NCRP Proc. 12:*94–116, 1991.

Boice, J. D., Jr., Blettner, M., Kleinerman, R. A., et al. Radiation dose and leukemia risk in patients treated for cancer of the cervix. *J. Natl. Cancer Inst. 79:*1295–1311, 1987.

Boice, J. D., Jr., Blettner, M., Kleinerman, R. A., et al. Radiation dose and breast cancer risk in patients treated for cancer of the cervix. *Int. J. Cancer 44:*7–16, 1989.

Boice, J. D., Jr., Day, N. E., Andersen, A., et al. Second cancers following radiation treatment for cervical cancer. An international collaboration among cancer registries. *J. Natl. Cancer Inst. 74:*955–975, 1985.

Boice, J. D., Jr., Engholm, G., Kleinerman, R. A., et al. Radiation dose and second cancer risk in patients treated for cancer of the cervix. *Radiat. Res. 116:*3–55, 1988.

Boice, J. D., Jr., Harvey, E., Blettner, M., et al. Contralateral breast cancer following radiotherapy for breast cancer. *N. Engl. J. Med.*, in press.

Boice, J. D., Jr., and Land, C. E. Adult leukemia following diagnostic x-rays? *Am. J. Public Health 69:*137–145, 1979.

Boice, J. D., Jr., and Land, C. E. Ionizing radiation. In: edited by D. Schottenfeld, J.F. Fraumeni, Jr. *Cancer Epidemiology and Prevention.* New York, WB Saunders, 1982, pp. 231–253.

Boice, J. D., Jr., Land, C. E., Shore, R. E., et al. Risk of breast cancer following low-dose radiation exposure. *Radiology 131:*589–597, 1979.

Boice, J. D., Jr., Morin, M. M., Glass, A. G., et al. Diagnostic x-ray procedures and risk of leukemia, lymphoma, and multiple myeloma. *J.A.M.A. 265:*1290–1294, 1991a.

Boice, J. D., Jr., Preston, D., Davis, F. G., and Monson, R. R. Frequent chest x-ray fluoroscopy and breast cancer incidence among tuberculosis patients in Massachusetts. *Radiat. Res. 125:*214–222, 1991b.

Boice, J. D., Jr., Rosenstein, M., and Trout, E. D. Estimation of breast doses and breast cancer risk associated with repeated fluoroscopic chest examinations of women with tuberculosis. *Radiat. Res. 73:*373–390, 1978.

Brachman, D. G., Hallahan, D. E., Beckett, M. A., et al. p53 gene mutations and abnormal retinoblastoma protein in radiation-induced human sarcomas. *Cancer Res. 51:*6393–6396, 1991.

Caldwell, G. G., Kelley, D., Zack, M., et al. Mortality and cancer frequency among military nuclear test (SMOKY) participants: 1957 through 1979. *J.A.M.A. 250:*620–624, 1983.

Checkoway, H., Mathew, R. M., Shy, C. M., et al. Radiation, work experience, and cause specific mortality among workers at an energy research laboratory. *Br. J. Ind. Med. 42:*525–533, 1985.

Chmelevsky, D., Kellerer, A. M., Land, C. E., et al. Time and dose dependency of bone-sarcomas in patients injected with radium-224. *Radiat. Environ. Biophys. 27:*103–114, 1988.

Conard, R. A. Late radiation effects in Marshall Islanders exposed to fallout twenty-eight years ago. In: Radiation carcinogenesis: epidemiology and biological significance, edited by J.D. Boice, Jr., J.F. and Fraumeni, Jr. New York, Raven Press, 1984, pp. 57–71.

Cook-Mozaffari, P. J., Ashwood, F. L., Vincent, T., et al. *Cancer Incidence and Mortality in the Vicinity of Nuclear Installations, England and Wales, 1959–1980,* OPCS Studies on Medical and Population Subjects 51. London, Her Majesty's Stationery Office, 1987.

Cook-Mozaffari, P., Darby, S., and Doll, R. Cancer near potential sites of nuclear installations. *Lancet 2:*1145–1147, 1989.

Curtis, R. E., and Boice, J. D. Jr. Second breast cancers after radiotherapy for Hodgkin's disease. N. Eng. J. Med. 319:244–246, 1988.

Curtis, R. E., Boice, J. D., Jr., Stovall, M., et al. Leukemia risk following radiotherapy for breast cancer. *J. Clin. Oncol. 7:*21–29, 1989.

Cuzick, J., and DeStavola, B. Multiple myeloma–a case-control study. *Br. J. Cancer 57:*516–520, 1988.

Darby, S. C., and Doll, R. Fallout, radiation doses near Dounreay, and childhood leukaemia. *Br. Med. J. 294:*603–607, 1987.

Darby, S. C., Doll, R., Gill, S. K., and Smith, P. G. Long term mortality after a single treatment course with x-rays in patients treated for ankylosing spondylitis. *Br. J. Cancer 55:*179–190, 1987.

Darby, S. C., Kendall, G. M., Fell, T. P., et al. A summary of mortality and incidence of cancer in men from the United Kingdom who participated in the United Kingdom's atmospheric nuclear weapon tests and experimental programmes. *Br. Med. J. 296:*332–338, 1988.

Davis, F. G., Boice, J. D., Jr., Hrubec, Z., and Monson, R. R. Lung cancer mortality in a radiation-exposed cohort of Massachusetts tuberculosis patients. *Cancer Res. 49:*6130–6136, 1989.

de Vathaire, F., François, P., Hill, C., et al. Role of radiotherapy and chemotherapy in the risk of second malignant neoplasms after cancer in childhood. *Br. J. Cancer 59:*792–796, 1989.

de Vathaire, F., François, P., Schweisguth, O., et al. Irradiated neuroblastoma in childhood as potential risk factor for subsequent thyroid tumor. *Lancet 2:*455, 1988.

Dobyns, B. M., Sheline, G. E., Workman, J. B., et al. Malignant and benign neoplasms of the thyroid in patients treated for hyperthyroidism: a report of the cooperative thyrotoxicosis therapy follow-up study. *J. Clin. Endocrinol. Metab. 38:*976–998, 1974.

Edmonds, C. J., and Smith, T. The long-term hazards of the treatment of thyroid cancer with radioiodine. *Br. J. Radiol. 59:*45–51, 1986.

Elkind, M. M., Benjamin, S. A., Sinclair, W. K., et al. Oncogenic mechanisms in radiation-induced cancer. *Cancer Res. 51:*2740–2747, 1991.

Evans, H. J. Gardner report. Leukaemia and radiation. *Nature 345:*16–17, 1990.

Evans, J. S., Wennberg, J. E., and McNeil, B. J. The influence of diagnostic radiography on the incidence of breast cancer and leukemia. *N. Engl. J. Med. 315:*810–815, 1986.

Favus, M. J., Schneider, A. B., Stachura, M. E., et al. Thyroid cancer occurring as a late consequence of head-and-neck irradiation. Evaluation of 1056 patients. *N. Engl. J. Med. 294:*1019–1025, 1976.

Fisher, B., Rockette, H., Fisher, E. R., et al. Leukemia in breast cancer patients following adjuvant chemotherapy or post-operative radiation: the NSABP experience. *J. Clin. Oncol. 3:*1640–1658, 1985.

Forman, D., Cook-Monzaffari, P., Darby, S., et al. Cancer near nuclear installations. *Nature 329:*499–505, 1987.

Fry, R. J. M., and Sinclair, W. K. New dosimetry of atomic bomb radiations. *Lancet 2:*845–848, 1987.

Fürst, C. J., Lundell, M., Holm, L.-E., and Silfversward, C. Cancer incidence after radiotherapy for skin hemangioma: a retrospective cohort study in Sweden. *J. Natl. Cancer Inst. 80:*1387–1392, 1988.

Gardner, M. J., Hall, A. J., Downs, S., and Terrell, J. D. Follow-up study of children born to mothers resident in Seascale, West Cumbria (birth cohort). *Br. Med. J. 295:*822–827, 1987.

Gardner, M. J., Snee, M. P., Hall, A. J., et al. Results of case-control study of leukaemia and lymphoma among young people near Sellafield nuclear plant in West Cumbria. *Br. Med. J. 300:*423–429, 1990.

Gilbert, E. S., Fry, S. A., Wiggs, L. D., et al. Analyses of combined mortality data on workers at the Hanford Site, Oak Ridge National Laboratory, and Rocky Flats Nuclear Weapons Plant. *Radiat. Res. 120:*19–35, 1989a.

Gilbert, E. S., Petersen, G. R., and Buchanan, J. A. Mortality of workers at the Hanford site: 1945–1981. *Health Phys. 56:*11–25, 1989b.

Ginzburg, H. M., and Reis, E. Consequences of the nuclear power plant accident at Chernobyl. *Public Health Rep. 106:*32–40, 1991.

Goldman, M. B., Maloof, F., Monson, R. R., et al. Radioactive iodine therapy and breast cancer: a follow-up study of hyperthyroid women. *Am. J. Epidemiol. 127:*969–980, 1988.

Goldsmith, R., Boice, J. D., Jr., Hrubec, Z., et al. Mortality and career radiation doses for workers at a commercial nuclear power plant: feasibility study. *Health Phys. 56:*139–150, 1989.

Greenberg, E. R., Rosner, B., Hennekens, C., et al. An investigation of bias in a study of nuclear shipyard workers. *Am. J. Epidemiol. 121:*301–308, 1985.

Greene, M. H., Young, R. C., Merrill, J. M., and DeVita, V. T. Evidence of a treatment dose response in acute nonlymphocytic leukemias which occur after therapy of non-Hodgkin's lymphoma. *Cancer Res. 43:*1891–1898, 1983.

Griem, M. L., Justman, J., and Weiss, L. The neoplastic potential of gastric irradiation. IV. Risk estimates. *Am. J. Clin. Oncol. 7:*675–677, 1984.

Hall, P., Holm, L.-E., Lundell, G., et al. Cancer risks in thyroid cancer patients. *Br. J. Cancer 64:*159–163, 1991.

Hamilton, T. E., van Belle, G., and LoGerfo, J. P. Thyroid neoplasia in Marshall Islanders exposed to nuclear fallout. *J.A.M.A. 258:*629–635, 1987.

Harris, C. C. Chemical and physical carcinogenesis: advances and perspectives for the 1990s. *Cancer Res. 51:*5023s–5044s, 1991.

Harvey, E. B., Boice, J. D., Jr., Honeyman, M., and Flannery, J. T. Prenatal x-ray exposure and childhood cancer in twins. *N. Engl. J. Med. 312:*541–545, 1985.

Hatch, M. C., Beyea, J., Nieves, J. W., and Susser, M. Cancer near the Three Mile Island nuclear plant: radiation emissions. *Am. J. Epidemiol. 132:*397–412, 1990.

Hempelmann, L. H. Risk of thyroid neoplasms after irradiation in childhood. *Science 160:*159–163, 1968.

Henshaw, D. L., Eatough, J. P., and Richardson, R. B. Radon as a causative factor in induction of myeloid leukaemia and other cancers. *Lancet 335:*1008–1012, 1990.

Hildreth, N. G., Shore, R. E., and Dvoretsky, P. M. The risk of breast cancer after irradiation of the thymus in infancy. *N. Engl. J. Med. 321:*1281–1284, 1989.

Hill, C., and Laplanche, A. Overall mortality and cancer mortality around French nuclear sites. *Nature 347:*755–757, 1990.

Hoffman, D. Effects of I-131 therapy in the United States. In: *Radiation Carcinogenesis: Epidemiology and Biological Significance,* edited by J.D. Boice, Jr., and J.F. Fraumeni, Jr. New York, Raven Press, 1984, pp. 273–280.

Hoffman, D. A., Lonstein, J. E., Morin, M. M., et al. Breast cancer in women with scoliosis exposed to multiple diagnostic x-rays. *J. Natl. Cancer Inst. 81:*1307–1312, 1989.

Hofmann, W., Katz, R., and Zhang, C. X. Lung cancer risk at low doses of alpha particles. *Health Phys. 51:*457–468, 1986.

Holm, L.-E., Hall, P., Wiklund, K. E., et al. Cancer risk after iodine-131 therapy for hyperthyroidism. *J. Natl. Cancer Inst. 83:*1072–1077, 1991.

Holm, L.-E., Wiklund, K. E., Lundell, G. E., et al. Cancer risk in population examined with diagnostic doses of 131-I. *J. Natl. Cancer Inst. 81:*302–306, 1989.

Holm, L.-E., Wiklund, K. E., Lundell, G. E., et al. Thyroid cancer after diagnostic doses of iodine-131. A retrospective cohort study. *J. Natl. Cancer Inst. 80:*1132–1138, 1988.

Hornung, R. W., and Meinhardt, T. J. Quantitative risk assessment of lung cancer in U.S. uranium miners. *Health Phys. 52:*417–430, 1987.

Howe, G. R. Risk of cancer mortality in populations living near nuclear facilities [editorial]. *J.A.M.A. 265:*1438–1439, 1991.

Howe, G. R., Nair, R. C., Newcombe, H. B., et al. Lung cancer mortality (1950–80) in relation to radon daughter exposure in a cohort of workers at the Eldorado Beaverlodge uranium mine. *J. Natl. Cancer Inst. 77:*357–362, 1986.

Howe, G. R., Nair, R. C., Newcombe, H. B., Miller, A. B., Burch, J. D., and Abbatt, J. D. Lung cancer mortality (1950–80) in relation to radon daughter exposure in a cohort of workers at the Eldorado Port Radium uranium mine: possible modification of risk by exposure rate. *J. Natl. Cancer Inst. 79:*1255–1260, 1987.

Hutchison, G. B., MacMahon, B., Jablon, S., and Land, C. E. Review of report by Mancuso, Stewart and Kneale of radiation exposure of Hanford workers. *Health Phys. 37:*207–220, 1979.

Ichimaru, M., Ohkita, T., and Ishimaru, T. Leukemia, multiple myeloma, and malignant lymphoma. In: *Cancer in Atomic Bomb Survivors,* GANN Monograph on Cancer Research No. 32, edited by I. Shigematsu and A. Kagan. Tokyo, Plenum Press, 1986, pp. 113–128.

Inskip, P. D., Monson, R. R., Wagoner, J. K., et al. Cancer mortality following radium treatment for uterine bleeding. *Radiat. Res. 123:*331–344, 1990a.

Inskip, P. D., Monson, R. R., Wagoner, J. K., et al. Leukemia following radiotherapy for uterine bleeding. *Radiat. Res. 122:*107–119, 1990b.

Jablon, S., Hrubec, Z., Boice, J. D., Jr., et al. Cancer in populations living near nuclear facilities. A survey of mortality nationwide and incidence in two states. *J.A.M.A. 265:*1403–1408, 1991.

Jablon, S., Hrubec, Z., Boice, J. D., Jr., and Stone, B. J. *Cancer in Populations Living Near Nuclear Facilities,* NIH Publ 90–874. Washington, DC, U.S. Government Printing Office, 1990.

Jablon, S., and Kato, H. Childhood cancer in relation to prenatal exposure to atomic-bomb radiation. *Lancet 2:*1000–1003, 1970.

Jablon, S., and Miller, R. W. Army technologists: 29-year follow-up for cause of death. *Radiology 126:*677–679, 1978.

Jessup, G. L., and Silverman, C. Personal usage of medical radiological procedures by radiologists, pathologists, and their families. *Am. J. Epidemiol. 114:*53–62, 1981.

Kinlen, L. Evidence for an infective cause for childhood leukaemia: comparison of a Scottish New Town with nuclear reprocessing sites in Britain. *Lancet 2:*1323–1327, 1988.

Kinlen, L. J., Clarke, K., and Hudson, C. Evidence from population mixing in British New Towns 1946–85 of an infective basis for childhood leukaemia. *Lancet 2:*577–582, 1990.

Knox, E. G., Stewart, A. M., Gilman, E. A., et al. Background radiation and childhood cancers. *J. Radiol. Prot. 8:*9–18, 1988.

Land, C. E. The hazards of fallout or of epidemiologic research? [editorial]. *N. Engl. J. Med. 300:*431–432, 1979.

Land, C. E. Estimating cancer risks from low doses of ionizing radiation. *Science 209:*1197–1203, 1980.

Land, C. E., McKay, F. W., and Machado, S. G. Childhood leukemia and fallout from the Nevada nuclear tests. *Science 223:*139–144, 1984.

Langlois, R. G., Bigbee, W. L., Kyoizumi, S., et al. Evidence for increased somatic cell mutations at the glycophorin A locus in atomic bomb survivors. *Science 236:*445–448, 1987.

Lewis, C. A., Smith, P. G., Stratton, I. M., et al. Estimated radiation doses to different organs among patients treated for ankylosing spondylitis with a single course of X rays. *Br. J. Radiol. 61:*212–220, 1988.

Lewis, E. B. Leukemia, multiple myeloma, and aplastic anemia in American radiologists. *Science 142:*1492–1494, 1963.

Lubin, J. H., and Boice, J. D., Jr. Estimating Rn-induced lung cancer in the United States. *Health Phys. 57:*417–427, 1989.

Lubin, J. H., Qiao, Y. L., Taylor, P. R., et al. Quantitative evaluation of the radon and lung cancer association in a case control study of Chinese tin miners. *Cancer Res. 50:*174–180, 1990a.

Lubin, J. H., Samet, J. M., and Weinberg, C. Design issues in epidemiologic studies of indoor exposures to Rn and risk of lung cancer. *Health Phys. 59:*807–817, 1990b.

Lundin, F. E., Jr., Wagoner, J. K., and Archer, V. E. *Radon Daughter Exposure and Respiratory*

Cancer: Quantitative and Temporal Aspects, NIOSH-NIEHS Joint Monogr. No. 1. Springfield, Va, National Technical Information Service, 1971.

Lyon, J. L., Klauber, M. R., Gardner, J. W., and Udall, K. S. Childhood leukemias associated with fallout from nuclear testing. *N. Engl. J. Med. 300:*397–402, 1979.

MacMahon, B. Some recent issues in low-exposure radiation epidemiology. *Environ. Health Perspect. 81:*131–135, 1989.

Mancuso, T. F., Stewart, A., and Kneale, G. Radiation exposure of Hanford workers dying from cancer and other causes. *Health Phys. 33:*369–385, 1977.

March, H. C. Leukemia in radiologists. *Radiology 43:*275–278, 1944.

Matanoski, G. M., Seltser, R., Sartwell, P. E., et al. The current mortality rates of radiologists and other physician specialists. Specific causes of death. *Am. J. Epidemiol. 101:*199–210, 1975.

Mays, C. W., and Spiess, H. Bone sarcomas in patients given radium-224. In: *Radiation Carcinogenesis: epidemiology and biological significance,* edited by J.D. Boice, Jr., and J.F. Fraumeni. New York, Raven Press, 1984, pp. 241–252.

McKinney, P. A., Alexander, F. E., Cartwright, R. A., and Parker, L. Parental occupations of children with leukaemia in West Cumbria, North Humberside, and Gateshead. *Br. Med. J. 302:*681–687, 1991.

Miller, A. B., Howe, G. R., Sherman, G. J., et al. Mortality from breast cancer after irradiation during fluoroscopic examinations on patients being treated for tuberculosis. *N. Engl. J. Med. 321:*1285–1289, 1989.

Modan, B., and Lilienfeld, A. M. Leukaemogenic effects of ionizing-irradiation treatment in polycythaemia. *Lancet 2:*439–441, 1964.

Mole, R. H. Antenatal irradiation and childhood cancer: causation or coincidence? *Br. J. Cancer 30:*199–208, 1974.

Monson, R. R., and MacMahon, B. Prenatal x-ray exposure and cancer in children. In: *Radiation Carcinogenesis: epidemiology and biological significance,* edited by J.D. Boice, Jr., and J.F. Fraumeni. New York, Raven Press, 1984, pp. 97–105.

Mulvihill, J. J., Myers, M. H., Connelly, R. R., et al. Cancer in offspring of long-term survivors of childhood and adolescent cancer. *Lancet 2:*813–817, 1987.

Najarian, T., and Colton, T. Mortality from leukemia and cancer in shipyard nuclear workers. *Lancet 1:*1018–1020, 1978.

National Academy of Sciences, Advisory Committee on the Biological Effects of Ionizing Radiations (The BEIR Report). *The Effects on Populations of Exposure to Low Levels of Ionizing Radiation.* Washington, DC, U.S. Government Printing Office, 1972.

National Academy of Sciences. *Health Risks of Radon and Other Internally Deposited Alpha-Emitters (BEIR IV Report).* Washington, DC, National Academy Press, 1988.

National Academy of Sciences. *Health Effects of Exposure to Low Levels of Ionizing Radiation (BEIR V).* Washington, DC, National Academy Press, 1990.

National Council on Radiation Protection and Measurements. *Influence of Dose and Its Distribution in Time on Dose-Response Relationships for Low-LET Radiations,* Report No. 64. Washington, DC, National Council on Radiation Protection and Measurements, 1980.

National Council on Radiation Protection and Measurements. *Evaluation of Occupational and Environmental Exposures to Radon and Radon Daughters in the United States,* Report No. 78. Bethesda, MD, National Council on Radiation Protection and Measurements, 1984.

National Council on Radiation Protection and Measurements. *Public Radiation Exposure from Nuclear Power Generation in the United States,* Report No. 92. Bethesda, MD, National Council on Radiation Protection and Measurements, 1987.

National Institutes of Health. *Report of the National Institutes of Health Ad Hoc Working Group to Develop Radioepidemiological Tables,* NIH Publication. No. 85–2748, U.S. DHHS. Washington, DC, Public Health Service, 1985.

Neel, J. V., Schull, W. J., Awa, A. A., et al. The children of parents exposed to atomic bombs: estimates of the genetic doubling dose of radiation for humans. *Am. J. Hum. Genet. 46:*1053–1072, 1990.

Nomura, T. Prenatal exposure to X-rays and chemicals induces heritable tumours and anomalies in mice. *Nature 296:*575–577, 1982.

Olsen, J. H., Andersson, M., and Boice, J. D., Jr. Danish epileptics given Thorotrast. In: *Risks from Radium and Thorotrast,* BIR Report 21, edited by D.M. Taylor, C.W. Mays, G.B. Gerber, and R.G. Thomas. London, British Institute of Radiology, 1989, pp. 116–118.

Pierce, D. A., and Vaeth, M. The shape of the cancer mortality dose-response curve for the A-bomb survivors. *Radiat. Res. 126:*36–42, 1991.

Pierce, D. A., Vaeth, M., and Preston, D. L. Analysis of time and age patterns in cancer risk for A-bomb survivors. *Radiat. Res. 126:*171–186, 1991.

Polednak, A. P. Bone cancer among female radium dial workers. Latency periods and incidence rates by time after exposure: brief communication. *J. Natl. Cancer Inst. 60:*77–82, 1978.

Pottern, L. M., Kaplan, M. M., Larsen, P. R., et al. Thyroid nodularity after childhood irradiation for lymphoid hyperplasia: a comparison of questionnaire and clinical findings. *J. Clin. Epidemiol.* 43:449–460, 1990.

Prentice, R. L., Kato, H., Yoshimoto, K., et al. Radiation exposure and thyroid cancer incidence among Hiroshima and Nagasaki residents. *Natl. Cancer Inst. Monogr. 62:*207–212, 1982.

Prentice, R. L., Yoshimoto, Y., and Mason, M. W. Relationship of cigarette smoking and radiation exposure to cancer mortality in Hiroshima and Nagasaki. *J. Natl. Cancer Inst. 70:*611–622, 1983.

Preston, D. L., and Pierce, D. A. The effect of changes in dosimetry on cancer mortality risk estimates in the atomic bomb survivors. *Radiat. Res. 114:*437–466, 1988.

Preston-Martin, S., Mack, W., and Henderson, B. E. Risk factors for gliomas and meningiomas in males in Los Angeles County. *Cancer Res. 49:*6137–6143, 1989a.

Preston-Martin, S., Paganini-Hill, A., Henderson, B. E., et al. Case-control study of intracranial meningiomas in women in Los Angeles County, California. *J. Natl. Cancer Inst. 65:*67–73, 1980.

Preston-Martin, S., Thomas, D. C., White, S. C., and Cohen, D. Prior exposure to medical and dental x-rays related to tumors of the parotid gland. *J. Natl. Cancer Inst. 80:*943–949, 1988.

Preston-Martin, S., Thomas, D. C., Yu, M. C., et al. Diagnostic radiography as a risk factor for chronic myeloid and monocytic leukemia (CML). *Br. J. Cancer 59:*639–644, 1989b.

Radford, E. P., and Renard, K. G. Lung cancer in Swedish iron miners exposed to low doses of radon 7. *N. Engl. J. Med.* 310:1485–1494, 1984.

Rallison, M. L., Dobyns, B. M., Keating, F. R., et al. Thyroid nodularity in children. *J.A.M.A. 233:*1069–1072, 1975.

Rallison ML., Lotz TM., Bishop M., et al. Cohort study of thyroid disease near the Nevada Test Site: a preliminary report. *Health Phys. 59:*739–746, 1990.

Rinsky, R. A., Melius, J. M., Hornung, R. W., et al. Case-control study of lung cancer in civilian employees at the Portsmouth Naval Shipyard, Kittery, Maine. *Am. J. Epidemiol. 127:*55–64, 1988.

Rinsky, R. A., Zumwalde, R. D., Waxweiler, R. J., et al. Cancer mortality at a naval shipyard. *Lancet 1:*231–235, 1981.

Robinette, C. D., Jablon, S., and Preston, T. L. *Mortality of Nuclear Weapons Test Participants.* Washington, DC, National Academy Press, 1985.

Ron, E., Modan, B., and Boice, J. D., Jr. Mortality after radiotherapy for ringworm of the scalp. *Am. J. Epidemiol 127:*713–725, 1988a.

Ron, E., Modan, B., Boice, J. D., Jr., et al. Tumors of the brain and nervous system following radiotherapy in childhood. *N. Engl. J. Med. 319:*1033–1039, 1988b.

Ron, E., Modan, B., Preston, D., et al. Thyroid neoplasia following low-dose radiation in childhood. *Radiat. Res. 120:*516–531, 1989.

Roscoe, R. J., Steenland, K., Halperin, W. E., et al. Lung cancer mortality among nonsmoking uranium miners exposed to radon daughters. *J.A.M.A. 262:*629–633, 1989.

Rowland, R. E., Stehney, A. F., and Lucas, H. F., Jr. Dose-response relationships for female radium dial workers. *Radiat. Res. 76:*368–383, 1978.

Rundo, J., Keane, A. T., Lucas, H. F., et al. Current (1984) status of study of ^{226}Ra and ^{228}Ra in humans at the Center for Human Radiobiology. *Strahlentherapie [Sonderb] 80:*14–21, 1986.

Saenger, E. L., Thoma, G. E., and Tompkins, E. A. Incidence of leukemia following treatment of hyperthyroidism. Preliminary report of the cooperative thyrotoxicosis therapy follow-up study. *J.A.M.A. 205:*855–862, 1968.

Samet, J. M. Radon and lung cancer. *J. Natl. Cancer Inst. 81:*745–757, 1989.

Schneider, A. B., Shore-Freedman, E., Ryo, U. Y., et al. Radiation-induced tumors of the head and neck following childhood irradiation. Prospective studies. *Medicine 64:*1–15, 1985.

Schoenberg, J. B., Klotz, J. B., Wilcox, H. B., Nicholls, G. P., Gil del Real, M. T., Stemhagen, A., and Mason, T. J. Case-control study of residential radon and lung cancer among New Jersey women. *Cancer Res. 50:*6520–6524, 1990.

Sevc, J., Kunz, E., Tomasek, L., et al. Cancer in man after exposure to Rn daughters. *Health Phys. 54:*27–46, 1988.

Sever, L. E., Gilbert, E. S., Hessol, N. A., and McIntyre, J. M. A case-control study of congenital malformations and occupational exposure to low-level ionizing radiation. *Am. J. Epidemiol. 127:*226–242, 1988a.

Sever, L. E., Hessol, N. A., Gilbert, E. S., and McIntyre, J. M. The prevalence at birth of congenital malformations in communities near the Hanford site. *Am. J. Epidemiol. 127:*243–254, 1988b.

Shimizu, Y., Kato, H., and Schull, W. J. Studies of the mortality of A-bomb survivors. 9. Mortality, 1950–1985: Part 2. Cancer mortality based on the recently revised doses (DS86). *Radiat. Res. 121:*120–141, 1990.

Shimizu, Y., Kato, H., Schull, W. J., et al. Studies of the mortality of A-bomb survivors. 9. Mortality, 1950–1985, Part 1. Comparison of risk coefficients for site-specific cancer mortality based on the DS86 and T65DR shielded kerma and organ doses. *Radiat. Res. 118:*502–524, 1989.

Shore, R. E., Albert, R. E., and Pasternack, B. S. Follow-up study of patients treated by x-ray epilation for tinea capitis: resurvey of post-treatment illness and mortality experience. *Arch. Environ. Health 31:*17–24, 1976.

Shore, R. E., Albert, R., Reed, M., et al. Skin cancer incidence among children irradiated for ringworm of the scalp. *Radiat. Res. 100:*192–204, 1984.

Shore, R. E., Hildreth, N., Woodard, E., et al. Breast cancer among women given x-ray therapy for acute postpartum mastitis. *J. Natl. Cancer Inst. 77:*689–696, 1986.

Shore, R. E., Woodward, E., Hildreth, N., et al. Thyroid tumors following thymus irradiation. *J. Natl. Cancer Inst. 74:*1177–1184, 1985.

Smith, P. G. Case-control studies of leukaemia clusters. *Br. Med. J. 302:*672–673, 1991.

Smith, P. G., and Doll R. Late effects of x irradiation in patients treated for metropathia haemorrhagica. *Br. J. Radiol. 49:*224–232, 1976.

Smith, P. G., and Doll, R. Mortality from cancer and all causes among British radiologists. *Br. J. Radiol. 54:*187–194, 1981.

Smith, P. G., and Doll, R. Mortality among patients with ankylosing spondylitis after a single treatment course with x-rays. *Br. Med. J. 284:*449–460, 1982.

Smith, P., and Douglas, A. J. Mortality of workers at the Sellafield plant of British Nuclear Fuels. *Br. Med. J. 293:*845–854, 1986.

Spiers, F. W. Background radiation and estimated risks from low-dose irradiation. *Health Phys. 37:*784–789, 1979.

Spiers, F. W., Lucas, H. F., Rundo, J., et al. Leukemia incidence in the U.S. dial painters. *Health Phys. 44:*65–72, 1983.

Spiess, H., Mays, C. W., and Chmelevsky, D. Malignancies in patients injected with radium 224. In: *Risks from Radium and Thorotrast,* BIR Report 21, edited by D.M. Taylor, C.W. Mays, and G. B. Gerber. London, British Institute of Radiology, 1989, pp. 7–12.

Stebbings, J. H., Lucas, H., and Stehney, A. Mortality from cancers of major sites in female radium dial workers. *Am. J. Ind. Med. 5:*435–459, 1984.

Stern, F. B., Waxweiler, R. A., Beaumont, J. J., et al. A case-control study of leukemia at a naval nuclear shipyard. *Am. J. Epidemiol. 123:*980–992, 1986.

Stevens, W., Thomas, D. C., Lyon, J. L., et al. Leukemia in Utah and radioactive fallout from the Nevada test site. A case-control study. *J.A.M.A. 264:*585–591, 1990.

Stewart, A. M., and Kneale, G. W. Radiation dose effects in relation of obstetric X-rays and childhood cancers. *Lancet 1:*1185–1188, 1970.

Stovall, M., Smith, S. A., and Rosenstein, M. Tissue doses from radiotherapy of cancer of the uterine cervix. *Med Phys. 16:*726–733, 1989.

Taylor, D. M., Mays, C. W., Gerber, G. B., and Thomas, R. G., eds. *Risks from Radium and Thorotrast*, BIR Report 21. London, British Institute of Radiology, 1989.

Tokunaga, M., Land, C. E., Yamamoto, T., et al. Incidence of female breast cancer among atomic bomb survivors, Hiroshima and Nagasaki, 1950–1980. *Radiat. Res. 112:*243–273, 1987.

Tucker, M. A., D'Angio, G. J., Boice, J. D., Jr., et al. Bone sarcomas linked to radiotherapy and chemotherapy in children. *N. Engl. J. Med. 317:*588–593, 1987a.

Tucker, M. A., Meadows, A. T., Boice, J. D., Jr., et al. Leukemia after therapy with alkylating agents for childhood cancer. *J. Natl. Cancer Inst. 78:*459–464, 1987b.

Tucker, M. A., Morris Jones, P. H., Boice, J. D., Jr., et al. Therapeutic radiation at a young age is linked to secondary thyroid cancer. *Cancer Res. 51:*2885–2888, 1991.

Ullrich, R. L., Jernigan, M. C., Satterfield, L. C., et al. Radiation carcinogenesis: time-dose relationships. *Radiat. Res. 111:*179–184, 1987.

United Nations Scientific Committee on the Effects of Atomic Radiation. *Genetic and Somatic Effects of Ionizing Radiation*, Publication E.86.IX.9. New York, United Nations, 1986.

United Nations Scientific Committee on the Effects of Atomic Radiation. *Sources, Effects, and Risks of Ionizing Radiation.* Publication E.88.IX.7. New York, United Nations, 1988.

Upton, A. C. Biological basis of radiological protection and its application to risk assessment. *Br. J. Radiol. 60:*1–16, 1987.

Upton, A. C. Carcinogenic effects of low-level ionizing radiation. *J. Natl. Cancer Inst. 82:*448–449, 1990.

Urquhart, J. D., Black, R. J., Muirhead, M. J., et al. Case-control study of leukaemia and non-Hodgkin's lymphoma in children in Caithness near the Dounreay nuclear installation. *Br. Med. J. 302:*687–692, 1991.

van Kaick, G., Wesch, H., Lührs, H., et al. The German Thorotrast Study–report on 20 years follow-up. In: *Risks from Radium and Thorotrast,* BIR Report 21, edited by D.M. Taylor, C.W. Mays, and G. B. Gerber. London, British Institute of Radiology, 1989, pp. 98–104.

Voelz, G. L., and Lawrence, J. N. P. A 42-year medical follow-up of Manhattan Project plutonium workers. *Health Phys. 61:*181–190, 1991.

Voelz, G. L., Wilkinson, G. S., Healy, J. W., et al. Mortality study of Los Alamos workers with higher exposures to plutonium. In: *Epidemiology Applied to Health Physics,* CONF-830101. Springfield, VA, National Technical Information Service, 1983, pp. 318–327.

Wakabayashi, T., Kato, H., Ikeda, T., and Schull, W. J. Studies of the mortality of A-bomb survivors, report 7. Part III. Increase of cancer in 1959–78, based on the Tumor Registry, Nagasaki. *Radiat. Res. 93:*112–146, 1983.

Wang, J. X., Inskip, P. D., Boice, J. D., Jr., Li, B. X., Zhang, J. Y., and Fraumeni, J. F., Jr. Cancer incidence among medical diagnostic X-ray workers in China, 1950 to 1985. *Int. J. Cancer 45:*889–895, 1990a.

Wang, Z., Boice, J. D., Jr., Wei, L., et al. Thyroid nodularity and chromosome aberrations among women in areas of high background radiation in China. *J. Natl. Cancer Inst. 82:*478–485, 1990b.

Wei, L. Health survey in high background radiation areas in China. *Science 209:*877–880, 1980.

Weichselbaum, R. R., Beckett, M. A., and Diamond, A. A. An important step in radiation carcinogenesis may be inactivation of cellular genes. *Int. J. Radiat. Oncol. Biol. Phy. 16:*277–282, 1989.

Wilkinson, G. S., Tietjen, G. L., Wiggs, L. D., et al. Mortality among plutonium and other radiation workers at a plutonium weapons facility. *Am. J. Epidemiol. 125:*231–250, 1987.

Wing, S., Shy, C. M., Wood, J. L., et al. Mortality among workers at Oak Ridge National Laboratory. Evidence of radiation effects in follow-up through 1984. *J.A.M.A. 265:*1397–1402, 1991.

Yamamoto, T., Kopecky, K. J., Fujikura, T., et al. Lung cancer incidence among Japanese A-bomb survivors, 1950–80. *J. Radiat. Res. 28:*156–171, 1987.

Yoshimoto, Y., Kato, H., and Schull, W. J. Risk of cancer among children exposed *in utero* to a-bomb radiations, 1950–84. *Lancet 2:*665–669, 1988.

Yoshimoto, Y., Neel, J. V., Schull, W. J., et al. *Frequency of Malignant Tumors during the First Two Decades of Life in the Offspring (F1) of Atomic Bomb Survivors,* RERF Technical Report 4–90. Hiroshima, Radiation Effects Research Foundation, 1990.

Chapter 7

Cancer Risk Estimates for External Radiation Exposure (BEIR V)

ARTHUR C. UPTON

ABSTRACT

Although neoplasms of many types can be induced by large doses of ionizing radiation, the extent to which the risk of any form of cancer may be increased by low-level irradiation is uncertain. The risks of cancer attributable to low-level irradiation can, therefore, be estimated only by extrapolation, based on assumptions about the relevant dose-incidence relationships. The estimates formulated by the BEIR V Committee were derived with the use of a modified multiplicative dose-response model, in which the risk of leukemia was assumed to increase as a linear quadratic function of the dose, and the combined risks of all other malignancies were assumed to increase linearly with the dose. The risks derived with these models, taking into account the revised A-bomb dosimetry and the additional years of follow-up of the A-bomb survivors, spondylitics, and other irradiated human populations, are appreciably larger than the International Commission on Radiological Protection (ICRP)-26 estimates and the "preferred" estimate of the BEIR III Committee. They are consistent, however, with estimates derived by the BEIR I Committee and 1988 United Nations Scientific Committee on the Effects of Atomic Radiation (UNSCEAR) with the use of analogous multiplicative, linear dose-incidence models. The extent to which the estimates should be reduced by a dose rate effectiveness factor (DREF) when applied to low-level, low linear energy transfer irradiation is a question that the BEIR V Committee did not resolve; however, the committee acknowledged that a value of 2 or more would not be unreasonable given that the DREFs for different neoplasms in laboratory animals range from 2–10.

INTRODUCTION

Information about the carcinogenic effects of radiation has been significantly amplified in recent years by large-scale epidemiological studies of Japanese A-bomb survivors; patients treated with radiation for ankylosing spondylitis, cancer of the uterine cervix, tinea capitis, and other medical conditions; workers exposed to radiation in various occupations; and populations residing in areas of high natural background radiation. Important new data have also come from studies of the carcinogenic effects of irradiation in laboratory animals and studies of radiation-induced neoplastic transformation of cells in culture. The new information, especially the increase in the carcinogenic risk per unit dose in A-bomb survivors implied by revisions in the dosimetry for the bombs, prompted recent updating of risk assessments by the United Nations Scientific Committee on the Effects of Atomic Radiation (UNSCEAR) (UNSCEAR 1988) and BEIR V Committee (National Academy of Sciences/National Research Council [NAS/NRC] 1990).

Although the carcinogenic effects of large doses of radiation are well documented, the choice of the appropriate dose-response model for use in estimating the effects of small doses remains uncertain, in part because the neoplasms induced by radiation characteristically take years or decades to develop, possess no features by which they can be distinguished individually from those induced by other causes, have generally been detectable only after relatively large doses (>0.5 Gy), and have varied with the type of neoplasm as well as with the age and sex of the exposed population (NAS/NRC 1990, UNSCEAR 1988). For purposes of risk assessment, the UNSCEAR and BEIR V committees have relied heavily on the data from the A-bomb survivors, owing to the comparatively large numbers of the survivors, the fact that they include individuals of both sexes and all ages, and the fact that they were exposed to a wide range of doses, extending from the lethal level down to the natural background radiation level.

DOSE-EFFECT RELATIONSHIP

Dose

Owing to revisions in the dosimetry of the A-bomb radiations and to the increase in numbers of excess cancer deaths observed on continuing follow-up of the survivors, particularly among those who were irradiated early in life, the carcinogenic effects of A-bomb radiation per unit dose are now estimated to be appreciably larger than heretofore (Preston and Pierce 1988). As evaluated with the new dosimetry, the contribution of neutrons to the total dose received by survivors no longer appears highly significant in either Hiroshima or Nagasaki (NAS/NRC 1990), with the result that there remains no adequate basis for estimating the relative biological effectiveness (RBE) of neutrons for carcinogenic effects in this population. In addition, the tissue dose of gamma radiation is now estimated to have been higher at Hiroshima and slightly lower at Nagasaki than heretofore, so that if an RBE of 10 or more is assigned to the neutrons, the estimated risk per unit dose equivalent is increased, depending on the depth of the tissue in question. Furthermore, the difference between Hiroshima and Nagasaki in overall risk per unit dose is reduced, no longer remaining statistically significant (NAS/NRC 1990).

During the most recent decade of follow-up, the cumulative radiation-induced excess in the number of deaths from cancer in A-bomb survivors has increased by more than 50%, from an estimated 205 deaths in 1975 to 340 deaths in 1985 (Shimizu et al. 1988, 1990). Furthermore, the increase has been roughly proportional to the increase in the baseline rate of cancer with aging (Fig. 7.1), while the relative risk of fatal cancer has remained comparatively constant over time

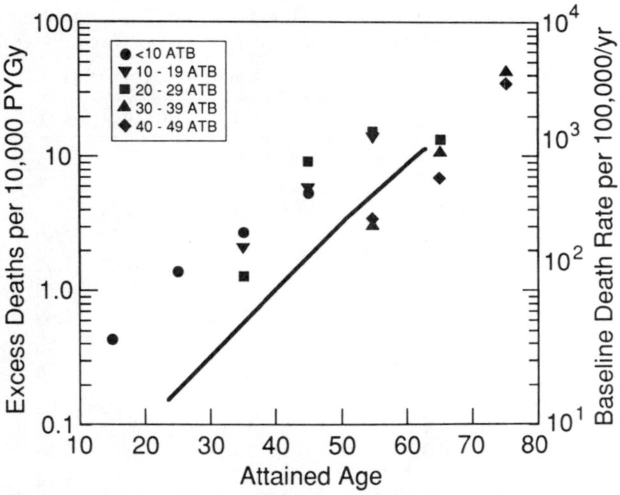

FIG. 7.1. Excess mortality in heavily irradiated (>1 Gy) cohorts of A-bomb survivors of different ages at the time of the bomb (ATB), in relation to attained age. The solid line represents the age-specific baseline cancer death rate for the population of Japan. (A-bomb survivor data from Shimizu et al. 1988; baseline cancer mortality rate data from Hirayama 1977). (From Upton AC. Radiobiology and radiation protection: the past century and prospects for the future. The Lauriston S. Taylor Lecture Series in Radiation Protection and Measurements, lecture no. 13. Washington, DC: National Council on Radiation Protection and Measurements, 1989.) Reprinted with permission of the National Council on Radiation Protection and Measurements.

(Table 7.1). Consequently, the overall cancer mortality data are now more compatible with the "relative" (or "multiplicative") risk projection model than with the "absolute" (or "additive") risk projection model, although neither model may be valid for cancer of a given type or site without appropriate qualification.

For total mortality from leukemia (excluding chronic lymphatic leukemia), the A-bomb survivor data conform best to a linear-quadratic dose-response function (Table 7.2, Fig. 7.2; NAS/NRC 1990), while for cancers other than leukemia, the overall dose-mortality relationship shows no significant departure from linearity in the range below 3 Gy (Table 7.2; NAS/NRC 1990). For solid cancers at certain sites (e.g., breast and thyroid) the data also are consistent with linearity, whereas for cancers at other sites (e.g., colon) the data appear to be more consistent with linear-quadratic or quadratic functions (Shimizu et al. 1988, 1990). It is noteworthy that chronic lymphocytic leukemia and Hodgkin's disease have not been observed to be induced by irradiation in the A-bomb survivors or in other human populations, in spite of their relatively high prevalence among hematological cancers as a group (NAS/NRC 1990, UNSCEAR 1988). These differences in dose-effect relationships for different types of cancer are not surprising in view of the diversity that is seen among neoplasms in laboratory animals (UNSCEAR 1977, 1986).

Additional evidence for the carcinogenicity of small doses, although not sufficient in itself to characterize the dose-effect relationship, is provided by the association between diagnostic x-irradiation in utero and the subsequent development of cancer in childhood (NAS/NRC 1990). Interpretation of the associa-

TABLE 7.1. RELATIVE AND ABSOLUTE RISKS OF DEATHS FROM CANCER IN A-BOMB SURVIVORS AT 1 GY (SHIELDED KERMA), 1950–1985, IN RELATION TO AGE AT IRRADIATION AND AGE AT DEATH[a]

	Age at Time of Death						
Age ATB[b]	0–19	20–29	30–39	40–49	50–59	60–69	70+
	Relative Risk at 1Gy						
Leukemia							
0–9	44.16	3.41	8.64	0.95			
10–19	54.74	—	2.45	1.02	0.82		
20–29		5.33	3.54	43.09	1.02	0.82	
30–39			0	24.05	10.58	1.47	3.89
40–49				0.83	3.82	0.82	3.10
50+					15.63	5.18	6.90
All ages	46.47	9.81	4.75	5.68	3.98	1.70	4.40
All cancers except leukemia							
0–9	(70.07)	5.89	1.96	1.86			
10–19	(40.90)	(0.82)	1.66	1.59	1.68		
20–29			(1.38)	2.09	1.74	1.37	
30–39			(0.84)	(1.12)	1.11	1.23	1.48
40–49				(1.25)	(1.12)	1.13	1.33
50+					(2.58)	(0.95)	1.15
All ages	75.32	2.22	1.60	1.58	1.39	1.13	1.29
	Excess Deaths Per 10^4 PYGy						
Leukemia							
0–9	6.71	0.93	1.27	−0.01			
10–19	3.95	—	0.56	0.02	−0.06		
20–29		3.93	1.52	4.84	0.01	−0.28	
30–39			0	3.18	2.26	1.09	3.89
40–49				−0.35	3.07	−0.24	3.50
50+					4.31	3.84	5.12
All ages	6.48	2.17	1.16	1.88	1.54	1.09	4.24
All cancers except leukemia							
0–9	(0.43)	1.32	2.85	5.16			
10–19	(3.96)	(−0.12)	2.00	5.84	13.91		
20–29			(1.39)	9.40	15.71	14.33	
30–39			(−1.32)	(1.33)	3.16	11.00	41.01
40–49				(2.48)	(3.37)	7.31	37.30
50+					(35.29)	(−2.88)	17.21
All ages	0.79	0.54	1.98	5.35	9.62	6.85	30.53

[a] Adapted from Shimizu Y, Kato H, Schull WJ. Life span study report 11, part II: cancer mortality in the years 1950–1985 based on the recently revised doses, DS86. RERF TR 5-88. Hiroshima, Japan: RERF, 1988; Shimizu Y, Kato H, Schull WJ. Studies of the mortality of a-bomb survivors. 9. Mortality, 1950–1985: part 2. Cancer mortality based on the recently revised doses (DS86). Radiat Res 1990;121:120–141.

[b] ATB, at the time of the bomb.

tion is complicated, however, by the absence of an excess of childhood cancer in prenatally irradiated A-bomb survivors (NAS/NRC 1990, UNSCEAR 1986).

The carcinogenicity of small doses is also indicated by the excess of thyroid cancer associated with therapeutic x-irradiation in childhood, which has been interpreted to increase linearly with the dose (Fig. 7.3) and which has been observed at doses of less than 100 mSv in children treated with x-rays to the

TABLE 7.2. EXCESS RELATIVE RISK OF DEATH FROM CANCER IN A-BOMB SURVIVORS PER GY OF ORGAN ABSORBED DOSE, 1950–1985[a]

					Equality of Excess RR[b]		
Dose (Gy)	0–6	<.20	<.50	<1.0	<.50	.50+	Test
Site of Cancer							
Leukemia	5.21[f]	−.12	2.40[d]	3.96[f]	2.44	5.53	[d]
All cancers except leukemia	.41[f]	.54	.38[d]	.46[f]	.34	.42	
Stomach	.27[f]	.17	.45[c]	.41[e]	.45	.26	
Lung	.63[f]	.17	1.09[d]	.83[e]	1.06	.60	
Female breast	1.19[f]	.21	.88	1.78[e]	.82	1.21	
Colon	.85[f]	−2.95[c]	−.53	−.10	−.52	.98	[c]

[a] Adapted from Shimizu Y, Kato H, Schull WJ. Life span study report 11, part II: cancer mortality in the years 1950–1985 based on the recently revised doses, DS86. RERF TR 5-88. Hiroshima, Japan: RERF, 1988; Shimizu Y, Kato H, Schull WJ. Studies of the mortality of a-bomb survivors. 9. Mortality, 1950–1985: part 2. Cancer mortality based on the recently revised doses (DS86). Radiat Res 1990;121:120–141.

[b] RR, Relative risk.

[c] $p < .10$.

[d] $p < .05$.

[e] $p < .01$.

[f] $p < .001$.

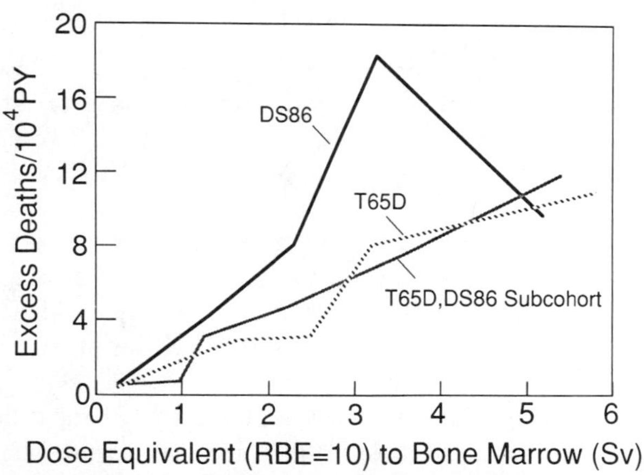

FIG. 7.2. Dose-mortality relationship for mortality from leukemia (excluding chronic lymphocytic leukemia [CLL]) in A-bomb survivors, 1950–1985. (From Preston DL, Pierce DA. The effect of changes in dosimetry on cancer mortality risk estimates in the atomic bomb survivors. Radiat Res 1988;114:437–466.) Reprinted with permission of Academic Press Inc. and the authors.

scalp for tinea capitis (Modan et al. 1989).

The potential carcinogenicity of radiation at low doses is also suggested by experiments with laboratory animals and cultured cells (UNSCEAR 1986), which imply that the frequency of tumor-initiating effects may increase as a linear nonthreshold function of the radiation dose (Upton 1987).

The above data notwithstanding, the dose-incidence curve remains highly

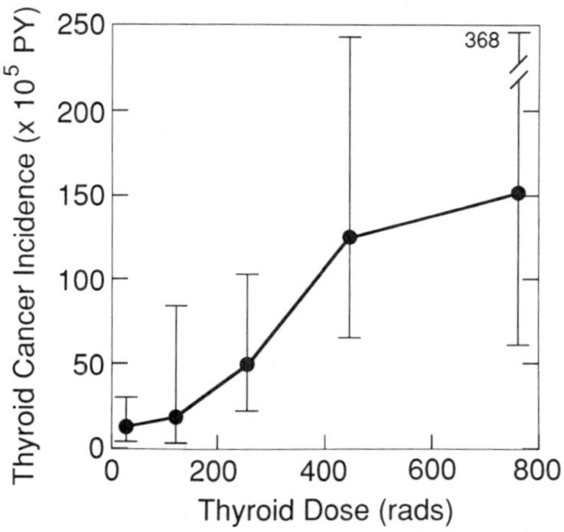

FIG. 7.3. Thyroid cancer incidence in relation to thyroid dose, adjusted for sex, ethnicity, and interval since irradiation. The vertical bars denote the 90% confidence limits. (From Shore RE, Woodard E, Hildreth N, et al. Thyroid tumors following thymus irradiation. J Natl Cancer Inst 1985;74:1177–1184.) Reprinted with permission of the authors.

uncertain at low doses, and the possible existence of a threshold in the mSv range cannot be excluded (NAS/NRC 1990).

DOSE RATE

In laboratory animals, the carcinogenic effectiveness of low linear energy transfer (LET) radiation in the 0.5–3 Gy dose range characteristically decreases with fractionation or protraction, whereas that of high-LET radiation remains unchanged or may even increase (NAS/NRC 1990, NCRP 1980, UNSCEAR 1986).

Comparative data for humans are lacking, except with respect to carcinogenic effects on the female breast, which appear to be similar in magnitude per unit dose whether the dose is delivered acutely or in many small increments from repeated fluoroscopic examinations of the chest (Miller et al. 1989; NAS/NRC 1990; Shore et al. 1986; UNSCEAR 1986, 1988).

Although patients given diagnostic doses of iodine-131 have been observed to develop far less thyroid cancer (Holm et al. 1988) than would be predicted from the effects of external x-radiation (Shore et al. 1985), if any, it is conceivable that the discrepancy may be attributable in part to spatial differences in dose distribution (National Council on Radiation Protection and Measurements [NCRP] 1985) since the carcinogenic effects of therapeutic x-irradiation have not been observed to diminish on fractionation (Shore et al. 1985).

LINEAR ENERGY TRANSFER

In laboratory animals, the carcinogenic effects of radiation vary as a function of LET (NAS/NRC 1990, UNSCEAR 1986). Depending on the neoplasm and the conditions of irradiation, the RBE of high-LET radiation typically increases with decreasing dose and dose rate, generally ranging from 2–30 for duration-of-life exposure, with substantially higher RBEs reported in certain instances (International Commission on Radiological Protection [ICRU] 1986).

In humans, data for external radiation are lacking, but the RBEs of internally emitted alpha particles for carcinogenic effects on the skeleton and the lung are consistent with values in the range of 20, as in laboratory animals (ICRU 1986; NAS/NRC 1988, 1990).

Age at Irradiation

For a number of different types of cancer, susceptibility varies in relation to age at the time of irradiation. In the case of leukemia (Fig. 7.4), the absolute risk appears to be higher in prenatal life, infancy, childhood, and late adult life than in adolescence or early adult life; the relative risk, on the other hand, is comparatively constant throughout adult life although appreciably lower than in childhood (NAS/NRC 1990, UNSCEAR 1986, 1988).

For cancer of the thyroid gland, susceptibility appears to be 2–3 times higher in infancy and childhood than in adult life (Shore et al. 1985). Likewise, susceptibility to cancer of the female breast appears to be higher in childhood and adolescence than later in life, decreasing markedly after the menopause (Shimizu et al. 1988, 1990, Tokunaga et al. 1984).

For cancers of other sites, the available data are not sufficient to define the relationship between susceptibility and age at the time of irradiation (NAS/NRC 1990, UNSCEAR 1988); however, it is noteworthy that the total excess of adult cancers in A-bomb survivors who were irradiated prenatally appears to be similar in magnitude to the excess in those who were irradiated during childhood (Table 7.3).

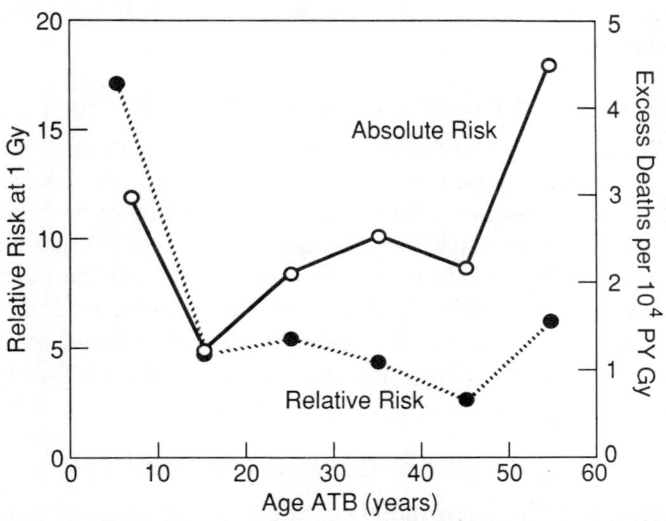

FIG. 7.4. Relative versus absolute risks of leukemia in A-bomb survivors 1950–1985, in relation to age at the time of the bombings (ATB). (From Shimizu Y, Kato H, Schull WJ. Studies of the mortality of A-bomb survivors. 9. Mortality, 1950–1985. Part 2. Cancer mortality based on the recently revised doses (DS86). Radiat Res 1990;121:120–141.) Reprinted with permission of Academic Press Inc. and the authors.

TABLE 7.3. CANCER RISKS IN A-BOMB SURVIVORS IRRADIATED PRENATALLY, AS COMPARED WITH THOSE EXPOSED AT 0–9 YEARS OF AGE[a, b]

	Irradiated In Utero (1950–84) DS86 Uterus Dose	Age 0–9 ATB (1950–85) DS86 Tissue Kerma	
No. of cancers	18 (2)[c]	142 (31)[c]	
RR at 1 Gy	All cancers 3.77 (1.1, 13.5)[d]	All cancers	3.97 (2.9, 5.4)[d]
		Leukemias	17.25 (9.3, 38.9)
		Other cancers	2.23 (1.6, 3.4)
Excess Risk/10^4 Person-Year-Gy	All cancers 6.57 (0.47, 14.5)	All cancers	5.47 (3.54, 7.17)
		Leukemias	2.93 (2.23, 3.60)
		Other cancers	2.27 (1.11, 3.65)

[a] From Yoshimoto Y, Kato H, Schull WJ. Risk of cancer among in utero children exposed to A-bomb radiation: 1950–1984. Lancet 1988;2:665–669. Reprinted with permission of The Lancet Ltd. and the authors.

[b] ATB, at the time of the bomb; RR, relative risk.

[c] No. of cases of leukemia.

[d] 90% confidence interval.

SEX

The induction of breast cancer by irradiation has been documented conclusively thus far only in women (Shore et al. 1986), which is not surprising in view of the fact that development of the disease is known to depend heavily on estrogenic stimulation of the mammary gland. Radiation-induced thyroid cancer also occurs about three times more commonly in females than in males, although the relative risk is similar in both sexes (NCRP 1985). For cancers of other sites, the relative risks generally appear higher in females than in males, but the sex differences are less pronounced (Table 7.4).

CONSTITUTIONAL AND PHYSIOLOGICAL FACTORS

Susceptibility to the induction of some types of cancer is thought to be increased in certain inherited disorders, e.g., susceptibility to osteosarcoma in association with familial retinoblastoma (Knudson 1985), susceptibility to breast cancer in association with heterozygosity for the ataxia telangiectasia gene (Swift et al. 1987), and susceptibility to skin cancer in association with the nevoid basal cell carcinoma syndrome (Strong 1977). In general, however, apart from limited data for different species and strains of laboratory animals (Storer et al. 1988), the influence of constitutional and physiological factors on susceptibility to radiation carcinogenesis is not known.

EFFECTS OF OTHER COFACTORS

Various interactions between radiation and other agents—including synergistic, additive, and antagonistic interactions—have been observed experimentally, depending on the agents in question and the conditions of exposure (Fry and Ullrich 1986, UNSCEAR 1982). In cancer patients receiving combined therapy with drugs and radiation, the risks of second, treatment-induced cancers are

TABLE 7.4. SEX DIFFERENCES IN RELATIVE AND ABSOLUTE RISKS OF CANCER MORTALITY IN A-BOMB SURVIVORS, 1950–1985[a]

	Estimated RR[b] at 1 Gy (Shielded Kerma)			Excess Deaths per 10^4 PYGy		
Site of Cancer	Male	Female	M/F Test	Male	Female	M/F Test (M/F)
Leukemia	4.96	4.92	1.00	3.14	1.80	1.74[c]
All cancers except leukemia	1.17	1.44	0.81[d]	5.76	8.78	0.66
Esophagus	1.19	2.99	0.40[c]	0.30	0.40	0.75
Stomach	1.15	1.36	0.85	2.01	2.18	0.92
Colon	1.45	1.67	0.87	0.60	0.51	1.18
Lung	1.26	1.86	0.68[c]	1.07	1.47	0.73
Urinary tract	2.00	2.15	0.93	0.81	0.42	1.93
Multiple myeloma	5.29	2.32	2.28	0.23	0.21	1.10

[a] Adapted from Shimizu Y, Kato H, Schull WJ. Life span study report 11, part II: cancer mortality in the years 1950–1985 based on the recently revised doses (DS86. RERF TR 5-88. Hiroshima, Japan: RERF, 1988; Shimizu Y, Kato H, Schull WJ. Studies of the mortality of A-bomb survivors. 9. Mortality, 1950–1985: part 2. Cancer mortality based on the recently revised doses (DS86). Radiat Res 1990;121:120–141.

[b] RR, relative risk.

[c] $p < .05$.

[d] $p < .01$.

higher than in those treated for similar malignancies with radiation alone (Fry and Ullrich 1986). Similarly, the carcinogenic action of radiation on the skin appears to be enhanced by ultraviolet radiation (Albert and Shore 1986). The combined carcinogenic effects of cigarette smoking and radiation on the lung appear to differ, depending on the conditions of irradiation (Cross et al. 1982), appearing more than additive in uranium miners (Whittemore and MacMillin 1985) but not in A-bomb survivors (NAS/NRC 1988, Shimizu et al. 1988, 1990).

LATENT PERIOD OF RADIATION-CANCER

The average period intervening between irradiation and cancer varies with the type of malignancy. With respect to leukemia and osteosarcoma, the radiation-induced excess first becomes evident within 2–5 years after irradiation, peaks within the first decade, and gradually decreases thereafter. With solid tumors other than those of bone, the first cases do not become evident until 10 or more years after irradiation, following which they tend to increase in numbers with advancing age, at least for a time. In A-bomb survivors, the overall excess of mortality from cancer has increased with attained age, roughly in parallel with the baseline cancer rate (Fig. 7.1), so that the relative risk during adult life has tended to remain more or less constant, as noted above (Table 7.1). Conversely, in patients treated for ankylosing spondylitis by x-irradiation of the spine, mortality from cancer has been observed to be maximal during the second decade after irradiation, following which it has declined (Darby et al. 1987). Whether the difference between the spondylitics and the A-bomb survivors reflects differences in the relative proportions of cancers of different organs or the influence of other factors remains to be determined, as noted below.

Age at Expression of Radiation-Induced Cancer

In age distribution, the cancers induced by irradiation appear generally to parallel those of the same histological types in the general population. It is noteworthy, therefore, that radiation-induced cancers of the female breast resulting from irradiation in childhood do not become evident until the affected women reach the age at which breast cancer typically becomes prevalent in the general population. This pattern of age distribution implies that irradiation merely initiates carcinogenesis in the breast, full development of which is dependent on age-related hormonal stimulation.

Risk Coefficients

The relative risk of cancer mortality in the Japanese A-bomb survivors, over the follow-up period 1950–1985 (2.2 million person-years), has been estimated to approximate 1.39 (1.32–1.46) per Gy (shielded kerma), corresponding to an absolute risk of 13.1 (10.4–15.9) excess deaths per 10^4 PYGy organ absorbed dose (Shimizu et al. 1988, 1990). In the x-irradiated spondylitics, by comparison, the relative risk of mortality from all cancers except carcinoma of the colon, averaged over a mean follow-up of 23.0 years (184,000 person-years), has been estimated to approximate 1.33, corresponding to 6.7 excess deaths per 10^4 PYGy (Darby et al. 1987; Smith and Doll 1981; UNSCEAR 1988). The difference between the two populations in excess cancer mortality per unit dose may conceivably be attributable to differences in (*a*) temporal and spatial distribution of the radiation dose, (*b*) age and sex structure of the population at risk, (*c*) duration of follow-up, (*d*) methods of cancer ascertainment, (*e*) nature of the reference population used for comparison, (*f*) constitutional differences in susceptibility, (*g*) subgroup selection, and (*h*) competing causes of death (e.g., Table 7.5).

Table 7.5. Salient Characteristics of the A-Bomb Survivor and Ankylosing Spondylitis Studies[a]

Characteristic	Atomic Bomb Survivor Study	Spondylitis Study
Nature of study	Prospective	Retrospective-prospective
Sample size	76,000	14,000
Sex composition	F = 59%	F = 17%
Age at irradiation (years)	0–>90	>15
Average follow-up (years)	28.8	23.0
Type of control	Internal	National rates
Type of dosimetry	Individual (DS86)	Individual for leukemia, 1/15 random sample elsewhere
Type of irradiation	Instantaneous, whole-body	Fractionated, non-uniform, partial-body
Dose distribution		
Mean dose (Gy)	0.24	1.9
Range of doses (Gy)	(0.01–6.0)	(0–8.06)
Person-years at risk	2,183,335	184,000

[a] Adapted from United Nations Scientific Committee on the Effects of Atomic Radiation. Sources, effects, and risks of ionizing radiation. Report to the general assembly with annexes. New York: United Nations, 1988.

Without a scientific basis for reconciling or combining the two risk estimates, both the UNSCEAR (UNSCEAR 1988) and the BEIR V committees (NAS/NRC 1990) considered the mortality experience of the A-bomb survivors to be more appropriate for projecting risk estimates applicable to the general population; however, present data from the follow-up of the A-bomb survivors cover only the first 40 years after irradiation, with the result that those survivors who were irradiated in childhood are just now attaining the age at which cancer becomes prevalent in the general population. It is not clear, therefore, how their cancer mortality will ultimately compare with that of cohorts irradiated at older ages. The most recent data suggest that the excess relative risk of overall cancer mortality has varied less with attained age than has the absolute risk, at least during adult life (Table 7.1). To this extent, therefore, the data are more consistent with a relative (or multiplicative) risk projection model than with an absolute (or additive) risk projection model, as noted above.

According to both types of models, the computation of lifetime risk following exposure to dose $\underline{D}$ at age $\underline{a}_o$ depends on q(a), the usual rate of death at age $\underline{a}$ for someone who has reached that age (for $a > a_o$), and δ $(a; a_o,D)$, the excess rate of death at that age resulting from exposure to dose $\underline{D}$. Both $\underline{q}$ and $\underline{\delta}$ can be expressed as cause-specific sums:

$$q(a) = q_1(a) + \ldots\ldots + q_k(a)$$
$$\delta(a;a_o,D) = \delta_1(a;a_o,D) + \ldots\ldots + \delta_k(a;a_o,D).$$

The probability of survival to age $\underline{a}$, given exposure to dose D at age a_o, is $L(a;a_o,D)$. That quantity is calculated from the many values of $q`$ $(a`;a_oD)$ for $a`$ $\leq a$, where

$$q`(a;a_o,D) = q(a) + \delta(a;a_o,D),$$

where $L(a + 1;a_oD) = L(a;a_oD) \times (1\text{-}q`(a;a_oD)$ for $a = 0,1,2\ldots$, and where $L(O;a_{o1}D) = 1$. Death from cause $\underline{i}$ at age $\underline{a}$ resulting from the exposure is proportional to $\delta_i(a;a_o,D) \times L(a;a_o,D)$.

In the additive risk projection model, the cause-specific excess risk coefficients δ_i depend on $\underline{a}$ only in terms of whether or not $\underline{a}$ exceeds $\underline{a}_o$ by at least a minimal latent period P_i:

$$\delta_i(a;a_o,D) = \begin{cases} A_i(a_oD) & \text{if } a > a_o + P_i \\ O & \text{if } a \leq a_o + P_i \end{cases}$$

In the multiplicative risk projection model, δ_i is a multiple of q:

$$\delta_i(a;a_o,D) = \begin{cases} M_i(a_o,D) \times q_i(a) & \text{if } a > a_o + P_i \\ O & \text{if } a \leq a_o + P_i \end{cases}$$

The coefficients A_i (a_o,D) and M_i (a_o,D), which depend only on dose $\underline{D}$ and age at exposure $\underline{a}_o$, for a given sex, are the so-called absolute and relative risk coefficients for the cause denoted by the index i.

For cancers of most sites, the rates increase with age, so that risks projected from the results of an incomplete follow-up tend to be substantially larger with the multiplicative model than with the additive model, the difference gradually disappearing with continued observation, as the follow-up nears completion. Lifetime risks projected with both models are now somewhat larger than heretofore, partly because of the dosimetric changes discussed above, but the estimates projected with the additive model have increased appreciably more than those projected with the multiplicative model (Table 7.6).

The above risk projection models applied to the data for A-bomb survivor cohorts of different ages imply that lifetime risks of radiation-induced cancer in those exposed during adult life are substantially lower than in those exposed during childhood or adolescence (Tables 7.7, 7.8).

TABLE 7.6. EVOLVING ESTIMATES OF EXCESS CUMULATIVE LIFETIME MORTALITY FROM CANCER, ALL TYPES COMBINED, ATTRIBUTABLE TO 1 GY ACUTE WHOLE-BODY LOW-LET IRRADIATION

Source of Estimate	Additive Risk Projection Model	Multiplicative Risk Projection Model
	(deaths per 10,000 persons)[a]	
BEIR I, 1972	120[b]	620[b]
UNSCEAR, 1977	250[b]	—
BEIR III, 1980	80[c]–250[b]	230[c]–500[b]
NUREG, 1985	290[c]	520[c]
UNSCEAR, 1988	400[b,d]–500[b,e]	700[b,e]–1100[b,d]
BEIR V, 1990	—	885[b,f]

[a] Values rounded; estimates for the general population, based largely on follow-up of Japanese A-bomb survivors, analyzed with T65D dosimetry prior to 1988 and with DS-86 dosimetry thereafter.

[b] Linear dose-incidence model.

[c] Estimate based on linear-quadratic dose-incidence model.

[d] Estimate for Japanese population, based on age-specific risk coefficients.

[e] Estimate for Japanese population, based on age-averaged risk coefficient.

[f] Estimate for U.S. population, based on modified multiplicative model (see footnotes to Table 7.8.).

TABLE 7.7. PROJECTED LIFETIME NUMBERS OF FATAL CANCERS AND YEARS OF LIFE LOST IN A POPULATION EXPOSED RAPIDLY TO 1 GY WHOLE-BODY LOW-LET RADIATION[a]

	Risk Projection Model	Excess Fatal Cancer[b]	Years of Life Lost[b]
Total population	Additive	400[c]–500[d]	9500[c]–12000[d]
	Multiplicative	700[d]–1100[c]	9500[d]–14000[c]
Working population (aged 25–64 years)	Additive	400[d]–600[c]	8800[d]–13300[c]
	Multiplicative	700[c]–800	8200[c]–9700[d]
Adult population (over 25 years)	Additive	500[d]	400[d]
	Multipliative	600[d]	6200[d]

[a] Adapted from United Nations Scientific Committee on the Effects of Atomic Radiation. Sources, effects, and risks of ionizing radation. Report to the general assembly with annexes. New York: United Nations, 1988.

[b] Values are for a population of 5000 males and 5000 females, based on cancer mortality rates for the population of Japan.

[c] Derived with age-specific risk coefficients.

[d] Derived with constant (age-averaged) risk coefficient.

TABLE 7.8. ESTIMATED EXCESS LIFETIME MORTALITY FROM CANCERS OF VARIOUS ORGAN SYSTEMS RESULTING FROM ACUTE EXPOSURE TO 0.1 GY WHOLE-BODY LOW-LET RADIATION, IN RELATION TO AGE AT EXPOSURE[a,b]

Males
(Deaths per 10^5)

Age at Exposure	Total	Leukemia	Nonleukemia[c]	Respiratory	Digestive	Other
5	1276	111	1165	17	361	787
15	1144	109	1035	54	369	612
25	921	36	885	124	389	372
35	566	62	504	243	28	233
45	600	108	492	353	22	117
55	616	166	450	393	15	42
65	481	191	290	272	11	7
75	258	165	93	90	5	—
85	110	—	96	14	17	—
Average[d]	760	110	650	190	170	300

Females
(Deaths per 10^5)

Age at Exposure	Total	Leukemia	Nonleukemia[c]	Breast	Respiratory	Digestive	Other
5	1532	75	1457	129	48	655	625
15	1566	72	1494	295	70	653	476
25	1178	29	1149	52	125	679	293
35	557	46	511	43	208	73	187
45	541	73	468	20	277	71	100
55	505	117	388	6	273	64	45
65	386	146	240	—	172	52	16
75	227	127	100	—	72	26	3
85	90	73	17	—	15	4	—
Average[d]	810	80	730	70	150	290	220

[a] Reprinted with permission from *Health Effects of Exposure to Low Levels of Ionizing Radiation: BEIR V*, c. 1990 by the National Academy of Sciences. Published by National Academy Press, Washington, DC.

[b] Based on a single exposure to radiation, and on a lifetable weighted average over each of the age groups listed, in a stationary population having U.S. mortality rates. The values tabulated represent the numbers of excess cancers in an irradiated, as compared with a nonirradiated, population; they do not include the numbers of persons dying from a radiation-induced cancer earlier than they would otherwise have died from a naturally occurring cancer. Use of the latter formulation, as in the UNSCEAR (1988) and BEIR III (1980) reports would increase the numbers by about 25%.

[c] Based on the sum of cancers of respiratory tract, digestive tract, breast, and other organs.

[d] Values rounded to nearest 10. The age-specific cancer risk $\gamma(d)$ was expressed as

$$\gamma(\mathrm{d}) = \gamma_0[1 + \mathrm{f(d)g}(\beta)].$$

where γ denotes the age-specific background risk of death to a specific cancer for an individual at a given age, which also depend upon the individual's sex and year of birth, f(d) represents a function of the dose, d (in sievert), which is a linear or linear quadratic function—i.e., $\mathrm{f(d)} = \alpha_1\mathrm{d}$ or $\mathrm{f(d)} = \alpha_2\mathrm{d} + \alpha_3\mathrm{d}^2$—and $\mathrm{g}(\beta)$ is the excess risk function, which may depend on sex, attained age, age at exposure (E), and time exposure (T).

For the leukemia model, the parameters were

$$\mathrm{f(d)} = \alpha_2\mathrm{d} + \alpha_3\mathrm{d}^2$$

$$\mathrm{g}(\beta) = \begin{cases} \exp[\beta_1\mathrm{I}(\mathrm{T} \le 15) + \beta_2\mathrm{I}(15 < \mathrm{T} < 25)] \text{ if } \mathrm{E} \le 20 \\ \exp[\beta_3\mathrm{I}(\mathrm{T} \le 25) + \beta_4\mathrm{I}(25 < \mathrm{T} < 30)] \text{ if } \mathrm{E} \le 20 \end{cases}$$

To the extent that the impact of a given cancer death depends on the age at which the affected person dies, the risks of radiation-induced cancer can also be expressed in terms of the years of life lost to the disease. On average, the loss of life expectancy per person attributable to the carcinogenic effects of 1 Gy rapid whole-body low-LET radiation in a population of all ages is estimated to be about 1 year, depending on whether the estimate is derived by the use of a multiplicative risk projection model or an additive risk projection model. In a population exposed continuously to whole-body radiation from age 18 to age 65, at a rate of 0.01 Sv per year, the loss of life expectancy from cancer is estimated to average slightly less than one half year per person, or about 16 years per excess cancer death (Table 7.9).

PRESENT NEW RISK ESTIMATES AS COMPARED WITH THOSE IN ICRP-26

The cumulative lifetime cancer risks projected in the most recent UNSCEAR (1988) and NAS/NRC (1990) reports are appreciably higher than those projected

---

where the indicator function $I(T \le 15)$ is defined as 1 if $T \le 15$ and 0 if $T > 15$. The estimated parameter values and their standard errors, in parentheses, are

$$\alpha_2 = 0.243(0.291),\ \alpha_3 = 0.271(0.314),$$
$$\beta_1 = 4.885(1.349),\ \beta_2 = 2.380(1.311),\ \beta_3 = 2.367(1.121),$$
$$\beta_4 = 1.638(1.321).$$

For cancer of the respiratory tract (ICD 160 − 163), the model parameters were as follows

$$f(d) = \alpha_1 d$$
$$g(\beta) = \exp[\beta_1 \mathrm{In}(T/20) + \beta_2 I(s)],$$

where $I(S) = 1$ if female, 0 if male, with $\alpha_1 = 0.636(0.291)$, $\beta_1 = 1.437(0.910)$, $\beta_2 = 0.711(0.610)$.

For breast cancer (female only), the model parameters were

$$f(d) = \alpha_1 d$$
$$g(\beta) = \begin{cases} \exp[\beta_1 + \beta_2 \ln(T/20 + \beta_3 \ln^2(T/20)] & \text{if } E < 15 \\ \exp[\beta_2 \ln(T/20) + \beta_3 \ln^2(T/20) + \beta_4 (E - 15)] & \text{if } E \ge 15, \end{cases}$$

where $\alpha_1 = 1.220(0.610)$, $\beta_1 = 1.385(0.554)$, $\beta_2 = -0.104(0.804)$, $\beta_3 = -2.212(1.376)$, $\beta_4 = -0.0628(0.0321)$.

For cancer of the digestive system (ICD 150 − 159), the model parameters were

$$f(d) = \alpha_1 d$$
$$g(\beta) = \exp[\beta_1 I(S) + \sigma E]$$

where $I(S)$ equals 1 for females and 0 for males and

$$\sigma E = \begin{cases} O & \text{if } E \le 25 \\ \beta_2(E - 25) & \text{if } 25 < E \le 35 \\ 10\beta_2 & \text{if } E > 35 \end{cases}$$

with $\alpha_1 = 0.809(9.327)$, $\beta_1 = 0.553(0.462)$, $\beta_2 = -0.198(0.0628)$.

For cancers other than those listed above (ICD 140 − 209 less those listed above), the model parameters were

$$f(d) = \alpha_1 d$$
$$g(\beta) = 1 \text{ if } E \le 10\ \exp[\beta_1(E - 10)] \text{ if } E > 10,$$

with $\alpha_1 = 1.220(0.519)$, $\beta_1 = -0.0464(0.0234)$.

TABLE 7.9. PROJECTED LIFETIME CANCER MORTALITY AND ASSOCIATED LOSS OF LIFE EXPECTANCY RESULTING FROM CONTINUOUS WHOLE-BODY IRRADIATION IN A POPULATION OF BOTH SEXES[a,b]

	Exposure Throughout Life (1mSv/yr)	Exposure from Age 18 to Age 65 (10 mSv/yr)
Excess cancer deaths		
No. per 10^4	56	320
% of normal	3	16
Loss of life expectancy (yr)		
Average per person exposed	0.2	0.5
Average per excess death	17	16

[a] Adapted from National Academy of Sciences, National Research Council. Committee on Biological Effects of Ionizing Radiation (BEIR V). Washington, DC: National Academy Presss, 1990.
[b] Calculations based on cancer and survival rates for the U.S. poulation and on use of the risk models presented in Table 7.10, which include an implicit DREF of about 2.0 for leukemia and a DREF of 1 for solid tumors.

TABLE 7.10. ESTIMATES OF LIFETIME CANCER RISKS BASED ON THE 1988 UNSCEAR AND BEIR V REPORTS, AS COMPARED WITH THOSE GIVEN IN ICRP-26

Organ at Risk	Cancer Deaths per 10^4 PSv	
	ICRP-26	UNSCEAR/BEIR[a]
Bone marrow	20	85
Bone	5	5
Lung	20	100
Thyroid	5	10
Breast	25	20
Subtotal	75	220
Remainder		
GI tract	—	150
Ovary	—	15
Bladder	—	30
Multiple myeloma	—	15
Skin	—	2
Other	—	65
Subtotal	50	282
TOTAL	125	500

[a] Rounded values, based on averges of the UNSCEAR and BEIR multiplicative projections derived with age-specific risk coefficients, divided by a DREF of 2.0 for compatibility with the estimate in ICRP-26, which were applicable to irradiation at low dose rates.

in ICRP-26 (ICRP 1977), even if reduced by a factor of 2–3 to allow for a decrease in the effectiveness of radiation at the low dose rates characteristic of occupational exposure (Table 7.10). Although neither the UNSCEAR nor the BEIR V Committee recommended a specific dose rate effectiveness factor (DREF) to use in adjusting risk estimates for exposures at low doses and low dose rates, both reports acknowledged evidence (NCRP 1980; UNSCEAR 1986) that the carcinogenic effectiveness of low-LET radiation in laboratory animals is generally lower by a factor of 2–10 at low doses (0.20 Gy) and low dose rates (<1 mGy min^{-1}) than at high doses (>3 Gy) and high dose rates (>100 mGy min^{-1}). On the basis of such evidence, both reports suggested that a DREF at the lower end of the above range would not be unreasonable. Since, however, the linear-quadratic dose-incidence model used for leukemia in the BEIR V report already introduced an implicit DREF of approximately 2.0 for this disease, no further adjustment for dose or dose rate in the risk estimates for leukemia was considered justifiable (NAS/NRC 1990).

Preparation of this report was supported in part by Grants ES 00260 and CA 13343 from the U.S. Public Health Service and Grant SIG 09 from the American Cancer Society. The author is grateful to Ms. Lynda Witte for assistance in the preparation of this manuscript.

REFERENCES

Albert, R. E., and Shore, R. E. Carcinogenic effects of radiation on the human skin. In: *Radiation Carcinogenesis*, edited by A. C. Upton, R. E. Albert, F. J. Burns, and R. E. Shore. New York, Elsevier, 1986, pp. 335–354.

Cross, F. T., et al. Carcinogenic effects of radon daughters, uranium ore dust and cigarette smoke in beagle dogs. *Health Phys. 42:*33–52, 1982.

Darby, S. C., Doll, R., Gill, S. K., et al. Long term mortality after a single treatment course with x-rays in patients treated for ankylosing spondylitis. *Br. J. Cancer 55:*179–190, 1987.

Fry, R. J. M., and Ullrich, R. L. Combined effects of radiation and other agents. In: *Radiation Carcinogenesis*, edited by A. C. Upton, R. E. Albert, F. J. Burns, and R. E. Shore. New York, Elsevier, 1986, pp. 437–454.

Hirayama, T., ed. *Comparative Epidemiology of Cancer in the U.S. and Japan. Mortality.* Tokyo: U.S.-Japan Cooperative Cancer Research Program, 1977.

Holm, L.-E., Wiklund, K. E., Lundell, G. E., Bergman, N. A., Bjelkengren, G., Aderquist, E. S., Ericsson, U.-B. C., Larsson, L.-G., Lidberg, M. E., Wicklund, H. V., and Boice, J. D., Jr. Thyroid cancer after diagnostic doses of iodine-131: a retrospective cohort study. *J. Natl. Cancer Inst. 80:* 1132–1138, 1988.

International Commission on Radiological Protection. *Recommendations of the International Commission on Radiological Protection,* ICRP Publication No. 26. Oxford, Pergamon Press, 1977.

International Commission on Radiation Units and Measurements. *The Quality Factor in Radiation Protection,* ICRU Report No. 40. Bethesda, MD, International Commission on Radiation Units and Measurements, 1986.

Knudson, A. G. Hereditary cancer, oncogenes, and multistep carcinogenesis. *Cancer Res. 45:* 1437–1443, 1985.

Miller, A. B., et al. Mortality from breast cancer after irradiation during fluoroscopic examinations in patients being treated for tuberculosis. *N. Engl. J. Med. 321:*1285–1289, 1989.

Modan, B., Alfandary, E., Chetrit, A., and Katz, L. Increased risk of breast cancer after low-dose irradiation. *Lancet 1:*629–630, 1989.

National Academy of Sciences, National Research Council Committee on Biological Effects of Ionizing Radiation (BEIR I). *The Effects on Populations of Exposure to Low Levels of Ionizing Radiation.* Washington, DC, Department of Health, Education, and Welfare, and U.S. Environmental Protection Agency, 1972.

National Academy of Sciences, National Research Council Committee on Biological Effects of Ionizing Radiation (BEIR III). *The Effects on Populations of Exposure to Low Levels of Ionizing Radiation.* Washington, DC, National Academic Press, 1980.

National Academy of Sciences, National Research Council. Health *Risks of Radon and Other Internally Deposited Alpha-Emitters. Advisory Committee on the Biological Effects of Ionizing Radiation (BEIR IV).* Washington, DC, National Academy Press, 1988.

National Academy of Sciences, National Research Council. *Committee on Biological Effects of Ionizing Radiation (BEIR V). Health Effects of Exposure to Low Levels of Ionizing Radiation,* Washington, DC, National Academy Press, 1990.

National Council on Radiation Protection and Measurements. *Influence of Dose and Its Distribution in Time on Dose-Response Relationships for Low-LET Radiations*, NCRP Report 64. Bethesda, MD, National Council on Radiation Protection and Measurements, 1980.

National Council on Radiation Protection and Measurements. *Induction of Thyroid Cancer by Ionizing Radiation*, NCRP Report 80. Bethesda, MD, National Council on Radiation Protection and Measurements, 1985.

Preston, D. L., and Pierce, D. A. The effect of changes in dosimetry on cancer mortality risk estimates in the atomic bomb survivors. *Radiat. Res. 114:*437–466, 1988.

Shimizu, Y., Kato, H., and Schull, W. J. *Life Span Study Report 11, Part II: Cancer Mortality in the Years 1950–1985 Based on the Recently Revised Doses (DS86)*, Radiation Effects Research Foundation TR 5-88. Hiroshima, Japan, Radiation Effects Research Foundation, 1988.

Shimizu, Y., Kato, H., and Schull, W. J. Studies of the mortality of A-bomb survivors. 9. Mortality, 1950–1985: Part 2. Cancer mortality based on the recently revised doses (DS86). *Radiat. Res. 121:*120–141, 1990.

Shore, R. E., Hildreth, N., Woodard, E. D., et al. Breast cancer among women given x-ray therapy for acute postpartum mastitis. *J. Natl. Cancer Inst. 77:*689–696, 1986.

Shore, R. E., Woodard, E., Hildreth, N., et al. Thyroid tumors following thymus irradiation. *J. Natl. Cancer Inst. 74:*1177–1184, 1985.

Smith, P. G., and Doll, R. Mortality from cancer and all causes among British radiologists. *Br. J. Radiol. 54:*187–194, 1981.

Storer, J. B., Mitchell, T. J., and Fry, R. J. M. Extrapolation of the relative risk of radiogenic neoplasms across mouse strains to man. *Radiat. Res. 114:*331–353, 1988.

Strong, L. C. Theories of pathogenesis: mutation and cancer. In: *Genetics of Human Cancer*, edited by J. J. Mulvihill, R. W. Miller, J. J. Fraumeni Jr. New York, Raven Press, 1977, pp. 401–414.

Swift, M., Reitnauer, P. J., Morrell, D., and Chase, C. L. Breast and other cancers in families with ataxia-telangiectasia. *N. Engl. J. Med. 316:*1289–1294, 1987.

Tokunaga, M., Land, C. E., Yamamoto, T., et al. Breast cancer among atomic bomb survivors. In: J. D. Boice and J. F. Fraumeni, eds. Radiation Carcinogenesis: Epidemiology and Biological Significance. New York: Raven Press, 1984:45–56.

United Nations Scientific Committee on the Effects of Atomic Radiation. *Sources and Effects of Ionizing Radiation. Report to the General Assembly with Annexes.* New York, United Nations, 1977.

United Nations Scientific Committee on the Effects of Atomic Radiation. *Ionizing Radiation: Sources and Biological Effects. Report to the General Assembly with Annexes.* New York, United Nations, 1982.

United Nations Scientific Committee on the Effects of Atomic Radiation. *Genetic and Somatic Effects of Ionizing Radiation. Report to the General Assembly with Annexes.* New York, United Nations, 1986.

United Nations Scientific Committee on the Effects of Atomic Radiation. *Sources, Effects, and Risks of Ionizing Radiation. Report to the General Assembly with Annexes.* New York, United Nations, 1988.

Upton, A. C. Biological basis of radiological protection and its application to risk assessment. *Br. J. Radiol. 6:*1–16, 1987.

U.S. Nuclear Regulatory Commission. *Health Effects Model for Nuclear Power Plant Accident Consequence Analysis*, U.S. Nuclear Regulatory Commission Contractor Report (Sandia No. SAND 85-7185) NUREG/CR-4214. Washington, DC, U.S. Government Printing Office, 1985.

Whittemore, A. S., and MacMillan, A. Cancer mortality among U.S. uranium miners: a reappraisal. *J. Natl. Cancer Inst. 71:*489–499, 1985.

Chapter 8

Estimates of Cancer Risk from Internally Deposited Radionuclides, with Emphasis on Alpha-Emitters: BEIR IV

W. A. MILLS

ABSTRACT

The National Academy of Sciences/National Research Council's BEIR IV report addresses health effects of exposure of human populations to internally deposited alpha-emitting isotopes, specifically, radon, polonium, radium, thorium, uranium, and the transuranic elements. Radionuclides and progeny of these elements are the major sources of human exposures to ionizing radiation and, except for the transuranics, are naturally occurring. Much of today's radiation protection effort involves risk assessments and control of exposures to the isotopes treated in BEIR IV. This article discusses briefly the scientific findings in BEIR IV and provides a summary of organ-specific coefficients appropriate for consideration in assessing risks of cancer mortality. Although not addressed in BEIR IV, risk coefficients for internally deposited radionuclides of iodine and strontium are also included, based on recent reports of the National Council on Radiation Protection and Measurements (NCRP).

INTRODUCTION

The intake of radionuclides by inhalation and ingestion and their subsequent deposition in bone and soft tissues results in about 70% of the average annual effective dose equivalent in soft tissues received from exposure to ionizing radiation by a member of the U.S. population. Of this 70%, radionuclides in the natural environment contribute about 65%, with the associated effective dose equivalent (effective dose in new International Commission on Radiological Protection [ICRP] terminology) (ICRP 90) to lung bronchial epithelial cells contributing about 55%. The remaining 5% or so of the dose from internal emitters results primarily from the use of radionuclides in nuclear medicine, with small additions from consumer products and even smaller additions from other sources. Table 8.1 shows the average per capita dose distribution for the U.S. population (NCRP 87).

An important perspective to bear in mind when considering the biological basis

TABLE 8.1. AVERAGE ANNUAL EFFECTIVE DOSE EQUIVALENT OF IONIZING RADIATIONS TO A MEMBER OF THE U.S. POPULATION[a]

Source	Dose Equivalent[b] (mSv)	Effective Dose Equivalent (mSv)	Effective Dose Equivalent (Percent)
Natural			
Radon[c]	24.0	2.0	55
Cosmic	0.27	0.27	8
Terrestrial	0.28	0.28	8
Internal	0.39	0.39	11
Total	—	3.0	82
Artificial			
Medical			
X-ray diagnosis	0.39	0.39	11
Nuclear medicine	0.14	0.14	4
Consumer products	0.10	0.10	3
Other			
Occupational	0.09	<0.01	<0.3
Nuclear fuel cycle	<0.01	<0.01	<0.03
Fallout	<0.01	<0.01	<0.03
Miscellaneous[d]	<0.01	<0.01	<0.03
Total artificial	—	0.63	18
Total natural and artificial	—	3.6	100

[a] From National Council on Radiation Protection and Measurements. Ionizing radiation exposure of the population of the United States. NCRP Report No. 93. Bethesda, MD: National Council on Radiation Protection and Measurements, 1987. Reprinted with permission of the National Council on Radiation Protection and Measurements.

[b] To soft tissues.

[c] Dose equivalent to bronchi from radon daughter products. The assumed weighting for the effective dose equivalent relative to whole-body exposure is 0.08.

[d] Department of Energy facilities, smelters, transportation, etc.

for radiation protection involving internal emitters is that unlike the case for some external radiation exposure conditions, exposures to radioactively contaminated materials in air or water do not occur in such a way that acute health effects are of concern. Thus the practice of protection is not directed toward prevention of large intakes of radionuclides but rather toward controlling the probability of a latent health effect from exposure to concentrations near natural background concentrations, specifically cancer. Potential hereditary defects or in utero effects are not of routine concern in radiation protection for internal emitters.

Attention herein is focused on the radioactive elements addressed in the 1988 National Academy of Sciences/Nuclear Regulatory Commission (NAS/NRC) publication entitled *Health Risks of Radon and Other Internally Deposited Alpha-Emitters: BEIR IV* (NAS/NRC 1988). Because of their special interest in radiation protection, brief comments are included on the health risks associated with radioactive iodine and strontium. The elements discussed in BEIR IV are radon (Rn), polonium (Po), radium (Ra), thorium (Th), uranium (U), and the transuranic elements, principally plutonium (Pu), with a major proportion devoted to the analysis for estimating the lung cancer risk from radon. In this paper these radionuclides are discussed in regard to their associated biological effects and the derivation of risk estimates, i.e., information used in performing risk assessments and making risk management decisions. No discussion is included that addresses environmental transport and metabolic behavior of these radionuclides, or the mechanics of calculating radiation dose to the tissue(s) at risk.

RADON

Regardless of whether the primary source is radon in air or radon in potable water, the risk of cancer due to exposure to radon is the risk of lung cancer following inhalation of radon progeny. (To assess the associated inhalation lung-cancer risk from radon in water, the most accepted transport ratio of radon in air to radon in water is 10^{-4} [Nazaroff et al. 1988]). The BEIR IV Committee considered two approaches for deriving inhalation risk estimates for radon in air: a dosimetric approach based on modeling the dose to target cells in the respiratory tract and an epidemiological approach based directly on the "substantial" body of available human data. The committee chose the epidemiological approach, analyzing four sets of underground miner data.

For radon progeny in air the BEIR IV Committee recommends an age-specific relative risk, time since exposure, model of the form:

BEIR IV: Relative Risk Model for Radon Progeny

$$r(a) = r_0(a)\,\{1 + 0.025(a)\,[W_1 + 0.5\,W_2]\}$$

where,
a = specific age of individual at risk;
$r(a)$ = age-specific lung-cancer mortality rate;
$r_0(a)$ = age-specific baseline lung-cancer mortality rate;
(a) = 1.2, when "a" is < 55 y,
1.0, when "a" is 55–65 y, and
0.4, when "a" is ≥ 65 y;
W_1 = working level months, WLM, incurred between 5 and 15 y before age "a";
W_2 = working level months, WLM, incurred 15 or more years before age "a."

In this general relative risk model, r(a) is the total lung cancer mortality rate from all causative agents, not just the rate due to radon exposure. Designating the lifetime probability of lung-cancer mortality in a nonexposed person to radon as R_0 and that of an exposed person as R_e, the ratio of lifetime risks R_e/R_0 is the proportional increased lifetime risk associated with exposure to radon progeny, and the excess risk is $R_e - R_0$. It is important to note that the excess relative risk due to radon in this model varies with time since exposure (TSE); thus cumulative exposure occurring 15 or more years prior to age (a) is discounted by 50%. This discounting of prior exposures is, in principle, consistent with the model adopted earlier by the National Council on Radiation Protection and Measurements (NCRP) which included a factor for diminishing risk from prior exposures with a half-time of 20 years (NCRP 1984). The NAS/NRC BEIR V Committee also included a decrease in risk over time since exposure in its models for leukemia, breast cancer, and cancers of the respiratory tract (NAS/NRC 1990).

In calculating R_e/R_0, the BEIR IV Committee considered confounding factors of sex and smoking status, assuming cigarette smoking to have a multiplicative interaction with radon exposure. No adjustments were made in baseline rates (R_0) for levels of "ambient radon progeny," concluding that "this correction would be small" compared with the overall uncertainty in the estimates. Interestingly, the committee included estimates for an exposure rate of 0.1 WLM/y, which represents a relatively small increase in the U.S. "ambient" average indoor exposure rate of 0.2 WLM/y.

Table 8.2 provides BEIR IV estimates of increased risk of lung cancer mortality due to radon exposure, according to sex and designation as smoker or nonsmoker. Smokers are assumed to be only cigarette smokers who start their habit at the age of 18 and continue at a steady-state consumption rate for life. For comparable smoking status, males have about a 2 times greater excess risk than females, and for both sexes, smokers have about a ten times greater excess risk than nonsmokers.

It is noteworthy that although the ratio of lifetime risks R_e/R_0 was the committee's preferred measure of increased risk, it considered the number of years lost because of radon exposure also as a measure of risk. The committee provided similar table entries of estimates for the differences between the expected years of life from birth for a "nonexposed" subject (L_o) and that of an exposed subject (L_e), where L_o is the average lifetime of nonsmokers. For a lifetime exposure range of 0.1 WLM/y to 10 WLM/y, values of $L_o - L_e$ range from about 3 days to 10 months in male nonsmokers, 1.6 years to 9 years in male smokers, 2 days to 7 months in female nonsmokers, and 10 months to 6 years in female smokers. According to the committee, the number of years of life lost by those who actually die of lung cancer is typically about 15 years for males and 18 years for females.

For convenience in transposing from "traditional" units for expressing radon concentrations (e.g., pCi/L) to SI-International System of units (e.g., Bq/m^3), selected values of exposure rate and corresponding R_e/R_0 ratios are shown in Table 8.3.

BEIR IV also provided lifetime risk coefficients for various ages at which exposure to radon progeny begins and ends, taking into account a 10-year latency period for lung cancer to be manifested. Ages at the start of exposure ranged from 0–60 years, and ages at the end of exposure ranged from 10–110 years. As an example, Table 8.4 shows values for males and females continuously exposed during the appropriate time period to a rate of 1 WLM/y.

Assuming an equal number of males and females in a given population with an unspecified smoking characterization, the BEIR IV nominal lifetime risk of lung-cancer mortality due to lifetime exposure to radon progeny is 350 cancer deaths per 10^6 person WLM.

However, this value is based on the BEIR IV Committee's assumption "that an exposure of 1 WLM in a home and exposure to 1 WLM in a mine have equivalent potency in causing lung cancer." Since the publication of BEIR IV, a

TABLE 8.2. INCREASE IN LIFETIME RISKS OF LUNG-CANCER DEATH RESULTING FROM EXPOSURE TO RADON DAUGHTERS[a,b]

Exposure Rate	Males						Females					
	Nonsmokers			Smokers			Nonsmokers			Smokers		
(WLM/yr)	R_e/R_0	R_e	R_e-R_0	R_e/R_0	R_e	R_e-R_0	R_e/R_0	R_e	R_e-R_0	R_e/R_0	R_e	R_e/R_0
0	1.0	0.0112	0.	1.0	0.123	0.	1.0	0.00603	0.	1.0	0.0582	0.
0.1	1.06	0.0118	0.0006	1.05	0.129	0.006	1.06	0.00637	0.00034	1.06	0.0614	0.0032
0.2	1.11	0.0124	0.0012	1.10	0.135	0.012	1.11	0.00672	0.00069	1.11	0.0646	0.0064
0.3	1.16	0.0131	0.0019	1.15	0.141	0.018	1.17	0.00706	0.00103	1.16	0.0678	0.0096
0.4	1.22	0.0137	0.0025	1.20	0.147	0.024	1.23	0.00741	0.00138	1.22	0.0710	0.0128
0.5	1.27	0.0143	0.0031	1.24	0.153	0.030	1.28	0.00775	0.00172	1.27	0.0742	0.0160
0.6	1.33	0.0149	0.0037	1.29	0.159	0.036	1.34	0.00809	0.00206	1.33	0.0773	0.0191
0.8	1.44	0.0161	0.0049	1.39	0.170	0.047	1.46	0.00878	0.00275	1.44	0.0836	0.0254
1.0	1.54	0.0173	0.0061	1.48	0.182	0.059	1.57	0.00946	0.00343	1.54	0.0898	0.0316
1.5	1.82	0.0204	0.0092	1.70	0.209	0.086	1.85	0.0112	0.00517	1.80	0.105	0.0468
2.0	2.08	0.0234	0.0122	1.91	0.235	0.112	2.14	0.0129	0.00687	2.06	0.120	0.0618
2.5	2.35	0.0264	0.0152	2.12	0.260	0.137	2.42	0.0146	0.00857	2.32	0.135	0.0768
3.0	2.62	0.0294	0.0182	2.31	0.284	0.161	2.70	0.0163	0.0103	2.56	0.149	0.0908
3.5	2.89	0.0324	0.0212	2.49	0.306	0.183	2.98	0.0180	0.0112	2.81	0.163	0.105
4.0	3.15	0.0354	0.0242	2.66	0.328	0.205	3.26	0.0197	0.0137	3.04	0.177	0.119
4.5	3.41	0.0383	0.0271	2.83	0.348	0.225	3.55	0.0214	0.0154	3.28	0.191	0.133
5.0	3.68	0.0413	0.0301	2.99	0.368	0.245	3.83	0.0231	0.0171	3.51	0.204	0.146
10.0	6.24	0.0700	0.0588	4.24	0.521	0.398	6.59	0.0398	0.0338	5.56	0.324	0.266

[a] Adapted with permission from *Health Risks of Radon and Other Internally Deposited Alpha-Emitters: BEIR IV*, © 1988 by the National Academy of Sciences. Published by National Academy Press, Washington, DC.

[b] R_0 = Baseline risk relative to persons of the same sex and smoking status. Value is not corrected for risk associated with average ambient radon-daughter exposure rate of 0.2 WLM/y; R_e = Total risk of lung-cancer mortality; R_e-R_0 = Excess risk of lung-cancer mortality associated with annual radon-daughter exposure rate.

TABLE 8.3. LIFETIME RISK OF LUNG CANCER MORTALITY FOR LIFETIME EXPOSURE TO RADON CONCENTRATIONS EXPRESSED IN TRADITIONAL AND SI UNITS

Ratio: $\frac{\text{Total Lung Cancer Risk } (R_e)}{\text{Baseline Lung Cancer Risk } (R_0)}$

Exposure[a]			Males		Females	
pCi L^{-1}	Bq m^{-3}[b]	Exposure Rate (WLM y^{-1})	Nonsmoker $R_0 = 0.0112$	Smoker $R_0 = 0.123$	Nonsmoker $R_0 = 0.00603$	Smoker $R_0 = 0.0582$
0	0	0[(c)]	1.00	1.00	1.00	1.00
0.7	25	0.1	1.06	1.05	1.06	1.06
2	74	0.3	1.16	1.15	1.17	1.16
4	150	0.6	1.33	1.29	1.34	1.33
10	370	1.5	1.82	1.70	1.85	1.80
20	740	3.0	2.62	2.31	2.70	2.56
30	1100	4.5	3.41	2.83	3.55	3.28
67	2500	10.0	6.24	4.24	6.59	5.56

[a] Neglecting "ambient" radon background level. Conversion assumptions for Exposure Rate/Exposure: [R_nDP] to [R_n] equilibrium = 0.4 WL to [R_n = 0.01 WL/pCi L^{-1} year at 1 WL = 51.5 WLM/y Indoor "home" occupancy = 75%

[b] 1 pCi/L = 37 Bq/m^3.

[c] Included are indoor [R_n] = 1.5 pCi L^{-1} or 0.2 WLM y^{-1}.

NAS/NRC panel has examined this assumption in regard to dosimetric considerations and concluded that the dose per unit exposure delivered to target cells in the respiratory tract for those exposed in the home environment should be reduced by about 30% for adults of both sexes and by 20% or less for infants and children aged 10 or less (NAS/NRC 1991). Thus for a lifespan of 70 years, the reduction is about 30%, and for the home environment the reduced BEIR IV risk coefficient becomes about 250 cancer deaths per 10^6 person WLM.

It is clear that smoking status plays a key role in attempting to estimate an individual's risk from radon exposure. The principal cause of lung cancer in the United States is cigarette smoking. Annual lung cancer mortality rates in the United States track very well with U.S. cigarette consumption. Of the estimated 143,000 lung cancer deaths expected in 1991, 83%, or about 120,000, can be attributed to cigarette smoking (85% among men and 75% among women) (American Cancer Society 1991).

If one assumes a per capita radon progeny exposure rate of 0.2 WLM/y for a U.S. population of 250 million and the reduced BEIR IV risk coefficient, an estimated 12,000 lung cancer deaths in 1991 can be attributed to radon exposure. However, 10,000 of these cases would occur in smokers and only about 2000 in nonsmokers, including those at risk from "environmental tobacco smoke" in the home. The vast majority of these radon-attributable lung cancers are associated with radon-progeny exposures below the U.S. Environmental Protection Agency's indoor radon guide of 4 pCi/L.

Therefore, it is difficult to conclude that a nationwide indoor radon mitigation program will result in a significant reduction in U.S. lung cancer mortality rates. In addition, the cleanup of uranium tailings seems questionable since it is based on the unvalidated assumption that a radium activity of 5 pCi/g of soil could

TABLE 8.4. LIFETIME RISKS (R_e/R_0) FOR VARIOUS AGE-DEPENDENT RADON EXPOSURE PERIODS[a,b]

MALES
Exposure Rate = 1.00 WLM/yr

	Age (yr) Exposure Ends								
Age (yr) Started	10	20	30	40	50	60	70	80	110
0	1.076	1.151	1.226	1.305	1.391	1.460	1.496	1.509	1.512
10		1.076	1.151	1.231	1.318	1.388	1.424	1.437	1.440
20			1.076	1.157	1.244	1.314	1.351	1.364	1.367
30				1.081	1.169	1.240	1.276	1.290	1.293
40					1.088	1.160	1.197	1.211	1.213
50						1.072	1.109	1.123	1.126
60							1.038	1.052	1.054

FEMALES
Exposure Rate = 1.00 WLM/yr

	Age (yr) Exposure Ends								
Age (yr) Started	10	20	30	40	50	60	70	80	110
0	1.082	1.165	1.248	1.338	1.432	1.504	1.539	1.555	1.560
10		1.083	1.166	1.256	1.351	1.422	1.458	1.473	1.479
20			1.084	1.174	1.269	1.341	1.376	1.392	1.397
30				1.090	1.186	1.258	1.293	1.309	1.314
40					1.096	1.186	1.203	1.219	1.224
50						1.072	1.108	1.124	1.129
60							1.036	1.052	1.057

[a] Excerpted with permission from *Health Risks of Radon and Other Internally Deposited Alpha-Emitters: BEIR IV*, © 1988 by the National Academy of Sciences. Published by National Academy Press, Washington, D.C.

[b] Smoking status of population unspecified.

result in an indoor radon progeny level of 0.02WL or 4 pCi/L. While there is a sound biological basis for protection programs to control radiation exposures of workers employed in underground mines, there seems to be much less of a biological basis for controlling radon directly.

POLONIUM

There are over 25 isotopes of polonium, all of which are radioactive and most of which are alpha-emitters. The important radioisotopes of interest to radiation protection are polonium-210, polonium-214, and polonium-218, with the latter two being important as progeny from the decay of radon-222. ^{210}Po is a major contributor to the internal dose received from natural radionuclides in the body, the other isotope being potassium-40. Radiation protection interests in ^{210}Po result from its use in experimental work, as a source of neutrons from the (α, n) reaction, and in consumer applications, such as static-eliminator devices. Polonium in tobacco has been implicated as a contributing factor in the induction of lung cancer in smokers.

No direct evidence of health effects from exposure to polonium exists. Therefore, estimates applicable to risk assessments for radiation protection must be obtained indirectly from results with other alpha emitters. Thus although it is less than a satisfactory approach, the BEIR IV report concluded that the toxicity of polonium is comparable to that of plutonium-239 and about five times that of radium-226. Over very long exposure times and very low doses, its toxicity approaches that of ^{226}Ra. This implies ^{210}Po tumor risk estimates of the order of 10^{-5}/cGy, primarily bone cancer.

RADIUM

Radium toxicity has been studied for more than a half-century and continues to be of great interest, not only because of its radioepidemiological significance but also because of its ubiquity in our environment. The history of radiation protection is intimately associated with epidemiological studies of radium toxicity.

Until the introduction of the concept of effective dose equivalent (now effective dose) by the ICRP in 1977, occupational limits in radiation protection for internal emitters were based on a level of safety associated with a ^{226}Ra body-burden equal to 0.1 μCi. The selection of this body-burden value was done because at or below this level no increase in bone cancer was observed in the extensive studies of radium dial painters. Maximum permissible body burdens (MPBB) and maximum permissible concentrations (MPC) in air or water for other internally deposited radionuclides were determined by dosimetric comparisons to this radium level of 0.1 μCi. Limits for the public were assumed to be 1/10 of the occupational limit.

ICRP Publication 26 recommended a different approach to controlling exposures resulting from internally deposited radionuclides. Under the ICRP's approach, the limits for internally deposited radionuclides are based on comparisons to total body risk associated with an acute (in time) exposure to low linear energy transfer (LET) radiation. Since effective dose is dependent on two quantitatively uncertain weighting factors, organ radiosensitivity and an alpha quality factor, this approach seems, at least for risk assessment, less sound scientifically than the earlier approach using body burden, concentrations, absorbed dose, and critical organs. For protection against radium risks, the earlier approach appears to be especially appropriate.

The biological endpoints associated with the ingestion of radium (the primary pathway for human intake) are bone sarcomas and facial carcinomas, specifically carcinomas of the paranasal sinuses and mastoids. Facial carcinomas are considered to be due, at least partially, to exposure to radon and its progeny, resulting from the decay of ^{226}Ra deposited in bone. It is noteworthy that except perhaps for the irradiation of bone sarcomas by the shorter half-life injected radium-224, leukemia has not been shown to be a discernible risk following radium ingestion.

Although the BEIR IV Committee offered a lifetime ^{224}Ra risk estimate for bone sarcoma of 2×10^{-4}/cGy, the committee, unfortunately, chose not to suggest risk coefficients for the radium nuclides of general interest in the practice of radiation protection, i.e., ^{226}Ra and radium-228. Studies of injected ^{224}Ra have limited application to radiation protection because of its short half-life, which limits exposure to bone surfaces and does not become a source of long-term exposure to bone.

The importance of the committee's extensive discussion of radium tumorigenesis relates strongly to various interpretations that can be placed on the observed dose-response relationships for bone cancer reported for the radium dial painters' ingestion of ^{226}Ra and ^{228}Ra. Figure 8.1 shows the dose-response relationships for a population subgroup of female radium dial workers exposed before 1930 as analyzed by three different analyses. While the abscissas are in units of skeletal dose, the ordinates are quite different. Evans's analysis includes both bone sarcomas and head carcinomas, and the Rowland et al. analysis gives a weight of 1.5 to the dose from ^{228}Ra (Evans et al. 1972, Rowland et al. 1978). The significance of Figure 8.1 is that while May's plot (Mays and Lloyd 1972) "leaves a strong visual impression of linearity . . . when subjected to statistical analysis, [it] is shown to be nonlinear with high probability" (National Research Council 1988). Thus regardless of how the data may be plotted, the existence of a threshold average-skeletal dose is not inconsistent with the findings.

Raabe et al. examined these findings of radium-induced bone tumors in humans in terms of appearance times and found an increase in median tumor appearance time with decreasing skeletal dose rate, suggesting a "practical" threshold (a concept introduced by Evans in the mid-1960s). On the basis of these findings, they suggested a threshold dose rate for human bone tumors of about .04 mGy/day and a cumulative skeletal dose of 80 cGy (Raabe et al. 1980). The BEIR IV Committee provided some support to this suggestion by concluding that "below an average skeletal dose of about 0.8 Gy, the chance of developing bone cancer from ^{226}Ra and ^{228}Ra during a normal lifetime is extremely small, possibly zero" (NAS/NRC 1988).

Schlenker et al. in their review of bone sarcomas occurring among radium-exposed females found "that the pattern of tumor risk within the skeleton does not reflect the pattern of radiation dose." They concluded that natural tumor susceptibility is "the primary correlate of site specific susceptibility to the induction of bone sarcoma by radium . . . (and) may even be the primary *determining* factor" (Schlenker et al. 1989). Noteworthy in their listing of bone cancer deaths in various cohorts studied is that the last bone sarcoma death was in 1988: An 83-year-old female dial painter who began employment at the age of 17 and had a cumulative skeletal dose of 8100 cGy. The lowest dose for a bone sarcoma death was 460 cGy.

Risk coefficients for radium derived from BEIR IV and based on skeletal absorbed dose are provided in Table 8.5.

THORIUM

Knowledge of thorium's biological effects in humans is limited to the findings from epidemiological investigations of patients intravascularly injected with

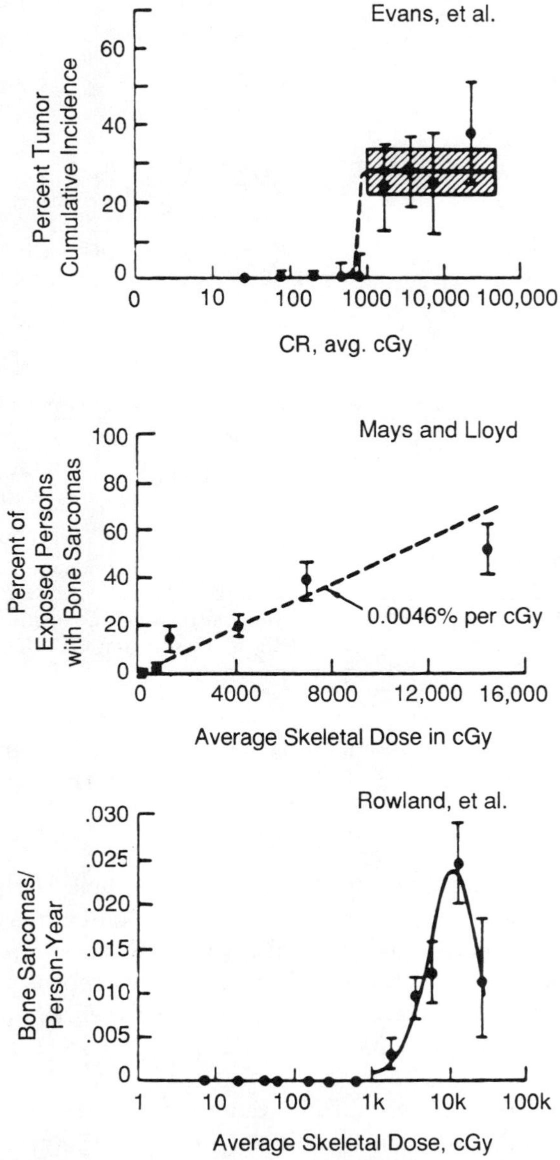

FIG. 8.1. ^{226}Ra and ^{228}Ra bone cancer relationships for dial painters.

Thorotrast (thorium-232 oxide [$^{232}ThO_2$]), a colloidal compound used in diagnostic radiology between 1928 and 1955. The liver, bone, and bone marrow were the tissues principally irradiated.

The BEIR IV Committee's excess cancer risk estimates and latency periods for $^{232}ThO_2$ are given in Table 8.6.

Considerable uncertainty exists in these estimates because of nonuniform tissue distribution, the chemistry of thorium, and incomplete follow-up in the

TABLE 8.5. RISK COEFFICIENTS AND LATENCY FOR RADIUM INGESTION

Radium-226		
Bone cancer[a]	37×10^{-6}/cGy (skeletal)	(5y latency)
Facial carcinoma	80×10^{-6}/cGy (skeletal)[b]	(10y latency)
Radium-228		
Bone cancer[a]	88×10^{-6}/cGy (skeletal)	(11y latency)
Radium-224		
Bone cancer	200×10^{-6}/cGy (skeletal)	(5y latency)

[a] Values from National Academy of Sciences/National Research Council. Health risks of radon and other internally deposited alpha-emitters: BEIR IV. Washington, DC: National Academy Press, 1988.

[b] A 50y period at risk is assumed.

TABLE 8.6. LIFETIME EXCESS CANCER RISKS FROM THOROTRAST[a]

Tissue	Risk Coefficient per 10^6 Person-cGy (Tissue)	Latency (Yr)
Liver	260–300	20
Bone Marrow (Leukemia)	50–60	5
Bone	55–120	10

[a] Values from National Academy of Sciences/National Research Council. Health risks of radon and other internally deposited alpha-emitters: BEIR IV. Washington, DC: National Academy Press, 1988.

epidemiological studies. As concluded by BEIR IV, "The extent to which these risk numbers apply to thorium radionuclides in other forms is unknown."

URANIUM

The major pathway for uranium transport to humans is through ingestion, principally of food, although water is an important source in some geographical areas. The uptake by the gastrointestinal tract is low, about 1% even for soluble materials. Although the skeleton is the major site of deposition, the biological half-life in bone is short, attaining equilibrium between intake and excretion. Other organ sites of depositions of natural uranium are the lung, liver, and kidney.

Health concerns are not limited to its radiation toxicity due to alpha particle decay but also include its chemical toxicity in inducing kidney damage, i.e., a nephrotoxic effect. The BEIR IV Committee concluded that it is premature to provide a risk estimate for the probability of developing renal damage from uranium but noted that estimates for a threshold concentration range from 3 to ≤0.1 ng/g, depending upon the choice of animal species and metabolic model.

Both animal and epidemiological studies support the conclusion that the induction of bone sarcoma is the most probable health effect following human exposure to uranium, especially to high-specific-activity uranium such as reactor-produced uranium-232. There is a suggestion of risk of lung cancer in humans from chronic inhalation of uranium dust in the workplace; however, the committee concluded that quantification of this risk "cannot be determined from published epidemiological studies because of confounding factors and the limited power of the surveys."

The likelihood of bone sarcomas being induced in the general population from exposure to environmental natural uranium concentrations is "virtually zero" when the estimate is based on a quadratic dose-response relationship. Even assuming a linear dose-response relationship and a constant nonoccupational intake of 1 pCi/day (estimated average U.S. daily intake from food and water consumption), the risk is relatively low. Mays et al. reported a lifetime risk of bone sarcoma = 1.5×10^{-6}/pCi day^{-1} or about 1×10^{-5}/cGy (Mays et al. 1985).

However, perhaps more importantly, the BEIR IV Committee concluded that "exposure to natural uranium is unlikely to be a significant health risk in the population and may well have no measurable effect."

TRANSURANIC ELEMENTS

Transuranic (TRU) elements are actinides with atomic numbers from 93 (neptunium) through 103 (lawrencium). Most have isotopes with half-lives of the order of minutes or less and thus are not of substantive health concern. The nuclides of greater interest in radiation protection are neptunium-237, plutonium-238, plutonium-239, americium-241, and curium-244. The most studied of these elements in regard to biological effects is plutonium, especially plutonium in the form of plutonium oxide ($^{239}PuO_2$), which is highly insoluble.

Human exposures to plutonium have not provided a data base from which to directly derive cancer risk estimates for this important element. From the U.S. Transuranic Registry and other studies of plutonium workers, it has been shown that plutonium concentrates in the lungs, liver, and bone, with only animal studies providing evidence of an association with cancer induction. Animal studies that compare tumorigenesis from radium and plutonium and findings from the human radium studies provide the basis for estimates of a bone cancer risk. A risk coefficient for lung cancer induced by plutonium inhalation was noted by the BEIR IV Committee based on its estimates of lung cancer risk for radon. Similarly, the risk of liver cancer was noted from its Thorotrast data analysis. In view of this approach, estimates of risks of lung and liver cancers are especially uncertain.

Table 8.7 shows the committee's organ risk estimates for transuranic elements based on the above indirect approach.

TABLE 8.7. TRANSURANIC ELEMENTS RISK ESTIMATES[a,b]

ORGAN	Cancer Deaths per 10^6 Person·cGy	Human Data Base
Lung	500	Radon progeny
Bone	80–1100	Radium
Liver	300	Thorotrast

[a] Values from National Academy of Sciences/National Research Council. Health risks of radon and other internally deposited alpha-emitters: BEIR IV. Washington, DC: National Academy Press, 1988.

[b] Based on 250 cases per 10^6 person·WLM.

IODINE AND STRONTIUM

Although the major focus of this presentation is on cancer risks from internally deposited alpha particle emitters, cancer risks following exposures to radioactive iodine and strontium are important in the practice of radiation protection. These are briefly discussed below.

Risk estimation for radiation-induced thyroid cancer is based primarily on studies of children exposed to x-rays for the treatment of a variety of benign diseases such as thymic enlargement and tinea capitis. Data from experimental animal studies and limited human data (e.g., Marshall Islanders exposed to weapons testing fallout) suggest, however, that iodine-131 is less of a carcinogen than x-rays, probably due to differences in dose rate and distribution of dose within the thyroid (NCRP 1985). The NCRP recommends that the organ dose received from ^{131}I be considered no more that one-third as effective as the organ dose received from external radiation. Its model for thyroid cancer risk estimation allows also for sex and age differences.

Table 8.8 shows NCRP's recommended risk estimates for mean thyroidal doses in the ranges of 6–1500 cGy. A maximum of 10% of "clinically evident radiation-induced thyroid cancers" are expected to be fatal.

NCRP recently published a comprehensive review of strontium radiobiology (NCRP 1991). Animal studies suggested that the organs at risk from internally deposited strontium, primarily strontium-90, are bone and blood-forming tissues, resulting in the risk of bone sarcomas and leukemia. However, these effects have been observed only at doses of thousands of cGy, and the nonlinear dose-response curve is similar to that observed for the radium dial painters. The NCRP concluded that "the dose-response relationships for ^{90}Sr-induced bone sarcomas in animal studies are in concordance with but do not prove the existence of a threshold." Based on comparisons to ^{226}Ra human data on bone sarcomas and the Japanese atomic bomb survivors' leukemia data, the NCRP suggested the risk estimates, and value ranges, as applicable to populations exposed to radiostrontium at low doses and dose rates. These estimates are shown in Table 8.9.

TABLE 8.8. THYROID CANCER (EXCESS LIFETIME RISK PER 10^6 PERSONS·cGY[a,b])

Exposure Age of Population[c]	Source of Irradiation	Total Cases[d]		Deaths[d]	
		Male	Female	Male	Female
≤ 18y	I-125, I-131	5	10	0.5	1
	External x or γ rays, I-132, I-133, I-135	10	30	1	3
> 18y	I-125, I-131	3	7	0.3	0.7
	External x or γ rays, I-132, I-133, I-135	8	20	0.8	2

[a] Adapted from National Council on Radiation Protection and Measurements. Induction of thyroid cancer by ionizing radiation, NCRP report no. 80. Bethesda, MD: National Council on Radiation Protection and Measurements, 1985.

[b] Thyroid doses from 6 to 1500 cGy; 5y minimum latency.

[c] Median ages: ≤ 18y 9.2 M, 9.3 F
> 18y 38.6 M, 41.4 F

[d] Rounded values.

TABLE 8.9. CANCER RISK FROM INTERNALLY DEPOSITED STRONTIUM

		Range
Bone sarcoma	1 case/10^6 person·cGy	0–500
Leukemia	3–6 cases/10^6 person·cGy	0–1600

TABLE 8.10. RISK COEFFICIENTS FOR INTERNALLY-DEPOSITED RADIONUCLIDES (TISSUE/ORGAN DOSE)

Radon	
Lung	350 cancer deaths per 10^6 person·WLM (underground miners) 250 cancer deaths per 10^6 person·WLM (modified for general population)
Polonium	
^{210}Po	order of 1 cancer death per 10^5 tissue cGy
Radium	
^{224}Ra Bone	200 cancer deaths per 10^6 tissue cGy (skeletal)
^{226}Ra Bone, facial	100 cancer deaths per 10^6 tissue (skeletal) cGy
^{228}Ra Bone	90 cancer deaths per 10^6 tissue (skeletal) cGy
Thorium	
$^{232}ThO_2$	
Liver	300 cancer deaths per 10^6 tissue cGy
Bone Marrow	50 leukemia deaths per 10^6 tissue cGy
Bone	90 cancer deaths per 10^6 tissue cGy
Uranium	
Bone	1 cancer death per 10^6 persons·1pCi/d (nephrotoxicity threshold 3 to ≤ 1 μg/g)
Transuranic elements	
Bone	700 cancer deaths per 10^6 person·cGy
Bone	80–1100 cancer deaths per 10^6 persons·cGy
Liver	300 cancer deaths per 10^6 person·cGy
Iodine	
Thyroid	1 cancer death per 10^6 person·cGy
Strontium	
Bone	1 cancer death per 10^6 person·cGy
Bone Marrow	3 cancer deaths per 10^6 person·cGy

Table 8.10 summarizes risk coefficients for the internally deposited radionuclides discussed in this paper.

SUMMARY

Except for human data on the ingestion of ^{226}Ra and ^{228}Ra and the intravenous injection of ^{224}Ra, the inhalation of radon progeny in a confounding underground mining environment, the thyroid uptake by the Marshall Islanders and medical patients of iodine, and the special case of patients treated with Thorotrast (^{232}Th), there are few little human data from which to directly assess the risk of internally deposited radionuclides. During World War II comparison to the safe body burden of 0.1 μCi of radium was the biological basis for limiting the amount of internally deposited radionuclides in critical organs. Based on animal studies and improvements in quantifying metabolic behavior of radionuclides, the ICRP in 1967 expanded the concept of using dosimetry as the means of controlling radiation exposures. The concept of effective dose became the biological basis for radiation protection. This approach depends on weighting organ radiosensitivities for cancer induction and genetic consequences, based in large measure on the atomic bomb survivors who appear to have been exposed primarily to low-LET gamma rays. However, with large uncertainties involved in using a quality factor for alpha particles and external radiation–derived radiosensitivities, the biological basis of current protection limits for internally deposited radionuclides falls far short of scientific certainty.

Perhaps it is time to rethink whether effective dose, equivalent dose, or even absorbed dose have sufficient scientific merit to serve radiation protection as measures of health risk, especially for internal emitters. Do we know quality factors and individual organ radiosensitivities well enough to have confidence in calculations of equivalent and effective dose, quantities that we cannot measure directly? Perhaps in radiation protection, we have too hastily opted against using maximum body burdens, air and water concentrations and intakes, and the concept of a critical organ as measures of achieved protection. How important is the summation of external and internal effective doses in the practice of risk management? The management of other hazardous materials appears to be acceptably controlled by measurable quantities without summation of indexes of toxicity.

How much more are we likely to learn about the risk of internally deposited radionuclides that will enhance radiation protection? In my view, not much. Animal studies can continue to provide information on metabolic behavior of radionuclides, but additional data on tumorigenesis and hereditary defects are not forthcoming. As for human studies, the populations considered in developing the BEIR IV report are likely to be the only source of additional information. Continued studies of these exposed populations are important, however, since many of those in the cohorts are still living. Perhaps before the end of this century "time since exposure" will be a proven factor reflecting repair, and we can accept the "hypothesis" of a "practical threshold" at environmental levels for those radionuclides of concern in our daily lives as internal emitters. Also, perhaps society will reestablish biological knowledge as the basis of radiation protection, instead of mandating implementation of the optimizing concept "as low as reasonably achievable" (ALARA) as the "prudent public health measure" based on the nonthreshold linear hypothesis.

REFERENCES

American Cancer Society. *Facts and Figures–1991.* New York, American Cancer Society, 1991.

Evans, R.D., Keane, A.T., and Shanahan, M.M. Radiogenic effects in man of long-term skeletal alpha-irradiation. In: *Radiobiology of Plutonium,* edited by B.J. Stover and W.S.S. Jee. Salt Lake City, UT, The J.W. Press, 1972, pp. 431–468.

Mays, C.W., and Lloyd, R.D. Bone sarcoma incidence vs. alpha particle dose. In: *Radiobiology of Plutonium,* edited by B.J. Stover and W.S.S. Jee. Salt Lake City, UT, The J.W. Press, 1972, pp. 409–430.

Mays, C.W., Rowland, R.E., and Stehney, A.F. Cancer risk from the lifetime intake of radium and uranium isotopes. *Health Phys. 48:*635–647, 1985.

National Academy of Sciences/National Research Council. *Health Risks of Radon and Other Internally Deposited Alpha-Emitters: BEIR IV.* Washington, DC, National Academy Press, 1988.

National Academy of Sciences/National Research Council. *Health Effects of Exposure to Low Levels of Ionizing Radiation: BEIR V.* Washington, DC, National Academy Press, 1990.

National Academy of Sciences/National Research Council. *Comparative Dosimetry of Radon in Mines and Homes.* Washington, DC, National Academy Press, 1991.

National Council on Radiation Protection and Measurements. *Evaluation of Occupational and Environmental Exposures to Radon and Radon Daughters in the United States,* NCRP Report No. 78. Bethesda, MD, National Council on Radiation Protection and Measurements, 1984.

National Council on Radiation Protection and Measurements. *Induction of Thyroid Cancer by Ionizing Radiation,* NCRP Report No. 80. Bethesda, MD, National Council on Radiation Protection and Measurements, 1985.

National Council on Radiation Protection and Measurements. *Ionizing Radiation Exposure of the Population of the United States,* NCRP Report No. 93. Bethesda, MD, National Council on Radiation Protection and Measurements, 1987.

National Council on Radiation Protection and Measurements. *Some Aspects of Strontium Radiobiology,* NCRP Report No. 110. Bethesda, MD, National Council on Radiation Protection and Measurements, 1991.

Nazaroff, W.N., Doyle, S.M., Nero, A.V., Jr., and Sexto, R.G. Radon entry via potable water. In: *Radon and its Decay Products in Indoor Air,* edited by W.N. Nazaroff and A.V. Nero, Jr. New York, John Wiley & Sons, 1988, pp. 131–155.

Raabe, O.G., Book, S.A., and Parks, N.J. Bone cancer from radium: canine dose response explains data for mice and humans. *Science 208:*61, 1980.

Rowland, R.E., Stehney, A.F., and Lucas, H.F., Jr. Dose-response relationships for female radium dial workers. *Radiat. Res. 76:*368, 1978.

Schlenker, R.A., Keane, A.T., and Unitt, K.K. Radium 226 and 228 in humans. In: *Risks from Radium and Thorotrast,* edited by D.C. Taylor, C.W. Mays, G.B. Gerber, and R.G. Thomas. BIR Report No. 21. London, British Institute of Radiology, 1989, pp. 55–63.

Chapter 9

Risk Assessment in Radiation Protection

CHARLES B. MEINHOLD

ABSTRACT

As result of newly acquired information, primarily from the Japanese survivors, the International Commission on Radiological Protection (ICRP) has issued and the National Council on Radiation Protection and Measurements (NCRP) will issue new recommendations. Several updated concepts, including the proposed risk projection model, the basis for selecting a dose limit, and a new approach accounting for different relative biological effectiveness (RBE) values are discussed. The ICRP's distinction between practices that will lead to increased exposure and intervention in which the only objective is to reduce exposure is presented.

RISK ASSESSMENT IN RADIATION PROTECTION

The ICRP began the revision of its Publication 26 in 1985, reacting primarily to the new biological information available from the Radiation Effects Research Foundation (RERF) (ICRP 1977). They also expressed the desire to improve the presentation of the recommendations and to maintain as much stability as they possibly could. This revision has now been published as ICRP Publication 60 (ICRP 1991). The NCRP incorporated preliminary information from the 1986 Japanese survivor data in its Report 91, published in 1987 (NCRP 1987), while still basing its recommended dose limits on the 1977 data of the United Nations Scientific Committee on the Effects of Atomic Radiation (UNSCEAR 1977). This revision is now in final preparation, based primarily on risk estimates provided by its Committee 1-2, chaired by Michael Fry.

In its 1977 recommendations, the ICRP used the fatal accident rate in "safe" industries as a measure of acceptable risk (ICRP 1977). The data for the period 1957 to 1988 indicate, however, the fatality rate in industries is being reduced at a rate of about 2% per year.

In a recent RERF commentary by Pierce, he has shown how apparent the cancer incidence increases are for those survivors who receive exposures above 0.5 Gy (Pierce 1989). That is, of the 2000 survivors who received shielded doses between 1 and 2 Gy, there were nearly twice as many nonleukemia cancers found as would have been expected and nearly 10 times as many leukemia cases.

Once the number of excess cases of cancer in the survivors has been determined, the problem of estimating the number of cancers to be seen during their remaining lifetimes needs to be addressed. The problem is difficult since no population has been observed over its entire lifetime, the different cancers may have different attributes with regard to protraction, and different populations have different spontaneous cancer rates and distribution of cancer types.

Pierce, in his commentary mentioned above, presented these data as a function of the time interval between follow-up periods. This presentation suggests that the excess cancer does indeed follow a pattern similar to that of the spontaneous cases (Pierce 1989). Based on this and other information, both the ICRP and NCRP have adopted a multiplicative model.

Perhaps the easiest way to visualize the "increase in cancer risk associated with radiation exposure is to compare the excess lifetime mortality from all cancers as estimated by the UNSCEAR as given in Table 9.1.

Roughly, the data in the table suggest that due to the revised dosimetry (approximately 10%) and the increase in the number of excess cancers observed, we can deduce a factor of 2 increase from 1977 to 1988 and, again, about a factor of 2 between the additive and multiplicative model. This greatly oversimplifies the issue, but provides some prospective on why we are talking about an increase by about a factor of 4.

Both the ICRP and NCRP developed values for the dose and dose rate effectiveness factor since neither UNSCEAR nor the National Academy of Science's Biological Effects of Ionizing Radiation (BEIR) Committee were particularly helpful in this regard. ICRP's Committee 1 suggested a value of 2, which was adopted by the commission. NCRP's Committee 1-2 suggested a value between 2 and 3. Since they both started with the same incidence data at high doses and dose rates, the final risk values were not very different. The NCRP draft adopted the ICRP risk values. Based primarily on considerations of the time required to repair single-strand breaks in DNA, the ICRP has applied the dose and dose rate effectiveness factor whenever the dose is less than 0.2 Gy or the dose rate is less than 0.1 Gy^{-1} (ICRP 1991).

The fetal concern centered on the developing central nervous system with a suggestion that in the 8th–16th week, the risk is about 0.4 Sv^{-1} for severe mental retardation. In addition, measurements of IQ in the survivors have suggested that there is about 30% IQ loss per Sv. This may well be a major oversimplification of the total risk to the fetus. From the dosimetric point of view, the commission explained that as a result of the adoption of the linear hypothesis, it was properly concerned with the average dose in the organ or tissue rather than the point in tissue. Although the commission adopted a Q-L relationship with Q = 1 for all radiations below 10 keV and Q = 300 divided by the square root of L for linear-

TABLE 9.1. UNSCEAR RISK ESTIMATES

	Additive Model	Multiplicative Model
UNSCEAR 1977	2.5×10^{-2} Gy^{-1}	—
UNSCEAR 1988	4.0–5.0 $\times 10^{-2}$ Gy^{-1}	7.0–11.0 $\times 10^{-2}$ Gy^{-1}

energy transfer (LET) or radiation with a LET greater than 100 (that is, $Q = 30$ at $LET = 100$), they did not believe that the precision implied by this relationship is warranted nor that the Q value relates to the average dose. It was for this reason that the commission adopted the quantity w_R (radiation weighting factors). These values range from 1 for all photons to 20 for 100 keV to 2 MeV neutrons, which are to apply to the average absorbed dose at a function of the incident radiation energy and type. The equivalent dose (new name for the dose equivalent) for an organ or tissue is $H_{T,R} = w_R D_{T,R}$ or if there is more than one radiation quality or type,

$$H_T = \sum_R w_R D_{T,R}.$$

New tissue-weighting factors have been developed based on the relative cancer mortality in each tissue together with the fatality fraction of that particular cancer type and the length of life lost due to each cancer. The w_Ts are given in four groups, with w_Ts at 0.01, 0.05, 0.12 and 0.2. The effective dose (formerly called the effective dose equivalent) is given as

$$E = \sum w_R H_{T,R}.$$

The distinction between the application of w_R and w_T can be simplified if one remembers that w_R is independent of the tissue involved and w_T is independent of the radiation quality or type. The ICRP has suggested that the International Commission on Radiation Units and Measurements (ICRU) quantities of ambient dose equivalent, directional dose equivalent, individual dose equivalent penetrating, and individual dose equivalent superficial provide reasonable approximations of effective dose and equivalent dose to the skin when the 1990 Q-L relationship mentioned above is used. Of course, it is expected that these questions will be reexamined when the planned revision of ICRP Publication 51 takes place.

With regard to the basis for radiation protection limit development, the ICRP noted that the standards for safe industries are neither constant nor uniform worldwide and that the mortality data applies to the average while the dose limit applies to the individual (ICRP 1991).

The commission suggested that a risk of 10^{-3} is unacceptable, a risk of 10^{-4} is tolerable, and a risk of 10^{-5} is perhaps acceptable. A multiattribute approach employing mortality considerations such as lifetime risk, loss of life expectancy, and the probability of the age of attributable death and morbidity attributes related to nonfatal cancer and genetic effects were used as the basis for judging the risk limit. This resulted in the dose limit for occupational exposure of 100 mSv in 5 years with the condition there be no more than 50 mSv in any 1 year (ICRP 1991).

Since the recommendation on the dose limit to the members of the public has been recently reviewed both by the NCRP and ICRP, no change is suggested, that is, 1 mSv per year averaged over any 5 consecutive years.

The commission made a distinction between the introduction or continuation of a practice, which involves a conscious decision to introduce a source of exposure, and intervention, in which the source of exposure already exists. For practices, the now classical system of radiation protection is retained, that is, justification of the practice, optimization of protection, and dose limits to protect the individual. For intervention, the most important aspect is that the whole intervention process is basically dose reduction and that justification simply means that all you need is to do more good than harm. The dose limits are inappropriate since the dose averted will depend upon the justification given above. It may be necessary to ensure that nonstochastic or deterministic effects are prevented.

New annual limits on intake (ALIs) will be provided within the same volume as ICRP Publication 60 (ICRP 1991), probably as Publication 61, and will incorporate the new w_Ts and ICRP 30 (ICRP 1988) and 48 (ICRP 1986) metabolic models and will be based on 20 mSv per year. These values are tentatively adopted by the NCRP in its revision. A revision of Publication 30 is planned that will incorporate the ICRP 56 (ICRP 1989) bone metabolism models, new respiratory tract models, and new reference man values.

In primary committed effective dose calculations, the remainder tissue is all the tissue of the body in which cancer might develop. Excluded are mineral bone, connective tissue, fat, contents of the stomach, and colon, etc. There is one further rule which requires that if any remainder tissue equivalent dose is larger than that in the twelve listed organs, one half the remainder w_T should be applied to that tissue.

REFERENCES

International Commission on Radiological Protection. *Recommendations of the International Commission on Radiological Protection*, ICRP publication 26. Oxford, Pergamon Press, 1977.

International Commission on Radiological Protection. *The Plutonium and Related Elements*, publication no. 48. Oxford, Pergamon Press, 1986.

International Commission on Radiological Protection. *Limits for the Intake of Radionuclides by Workers*, publication no. 30. Oxford, Pergamon Press, 1988.

International Commission on Radiological Protection. *Age-Dependent Doses to Members of the Public from Intake of Radionuclides*, publication no. 56 Oxford, Pergamon Press, 1989.

International Commission on Radiological Protection. *1990 Recommendations of the International Commission on Radiological Protection*, publication no. 60. Oxford, Pergamon Press, 1991.

National Council on Radiation Protection and Measurements. *Recommendations on limits for exposure to ionizing radiation*, report no. 91. Bethesda, MD, National Council on Radiation Protection and Measurements, 1987.

Pierce, D. A. *An Overview of the Cancer Mortality Data on the Atomic Bomb Survivors*, RERF Commentary and Review Series, RERF CR 1–89. Hiroshima, Radiation Effects Research Foundation, 1989.

United Nations Scientific Committee on the Effects of Atomic Radiation. *Sources and Effects of Ionizing Radiation. 1977 Report to the General Assembly with Annexes.* New York, United Nations, 1977.

United Nations Scientific Committee on the Effects of Atomic Radiation. *Sources, Effects and Risks of Ionizing Radiation. 1988 Report to the General Assembly with Annexes.* New York, United Nations, 1988.

Chapter 10

The Radioepidemiological Tables in Radiation Litigation

RALPH H. JOHNSON

ABSTRACT

The applications, limitations, and potential misuses of the National Institutes of Health radioepidemiological tables in litigation regarding claims of radiation-induced disease are discussed.

INTRODUCTION

The philosopher Bertrand Russell once remarked, "The triumphs of science are due to the substitution of observation and inference for authority." The legal resolution of the causation issue in radiation-induced disease claims is necessarily dependent upon epidemiological observation and inference: "There is no factual alternative to an epidemiologic approach when trying to relate radiation exposure to a given cancer" (Meinhold 1986). In 1983 Congress passed Public Law 97–414, directing the Secretary of Health and Human Services to review the epidemiological evidence regarding radiation-disease and devise radioepidemiological tables showing the probability that certain cancers resulted from prior exposure to ionizing radiation. This paper discusses the role of those tables in radiation litigation.

THE ORIGINAL PURPOSE FOR THE TABLES

The radioepidemiological tables were intended by their congressional sponsors to be a scientific basis for a legislative scheme to compensate residents in the vicinity of the Nevada Test Site, who allegedly were exposed to radioactive fallout and contracted radiogenic disease. Compensation was to be made according to "standards of proof that [were] more generous to claimants than those standards generally available under provisions of tort law or workers compensation law," and the claimants were to be given "the benefit of the doubt" in the assessment of their radiation doses (131 Congressional Record S15568–01, Nov. 14, 1985). The President, however, instructed the Secretary of Health and Human Services as follows:

> There is as yet no consensus among radiation experts in relating human cancers and exposure to low levels of radiation. Yet, [the legislation] mandates that

> probability of causation tables be calculated for even very small dose levels. Accordingly, I am directing the Secretary of Health and Human Services to complete the tables to the extent that may be possible and scientifically responsible, in light of the analysis also mandated by [the legislation], which requires him to "assess the credibility, validity, and degree of uncertainty associated with such tables." (National Institutes of Health [NIH] 1985)

Similarly, the NIH committee responsible for devising the tables observed:

> Although we know more about the effects of ionizing radiation than any other carcinogen, and much of the knowledge is quantitative in form, our present knowledge is far from complete, and the tables can be regarded as no more than a guide to causation in the particular case. (NIH 1985)

The resultant NIH tables, contrary to the apparent congressional expectation, indicated that the probability that cancer among residents in the vicinity of the Nevada Test Site was attributable to radioactive fallout exposure was minimal (National Research Council [NRC] 1984). Accordingly, in 1990 the Radiation Exposure Compensation Act, Public Law 101–426, was passed to compensate claimants without reference to the NIH tables. Instead, awards were to be made if a claimant lived in particular counties, during certain time spans, and contracted specific diseases within defined periods after exposure. Thus while the tables' original purpose was extinguished by congressional fiat, the tables, a comprehensive study by distinguished scientists, were available for use in radiation litigation. This use was anticipated by the NIH committee:

> Even an imperfect guide that summarizes the relevant literature in such a way that it can be brought to bear on the individual case should be helpful in determining whether a prior exposure to radiation was a significant factor. (NIH 1985)

THE ATMOSPHERIC TESTING LITIGATION

One of the most significant radiation cases, *Allen v. United States* (588 F. Supp. 247 [D. Utah 1984], *rev'd* 816 F.2d 1417 [10th Cir. 1987], *cert. denied*, 108 S. Ct. 694 [1988]), was tried before the NIH tables were prepared, yet the case provides factual circumstances useful for considering the tables' application. The *Allen* case involved several hundred plaintiffs who lived downwind from the Nevada Test Site and contended that radioactive fallout from the atmospheric testing of nuclear weapons at the site had deposited on their communities, causing many different types of cancer. Plaintiffs alleged that the federal government had negligently failed to warn the off-site public regarding the cancer risks associated with the fallout and had given false assurances of safety. As part of the failure-to-warn claim, plaintiffs asserted that the government had not advised the residents of simple precautions, such as remaining indoors during fallout episodes, which would have reduced their exposures. Plaintiffs causally related the lack of warnings with their exposures by testifying that they would have abandoned the downwind area had they been accurately apprised of the risk.

The court's conclusion that the United States' failure to warn constituted negligence was based on its conclusion that

> [A] reasonable person, exercising great care in light of the best of available scientific knowledge, would err on the side of caution by assuming no "safe" threshold exposure to atomic radiation . . . and by determining that every practicable step be taken to minimize unnecessary radiation exposure. (*Allen*, 588 F. Supp. at 362)

If the NIH tables had been accepted by the judge in the *Allen* case as the appropriate method for deciding causation, the calculation of the probability of causation value for, for example, a breast cancer victim could have been summarized as follows:

1. The probability of causation formula is $PC = R/(1 + R)$ where R is the relative excess.
2. R is the product of three functions–the person's relevant tissue dose (D) in cGy, the influence of exposure age and diagnosis age (T), and the relative excess of breast cancer for the person's exposure age (K). The values for T and K are found in the radioepidemiological tables at pages 244 and 248, respectively.
3. The doses calculated for the breast cancer victim by the defendant's and plaintiffs' experts are set forth in the court's opinion (*Allen*, 588 F. Supp. at 427). The defendant's estimate was 0.38 cGy; the plaintiffs' estimate dose was 36.7 cGy. The person was born in 1941 and was diagnosed as having breast cancer in 1978. For purposes of this calculation it is assumed that she received her exposure acutely in 1953.
4. The probability of causation value based on the defendant's estimated dose is approximately 0.5%; the value using the plaintiffs' dose is 31.6%.
5. If the 1990 National Research Council (BEIR V) breast cancer risk data had been used to calculate the breast cancer probability of causation, the value based upon the defendant's estimate would have been between 0.2% and 0.3%; the best estimate being 0.3%. Using the plaintiffs' estimate, the value would have been between 14.9% and 21.7%, the best estimate being 18.0% (NRC 1990).

It is quite unlikely, however, that the judge would have used the NIH tables as a basis for his decision, even if they had been available. The reason is that dose estimates are an essential variable in the probability of causation formula, and the judge expressed substantial skepticism that reliable doses could be reconstructed for the plaintiffs:

> Even under controlled laboratory conditions, accurate determination of absorbed dosages of radiation in human tissue is problematical. Usually where the question of radiation-induced cancer or leukemia is raised, dose must be estimated long after the fact of actual exposure. Accurate reconstruction of dose proves difficult even when a good deal is known from measurements about the source, the times and extent of exposure, effect of shielding, and other factors. Where measurements are scarce or non-existent, dose estimation is exceedingly difficult. (*Allen*, 588 F. Supp. at 425)

The plaintiffs' doses were estimated using what the judge described as "reverse engineering"–plaintiffs' expert "retrospectively estimate[d] the dose that would account for the observed increase" of disease detected in a telephone survey (*Allen*, 588 F. Supp. at 427). The court referred to this procedure as "novel" and "certainly not controlling" (*Allen*, 588 F. Supp. at 428).

Although the federal government's program to reconstruct the plaintiffs' doses, the "Off-Site Radiation Exposure Review Project," was one of the most comprehensive retrospective dosimetry exercises ever undertaken (Church et al. 1990), its findings were not fully accepted by the court:

> [The Government] produced evidence concerning the current Off-Site Radiation Exposure Review Project . . . to reconstruct external and internal dose estimates using the few measurements available in conjunction with theoretical models and data from other sources and events. . . . This court has carefully reviewed the testimony and documentary evidence offered as relating to dosimetry. Even as much recent work as has been done still leaves the question of radiation exposure of individual plaintiffs very much in doubt. Even upon the completion of current studies . . . the exact dose of radiation delivered to individuals or to the group will never be known with precision. (*Allen*, 588 F. Supp. at 426)

Later, in his opinion, the court added:

> The Government implicitly and explicitly urges the conclusion that its estimates of fallout exposure to the plaintiffs–educated guesses, really–represent the probable upper limits of the radiation dosage and corollary risk of cancer and leukemia actually experienced by the plaintiffs. Months of review of the record and exhibits in this action compels this court to find otherwise. Rather than serving as the ceiling of retrospective dosimetry for many of the plaintiffs, the exposure estimates currently offered by the Government should likely be deemed to be minimum figures for use in risk estimation. While without question some individuals received less exposure than is currently estimated, it is extremely likely that many people received more exposure to alpha, beta and gamma radiation both externally and internally than more conservative guesses have indicated. (*Allen*, 588 F. Supp. at 427)

Thus rather than require the plaintiffs to prove that a specific radiation dose caused a specific disease, the court created a new presumption of causation predicated on the defendant's negligence:

> Where a defendant who negligently creates a radiological hazard which puts an identifiable population group at increased risk, and a member of that group at risk develops a biological condition which is consistent with having been caused by the hazard to which he has been negligently subjected, such consistency having been demonstrated by substantial, appropriate, persuasive and connecting factors, a fact finder may reasonably conclude that the hazard caused the condition absent persuasive proof to the contrary offered by the defendant. (*Allen*, 588 F. Supp. at 415)

The court then relied primarily upon the 1980 National Research Council (BEIR III) report to determine which types of cancer were "consistent with having been caused by the (radiological) hazard." The three cancer types found by the court to be caused by fallout exposures–leukemia, thyroid, and breast cancer–were the only cancers identified in the BEIR III report as having a "very high" or "high" "sensitivity to radiation induction" (NRC 1980). Other "connecting factors" considered by the court included whether plaintiffs were residing in the affected communities during the testing period, the general levels of radiation measured in the vicinity of plaintiffs' homes, exposure to other carcinogens, the latency periods for specific plaintiffs' cancers, and the plaintiffs' sex and age at the time of exposure and diagnosis (*Allen*, 588 F. Supp. at 428–443).

The case control studies performed by the University of Utah (Stevens et al. 1990) to assess the relationship between leukemia in Utah and fallout from the test site provide equivocal support for the *Allen* decision:

> In view of the consistency of our findings with other literature and the lack of other plausible explanations, we conclude that the excess of acute leukemias in southwestern Utah is probably not due to chance and may be attributable to fallout radiation. However, the estimated number of cases attributable to fallout in this region is small, and these cases are indistinguishable from those caused by other factors. If a linear, dose-response hypothesis is correct, there would also have been a substantial number of cases attributable to fallout in the rest of the state because of its much larger population. In relative terms, the predicted increased risk outside of southwestern Utah is undetectable by epidemiologic means, and the small but nonsignificant increases we found in the intermediate-dose group were consistent with this prediction. (Reprinted from Stevens et al., *Leukemia in Utah and Radioactive Fallout From the Nevada Test Site.* JAMA 264:585–91;1990. Copyright 1990, American Medical Association.)

THE RADIUM AIRCRAFT INSTRUMENT DIAL LITIGATION

Contemporaneous with the *Allen* decision, Judge Patrick Kelly issued his decision in *Johnston v. United States* (597 F. Supp. 374 [D. Kan. 1984]). He encountered the same disparity in the parties' dose estimates that occurred in the *Allen* case:

> The question remains of the exact dosage these plaintiffs received while employed at AID [an aircraft instrument company that repaired some radioactive instrument dials], and whether that total dose should be considered hazardous. As will be explored later in this opinion, the answer depends upon just what assumptions are made by the expert witnesses who have attempted to "reconstruct" a precise dose for these plaintiffs. Defendant's experts generally testified that the doses were 7 rem or less, while plaintiffs' experts generally testified that the doses were hundreds or thousands of rems. Obviously, this is a huge unbridgeable gap. If the defendant's experts are correct, the doses these plaintiffs received at AID are less than the doses the average person will get in his lifetime from dirt and x-rays. If plaintiffs' experts are correct, then the doses these plaintiffs received at AID are so large that it is scientifically reasonable to consider them likely to cause harm and to label them "hazardous" or even "lethal." (*Johnston*, 597 F. Supp. at 390)

The court then insightfully explained how a probability of causation calculation could be manipulated:

> [A probability of causation formula] entices the expert witness to calculate backwards. If the witness desires to have a final result of over 50%, i.e., more probable than not, that can easily be accomplished by playing with the input numbers until

the output is what that witness wanted. For example, if one increases the risk [coefficient], the output is a higher percentage. If that still does not yield over 50%, then one can simply find ways to increase the dose estimate until the magic figure is reached.

These comments should not be taken to mean that a probability of causation calculation is without any value. It can be helpful as a supporting argument for other evidence, but an informed court should realize its limitations and assure that the input data represent reasonably valid assumptions and reflect scientific consensus because the result will magnify any error or uncertainty in the input data. As learned here, a court must consider statistical risk calculations with a healthy degree of skepticism and must not rely upon this method to provide an easy answer to a complex problem. (*Johnston*, 597 F. Supp. at 394)

THE BANE LITIGATION

The danger of relying upon probability of causation tables alone, without reference to the particular circumstances of each claimant's case, is underscored by another recently dismissed case, *Bane v. Babock & Wilcox Co.* (Civil Action No. CA-85–137 [E.D. Va. 1987]. In that case, the plaintiff's decedent, who did not drink or smoke, died of a gastroesophageal adenocarcinoma (a cancer that occurred at the junction of the stomach and esophagus). As a health physicist at a shipyard where nuclear-powered submarines were refueled, he had received a relatively small film badge dose–6.2 cGy. The calculated probability of causation was less than 1% (Fabrikant 1987a). Thus if the recorded film badge dose would have been accepted by the court, the tables would have supported a finding that the decedent's disease was probably unrelated to his work exposure. The film badge dose, however, was disputed by the plaintiff's experts, who contended that the decedent had received a much higher, though unquantified, dose. If the plaintiffs had been able to prove that their decedent's dose had significantly exceeded 100 cGy, the tables would have supported the plaintiffs' causation claims (NIH 1985).

The decedent's medical history, however, was reviewed for other possible causes of the decedent's disease. The medical records revealed that the decedent had a chronic hiatal hernia. Hiatal hernias are a primary cause of the ulceration of the normal esophageal epithelium by gastric acid from the stomach. Columnar-lined epithelium or Barrett's esophagus is caused by the ulceration of the normal esophageal squamous epithelium by gastric acid from the stomach. Pathological studies of the decedent's esophageal tissue demonstrated that he had a columnar-lined epithelium or Barrett's esophagus. Adenocarcinoma of the esophagus or gastroesophageal region is causally related to columnar-lined epithelium or Barrett's esophagus (Schein 1987).

In short, gastric acid associated with the decedent's hiatal hernia had caused an ulceration of his esophageal tissue. The normal tissue was replaced with abnormal tissue–a condition known as Barrett's esophagus. Barrett's esophagus is a premalignant condition leading to gastroesophageal cancer. Once this information was made available to plaintiffs' attorneys and expert witnesses, the plaintiffs dismissed the case. If the plaintiff's experts' version of the decedent's dose history had prevailed and the court had relied exclusively upon the NIH tables (without knowing of the actual cause of the decedent's cancer), a scientifically irrational decision would have been made.

A MISUSE OF THE PROBABILITY OF CAUSATION FORMULA

The BEIR V Committee defined the probability of causation formula as "an estimate of the probability that a given radiation dose in the history of a patient was the cause, in some sense, of a subsequent malignant neoplasm that has actually occurred" (NRC 1990). The limitation of the formula's application to cases where radiogenic disease has actually occurred should be emphasized. The following hypothetical fact situation illustrates how the formula can be misused to cause confusion and fear in a case where no cancer has been diagnosed.

An expert witness estimates a 10-cGy dose equivalent to a person 20 years old, who has not been diagnosed as having any form of cancer. The expert witness then prepares a report that states, "If this person contracts leukemia in the future, the probability that the leukemia was caused by the radiation exposure is 78%. This calculation is based on the excess relative risk value of 3.637 for leukemia given at page 226 of the BEIR V Report. PC = 3.637/(1 + 3.637) = .78 = 78%."

To base a retrospective probability of causation computation on the assumption that a person will contract leukemia at some undefined time in the future without clearly disclosing the minimal likelihood of such a disease occurring constitutes a profound abuse of the probability of causation formula. It certainly must be anticipated that the typical claimant, lawyer, or judge reading the hypothetical expert's report will mistakenly infer that the chances the person will contract leukemia are 78%, unaware that current estimates of the lifetime risk of excess leukemia mortality per 100,000 exposed persons as a result of a single exposure to 10 cGy is 110 (NRC 1990).

Once such statistical sophistry has occurred, its misleading effect on nonscientists is difficult to correct. The assistance of an expert witness with substantial training and experience in radiation epidemiology is indispensable.

SUMMARY

Although the NIH tables did not confirm the congressional supposition that cancers in the vicinity of the Nevada Test Site were caused by fallout exposure, the tables nevertheless provide valuable information about cancer causation, particularly where reliable radiation dose information is available. One leading radiobiologist has observed that the tables

> address methods for evaluating prospective risk that are more advanced than any previously available approach for quantifying risk. . . . While there may be some scientific hesitancy in accepting an effort to employ retrospective risk analysis, it is already being done on a practical basis, both in industry and in the courts. Thus, the [tables] may be considered to be a step forward in helping to determine the scientific validity of causation in those cases of illness induced by an environmental carcinogen. (Fabrikant 1987b)

The same author also warned, however, that the sources of uncertainty (e.g., sampling variability in estimating cancer incidence and the estimation of cancer hazard rates at different dose levels) could render the confidence interval surrounding the probability of causation estimate "extremely broad, sufficient, some argue, to make it difficult and, in some cases, impossible, to draw reliable conclusions" (Fabrikant 1987b). The tables are, therefore, susceptible to abuse, especially where their limitations are not fully explained to the lay decision maker. The challenge for the radiobiologist is to teach the reliable inferences that can be made using the tables and to communicate plainly their inherent limitations.

REFERENCES

Church, B. W., Wheeler, D. L., Campbell, C. M., and Nutley, R. V. Overview of the Department of Energy's Off-Site Radiation Exposure Review Project (ORERP). *Health Phys. 59:*503–510, 1990.

Fabrikant, J. I. *Affidavit in Support of Defendants' Motion for Summary Judgment. Bane v. Babcock & Wilcox Company,* Civil Action No. CA-85–137, U.S. District Court for the Eastern District of Virginia, 1987a.

Fabrikant, J. I. Probability of causation: implications for radiological protection and dose limitation. Paper read at Annual Conference, Canadian Radiation Protection Association, Saskatoon, Saskatchewan, May 29, 1987b.

Meinhold, C. The impact of probability of causation on the radiation protection program. *Health Phys. 55:*375–377, 1986.

National Institutes of Health. *Report of the National Institutes of Health Ad Hoc Working Group to Develop Radioepidemiological Tables,* NIH Publication No. 85–2748. Washington, DC, U.S. Government Printing Office, 1985, pp. 79, 209 (Fig. X-5).

National Research Council. *The Effects on Populations of Exposure to Low Levels of Ionizing Radiation: 1980, BEIR III Report.* Washington, DC, National Academy Press, 1980, pp. 266–267 (Table A-1).

National Research Council. *Assigned Share for Radiation as a Cause of Cancer: Review of the Radioepidemiological Tables Assigning Probability of Causation.* Washington, DC, National Academy Press, 1984, p. 129.

National Research Council. *Health Effects of Exposure to Low Levels of Ionizing Radiation, BEIR V Report.* Washington, DC, National Academy Press, 1990, pp. 226 (Table 4F-1), 172 (Table 4–2).

Schein, P. S. *Affidavit in Support of Defendants' Motion for Summary Judgment. Bane v. Babcock & Wilcox Company,* Civil Action No. CA-85–137, U.S. District Court for the Eastern District of Virginia, 1987.

Stevens, W., Thomas, D.C., Lyon, J.L., et al. Leukemia in Utah and radioactive fallout from the Nevada test site. *JAMA. 264:*585–591, 1990.

Section III

Whole Body Radiation Exposure and Effects on Specific Tissues and Organs

Chapter 11

Biology of Radiosensitive Tissues and Organs

TIMOTHY J. JORGENSEN

ABSTRACT

Tissues and organs with relatively high radiosensitivities are of interest to the health physicist. They represent the dose-limiting targets for determination of maximum permissible radiation doses. Several different human organ systems are radiosensitive. The skin, testes, ovaries, and lens of the eye are considered here. The physiology, cell biology, radiation effects, and dose response for each organ are reviewed. The biological endpoints considered are skin cancer, sterility, and cataracts. Although the endpoints differ, these organs are biologically similar in that they all contain cells that are continually differentiating. This continuous cell replication and turnover is the major factor contributing to radiosensitivity. Basic principles of stem cell kinetics are reviewed, with specific characteristics for each particular organ. Unique aspects of each organ and special dose considerations are explained. The effects of dose fractionation, linear energy transfer, and time of exposure are considered.

INTRODUCTION

The radiosensitivity of different mammalian cells is typically measured by survival curve analysis. Survival curves for a wide variety of cell types measured in vitro and in vivo show remarkable similarities, both in shape of the dose-response curve and in the mean lethal dose (D_0). Even though cells are the building blocks of tissues and organs and are similar in radiation sensitivity, organs vary greatly in their relative radiosensitivities. This apparent paradox is a reflection of the different physiological states of cells within organs and the relative contribution of proliferating cells to an organ's overall function. Also, radiosensitivity differs with the endpoint that must be considered and the effect of that endpoint on the functioning of the whole body.

In this chapter, I will focus on several organs and tissues that are considered to be at high risk, from a health physics point of view, and discuss the biological basis for the radiosensitivity. For each organ I will review the relevant physiology, histology, and cell biology. I will discuss the unique aspects of the organ, the significant effects of ionizing radiation, and the considerations in determining safe dose levels.

SKIN

Physiology and Histology

A major function of the skin is to provide a barrier that shields internal organs from harsh environmental conditions and bacterial infections. However, the skin also performs other important functions in maintaining body homeostasis, such as thermoregulation and control of water loss. The skin also has special sensory functions, which allow the body to communicate with the environment. Some areas of skin have secondary specialized functions, such as hair production or chemical secretions.

Since it is comprised of different tissues that interact in performing functions, the skin is an organ in its own right. Compared to other body organs, the skin is quite large. In humans, skin constitutes about 3% of total body weight, which makes it larger than the liver or the brain (Hopewell 1990).

The structure of skin is stratified into different regions. The major subdivisions are the outer epidermis region and the inner dermis region separated by a basement membrane. The dermis is 1–2 mm thick and is made up mostly of interwoven elastin, collagen, and reticulin fibers, through which blood vessels and nerves pass. Fibroblasts are embedded within the dermis and are responsible for producing most of the fibers. The blood vessels and nerves do not penetrate the basement membrane into the epidermis. The cells of the epidermis must rely on diffusion through the basement membrane to obtain blood born nutrients (Hopewell 1990).

The cells closest to the basement membrane are the basal cells, which are cuboidal or columnar in shape and are anchored to the basement membrane by hemidesmosomes. Basal cells probably represent a mixture of stem cells for different lineages of epidermal cells. Radiation survival analysis suggests that despite about 20,000 basal cells per square millimeter of epidermis, there are only about 1500 proliferative units. This indicates that only 1,500 cells per square millimeter are clonogenic stem cells, which can repopulate tissue. The rest of the cells must already be committed to differentiation. The major cell type of the epidermis is the keratinocyte. Keratinocytes undergo terminal differentiation as they transit through the epidermis from the basement membrane to the skin surface. The transit time is estimated to be about 14 days. Other cells, such as melanocytes and Langerhans cells, are also found in the epidermis.

Cell Biology

The surface of the skin, the stratum corneum, is made up entirely of dead keratinocytes. They contain no nucleus or other organelles, and are merely empty keratin shells. These dead cells are eventually sloughed off, and underlying cells move up to take their place. Approximately 1 g of surface cells is shed each day. This process is not simply necrosis of cells as they move away from the basement membrane but rather a genetically programmed differentiation process. Differentiation is characterized by a sequential expression of biochemically distinct keratin molecules. The keratin composition of cells near the basement membrane is different from the cells near the surface. Even the loss of nuclear DNA is a programmed event that results from temporal expression of a specific DNA nuclease that digests the nuclear DNA into fragments of characteristic size.

Keratinocyte differentiation can be induced in tissue culture with calcium (Hennings et al. 1980). In fact, care must be taken to exclude even small amount of calcium if induction of differentiation is to be avoided. The failure of a supposedly normal keratinocyte to differentiate in response to calcium indicates aberrant growth regulation–a precondition for neoplastic transformation (Kawamura et al. 1985, Kilkenny et al. 1985). Such cells are immortal in that they fail to become senescent like normal cells. This provides a convenient assay of testing the carcinogenic potential of radiation and other agents since immortalized keratinocytes will grow to produce colonies in tissue culture, while normal cells will die out. Loss of contact inhibition and ability to grow in soft agar are other tissue culture assays thought to correlate with malignant transformation. Human cells are extremely difficult to transform in tissue culture with ionizing radiation, and most experimental work has been done in rodent systems. Keratinocytes are thought to be the target cell for squamous cell carcinomas–the most common type of skin cancer. Basal cells and melanocytes are the respective target cells for basal cell carcinomas and melanomas–the other two common skin cancers. Although all of these tumors have been linked to ultraviolet radiation exposure, only squamous cell and basal cell carcinomas have been shown to be induced by ionizing radiation (Shore 1990).

Destruction of Langerhans cells by ionizing radiation may also contribute to skin cancer induction, since Langerhans cells have functions thought to be involved in immune surveillance against transformed cells (Cole 1986). Langerhans cells represent 2%–4% of the epidermal cells, and the effect of radiation on these cells is the most poorly understood area of skin radiation biology. Langerhans cells differ from other epidermal cells in that they originate in the bone marrow and migrate to the epidermis. They are found at all skin depths, and the number of Langerhans cells is correlated with the thickness of the epidermis.

Unique Aspects

One of the major issues concerning radiation protection of the skin is the exact location of the target cells for carcinogenesis. Since the most vulnerable cells lie along the basal membrane, the question becomes the location of the basal membrane. Historically, radiation protection guidelines have used 70 μm as the mean depth of the basal layer; however, this depth actually varies in different body areas from 30–300 μm (Fry 1990).

Another issue is the determination of the area over which the radiation dose should be measured. For external beam irradiation over relatively large exposed areas, this is not a difficult problem. Exposure due to contamination with radioactive "hot particles" (i.e., radioactive particles less than 2 mm), however, produces significant complications. These particles can result in very high local nonuniform skin doses that are difficult to estimate (Chabot et al. 1988). The biological consequences of this type of nonuniform radiation exposure are not clear (Charles et al. 1988). The National Council on Radiation Protection and

Measurements (NCRP) has recently published a report dedicated to this particular issue (NCRP 1989).

One unique route of radiation exposure for the skin is the use of gaseous positron-emitting radioisotopes in nuclear medicine. The skin is considered a tissue at risk for calculation of maximum permissible concentration values of positron emitters in air (Piltingsrud et al. 1985).

Skin carcinogenesis may differ from the general principle that protracted exposure reduces cancer risk. Some investigators have found that increased dose fractionation of x-rays significantly increased the induction of squamous cell carcinomas in mice (Fry 1990). This area needs further study.

Lastly, ultraviolet (UV) radiation may act as a promotor for skin carcinogenesis in that UV exposure following ionizing radiation exposure acts synergistically to increase tumor induction (Fry 1990). The effect of UV promotion on cells initiated by ionizing radiation may be a general cellular phenomenon but is of practical importance only for skin, since skin is the only major tissue exposed to UV light. Different racial groups vary in their susceptibility to UV-induced nonmelanotic skin cancer as much as 50-fold (Shore 1990). Although racial differences in sensitivity to ionizing-radiation-induced cancer has not been determined, it is suspected that the degree of UV promotion of ionizing-radiation-induced tumors may vary among races as well.

Radiation Effects

The significance of skin as a critical organ for dose limitation was much more important in earlier years when low-energy x-rays were the major source of exposure. These x-rays had low penetration and resulted in relatively high skin doses and visually evident radiation damage. The most obvious and immediate effect of radiation damage to skin is erythema. This is the first biological effect of ionizing radiation ever to be recognized. In fact, it was used as a biological dosimeter of radiation dose. There are three periods or phases to radiation-induced erythema (Potten 1985). The first phase appears as reddening within a few hours of exposure, peaks at 24 hours, and then subsides after 2 or 3 days. This reaction has an apparent threshold of about 3 Gy, is only slightly dose dependent, and is primarily due to vascular changes such as capillary leakage and dilation of blood vessels. The second phase is a wave of erythema beginning at 8 days, peaking at 14, and subsiding after a month. This phase is primarily due to inflammatory reactions, including release of histamine-like compounds. The intensity of this second phase has the greatest degree of dose dependence. Finally, a third phase may occur beginning at 35 days and is largely dose independent. It is probably due mostly to loss of epidermal cells and a reduction in epidermal thickness.

The basal layer of skin is one of the few tissues that allows for in vivo radiation survival analysis. This has been achieved by irradiating test areas of skin with large doses of radiation to denude the skin of proliferating cells. Following irradiation the skin is repopulated from small nodules of dividing cells. These nodules are analogous to the colonies seen in tissue culture assays, and each nodule is thought to originate from a single stem cell. By determining the number of nodules per unit area of skin at different radiation doses, an in vivo survival curve can be generated. Using this approach a D_0 of 1.35 Gy has been determined (Withers, 1967), which is similar to estimates from in vitro experiments. If heavily irradiated skin sections are examined, very few degenerating cells are seen (Tubiana et al. 1990). The denuding of skin is due to sterilization of clonogenic cells rather than outright cell death. The time course for denuding the skin following irradiation is, therefore, linked to the period required for terminal differentiation of previously committed cells (about 20 days) and the failure of replacement by their clonogenic progenitors.

Of most concern, from a radiation protection standpoint, is radiation-induced skin cancer, because it is a stochastic effect with no dose threshold and has potentially lethal consequences (National Academy of Sciences [NAS] 1990). Skin cancer was the first type of cancer recognized to be caused by ionizing radiation, since early radiologists developed skin cancer in areas of chronic skin erythema. Despite this early association, good epidemiological studies were difficult since most skin cancers are not fatal and are not reported as often as other cancers. Nevertheless, several classic studies strongly suggest that skin may be particularly sensitive to radiation carcinogenesis (Shore 1990). Patients treated with ionizing radiation to treat tinea capitis, a parasitic disease of the scalp, developed skin cancers on their heads and necks with a dose-dependent incidence (Harley et al. 1983; Shore et al. 1984). A case-control study of children irradiated for enlarged thymus showed increased incidence of skin cancer in the irradiation field (Albert et al. 1986). Uranium miners developed skin cancers on the face, presumably due to deposition of uranium dust (Sevcoca et al. 1978). The skin tumors with increased incidence in these studies were all of the squamous cell and basal cell type. Squamous cell and basal cell carcinomas have also been experimentally produced in mice with ionizing radiation (Bowden et al. 1990; Jaffe et al. 1986, 1987).

The current estimates of skin cancer risk in humans is 7.1×10^{-6} excess cases/cm^2 person Gy (NAS 1990). However, there seems to be some interaction with ultraviolet radiation, making the risk for areas of skin also exposed to sunlight about five times higher. Also, recent studies suggest that fair-skinned people, who sunburn easily, may also be at higher risk for ionizing-radiation-induced skin cancer (Davis et al. 1989).

Dose Considerations

The doses that will induce erythema are very high by today's radiation protection standards. The presence of chronic cutaneous damage indicates tens of gray of dose at a rate of at least 0.5 cGy per day. This exceeds today's standards for radiation protection, which are based on carcinogenic risk, by several orders of magnitude. Currently, the maximum allowable occupational skin dose per year is 0.5 Sv, with a total lifetime limit of 20 Sv (NCRP 1987). There has recently been some discussion that radiation protection guidelines for the skin need to be reevaluated (Fry 1990). The current guidelines were established in 1977, before knowledge about synergism of carcinogenesis with UV exposure and possible increased risk of protracted exposure was appreciated. This new information suggests that total dose limits will need to be adjusted downward if the same level of absolute skin cancer risk is to be maintained.

TESTIS

Physiology and Histology

The testes are the male sexual organs responsible for sperm production. They are contained externally in a pocket of skin called the scrotum, which allows them to have a temperature below that of the body core. This lower temperature is required for development of the sperm. The testes are composed mostly of seminiferous tubules in which sperm maturation takes place. A basement membrane and connective tissue define the margins of the tube. Internally, there is an epithelial layer composed of stem cells and other progenitor cells of terminally differentiated sperm cells. Cells migrate from the basement membrane to the interior surface of the tubule during differentiation. The production of sperm by this differentiation process is called spermatogenesis. The fully mature sperm cells are released to the lumen of the tubule. They then move through the tubule due to the peristaltic contraction of the myoid cells within the collagenous matrix of the basement membrane. The sperm cells eventually pass through the rete testis into the excurrent ducts, from which they are ultimately expelled by ejaculation.

Between the seminiferous tubules are blood vessels and connective tissues (Meistrich 1986a). Also found in this interstitial area are Leydig cells. Leydig cells are steroid hormone secreting cells and are responsible for producing testosterone. These hormones act locally in the testis and other sex organ and distally to effect secondary sex characteristics and control libido.

Cell Biology

A very small percentage of all of the available stem cells are dividing at any one time (Meistrich 1986a). When they do divide, the division can serve one of two functions. The first is to maintain the number of stem cells, and the second is to produce cells committed to the differentiation process (Meistrich 1986b). These two functions are called *regeneration* and *repopulation*, respectively. Spermatogonia are daughters of the stem cells that have become committed to differentiate via the spermatogenic pathway. They are found along the basement membrane. Spermatogonia undergo a series of mitotic divisions to produce primary spermatocytes, which in turn undergo the first meiotic division to become secondary spermatocytes. Meiosis results in cells with a haploid genome, a characteristic that distinguishes spermatogenesis from somatic differentiation. A second meiotic division produces smaller spermatids, at which point cell division stops. Spermatids then undergo a loss of cytoplasm and the development of a tail and become spermatozoa (i.e., sperm cells). Sperm cells are released to the lumen of the tube and move out of the seminiferous tubules into the epididymis. In man, it takes about 8.5 weeks for a differentiated spermatogonium to become a mature sperm cell. The period of each stage of the differentiation process is rigidly fixed and is unaffected by radiation exposure. The cycling of the stem cells, however, can be altered by radiation, and stem cell kinetics do change with radiation exposure.

Sertoli cells are large cells that extend from the basal membrane up toward the lumen of the tube. They are believed to provide nourishment to the developing spermatids. They also produce androgen-binding protein and inhibin, which regulates the amount of follicle-stimulating hormone (FSH) hormone produced by the pituitary. These cells also arise from basal layer stem cells. Leydig cells are found in interstitial areas between the tubules and have large endoplasmic reticuli associated with their primary function of steroid hormone production. Leydig cells and Sertoli cells are not thought to divide once an individual reaches adulthood, which is consistent with their relative radioresistance.

Unique Aspects

Spermatogenesis is similar to keratinocyte differentiation in that a basal membrane of stem cells gives rise to cells that differentiate as they move to the surface. Spermatogenesis differs in that the terminally differentiated sperm cells are alive and, therefore, present an additional potential target for radiation action.

Radiation Effects

Radiation affects the rate of production of differentiating cells from the stem cells and the yield of sperm produced from the spermatogonium (Tubiana et al. 1990). Radiation survival of stem cells has an approximate D_0 of 3 Gy and an n value of 4. Their relative radioresistance is probably related to their very small proliferating fraction. The reduction in sperm is typically more pronounced than what can be accounted for purely on the basis of stem cell mortality. The primary effect on sperm production at lower doses is due more to depletion of the relatively sensitive differentiating spermatogonia. These cells usually survive their first mitotic division postirradiation but then die shortly after that. The spermatogonia are the most sensitive of the cells and can be considered the major cell target at low radiation doses. However, at low doses the more resistant stem cells are spared, and they can rapidly repopulate the spermatogonia. For this reason, even complete loss of spermatogonia will not cause permanent sterility. As doses increase and the reduction of stem cells becomes pronounced, division of the remaining stem cells serves to regenerate the stem cell compartment before spermatogonia are repopulated. Thus the length of azoospermia increases with increasing dose (Meistrich 1986b). Only with complete elimination of the stem cell compartment will permanent sterility occur.

A unique in vitro colony-counting assay has been used by some investigators to determine the number of stem cells that survive in irradiated testes (Lett et al. 1991). Thin cross-sections of the testis through the seminiferous tubules are made 5 weeks postirradiation and then examined microscopically. Each tubule in the field is scored for the presence or absence of spermatogenic cells. The presence of spermatogenic cells indicates at least one surviving stem cell in that region of the tubule. Tubules without spermatogenic cells are considered to be devoid of stem cells in that area.

Cells that have completed meiosis, such as spermatozoa and spermatids, are less sensitive to the lethal effects of radiation than spermatogonia. Due to the resistance of these cells nearing completion of the differentiation process, infertility does not usually appear until 6 weeks postirradiation.

It is interesting to note that sterilization of rams by irradiation of the testes provided the first evidence that dose fractionation could enhance cell killing in dividing versus nondividing tissues. Experiments performed in the 1920s showed that rams could not be sterilized with a single dose of radiation to the testes without seriously damaging surrounding tissue. Dose fractionation allowed sterilization with a minimum of complications. We now know that fractionation recruits nondividing stem cells into the cycling population as they regenerate and repopulate the lethally irradiated cells. These cycling cells have increased radiosensitivity, and this probably contributes somewhat to the fractionation effect seen. Another factor is that the stem cells of the testes appear to be less efficient at sublethal damage repair compared with other renewing normal tissues (Withers et al. 1974), allowing greater accumulation of damage during dose fractionation relative to the other tissues. It was postulated that the testes may model the growth kinetics of a human tumor, and dose fractionation was applied to tumor radiotherapy. The application of dose fractionation in tumor control has turned out to be a major advance in radiation oncology (Hall 1988).

Dose Considerations

Production of sperm can be affected by relatively low doses of radiation (Bianchi 1983, Fajardo 1982). A dose of 0.08 Gy can cause a transient mild reduction, and 0.2 Gy causes a pronounced reduction that can last several months (Tubiana et al. 1990). Absence of sperm for 1–2 years occurs after doses of 2 Gy, and subsequent sperm counts may plateau at a level below normal. The reason that recovery is not complete is not clear; however, it may be due to unrepaired sublethal damage within the surviving stem cells or damaged homeostatic mechanisms (Meistrich 1986b). These lower sperm counts may result in impaired fertility. It has been estimated for each two-fold reduction in sperm count in a population of men, another 4% of the population will become infertile (Meistrich 1986b). Doses of 6 Gy or higher usually produce complete sterilization with complete loss of sperm cells.

Radiation effects are usually restricted to spermatogenesis. Leydig cells are much more resistant and are usually not affected by these low doses. Gonadotrophin levels, secondary sex characteristics, and libido are, therefore, usually not affected at doses that cause complete sterility (Jaffe et al. 1988).

Radiation effects on spermatogenesis appear to be independent of dose rate. Dose fractionation does not have a sparing effect, and animal experiments have shown that dose rates as low as 1 cGy per day interfere with spermatogenesis.

The current annual maximum permissible dose for the human testis is 0.5 Sv (NCRP 1987).

OVARY

Physiology and Histology

The main organs of the female reproductive tract include the vagina, the uterus, the fallopian tubes, and the ovaries. The ovaries are the site of egg (ovum) development and storage. They are situated next to the uterus on the side of the pelvic cavity and are held in place by mesovaria and ovarian ligaments. The surface layer of the ovary is composed of a single layer of cells called the gonadal epithelium. Under this layer is the cortex, and within the cortex is the medulla. Although they perform different functions, there is no definitive line between the end of the cortical tissue and the beginning of the medulla. The most obvious ovarian structures are the follicles, each of which contains a single egg at various stages of development, and the corpora lutea, a yellow-colored body that develops from a postovulatory follicle (Guyton 1981).

The different structures of the ovary vary in function during the menstrual cycle, which usually encompasses 28 days. Gonadotropins from the pituitary stimulate follicles to grow in the ovaries. One of the follicles ovulates and releases an egg on the 14th day of the cycle. While the follicles are growing, they are releasing estrogen, a steroid hormone. After ovulation, the ruptured follicle develops into a corpus luteum, which releases large amount of estrogen and progesterone. Two weeks later the corpus luteum degenerates, blood levels of estrogen and progesterone drop, and menstruation begins. After this, the cycle begins again (Guyton 1981).

Cell Biology

The gonadocytes undergo a complicated and protracted form of development. During human embryogenesis these cells multiply and differentiate into oogonia and then primary oocytes. The primary oocytes enter into their first meiotic division, which proceeds to late prophase and stops. At this point the oocyte and the follicle surrounding it increase in size. The oocytes remain in this suspended state from the time of birth to the onset of puberty. The largest number of oocytes exist at the 30th week of gestation, at which time the developing ovaries contain about 7 million. Many of these degenerate, so there are about 2 million at birth. By puberty the number is only 300,000. Degeneration continues until menopause, when the oocytes are completely gone. Between puberty and menopause only about 400 oocytes actually ovulate. At each estrous cycle, a single oocyte finishes its first meiotic division and enters into its second meiotic division up to metaphase at which it remains until it is penetrated by a sperm cell. Typically, oocyte maturation alternates between the two ovaries with each estrous cycle. Ovarian development, oocyte maturation, and the estrous cycle are rigidly controlled by hormones. Estrogen and progesterone are produced by the ovaries themselves. Other reproductive hormones are produced by the hypothalamus and the anterior pituitary (Turner et al. 1976).

Unique Aspects

The oocytes are the target cells for radiation-induced sterility in females. The situation is different from that of spermatocytes in males in that these cells are not dividing. Despite the lack of division, oocytes are very radiosensitive. The D_0 for oocytes has been estimated to be 0.12 Gy (Bianchi 1983), making them the most sensitive cell type in the body. This may be due to higher sensitivity of the cell membrane to radiation and its increased importance as a target for radiation-induced lethality. Target theory suggests a target size closer to the cell membrane than the nucleus (Straume et al. 1987).

Radiation Effects

All ovarian follicles have equal sensitivity regardless of their maturation state. For this reason, sterilization effects of radiation are usually immediate and permanent. Also, since female hormonal secretions depend on follicle maturation, radiation sterilization is the equivalent of castration. For radiation doses that do not cause sterility, oocytes are five-fold more resistant than spermatocytes to mutagenesis. Many of the mutations that occur are dominant lethal mutations that result in embryo death at the time of implantation. The induction of dominant lethal mutations by ionizing radiation has been studied in mouse oocytes and was found to have a linear dose response (Goldstein 1987).

Dose Considerations

The dose resulting in sterilization/castration depends on the age of the woman. Five to seven Gy will usually sterilize a woman in her forties, while a woman in her twenties would have to receive 12–15 Gy before sterilization would occur. Women receiving doses from 3–4 Gy may experience long periods of amenorrhea and then recover fertility. A high dose of 20–30 Gy may result in high levels of gonadotropins, as the body tries to stimulate the dysfunctional ovaries into estrogen production. This situation is a sign of high dose exposure and irreversible ovarian damage (Tubiana et al. 1990).

Although it is difficult to assess the genetic risks associated with sublethal irradiation of oocytes, the risks are considered quite low. It is estimated that a dose exceeding 1 Gy must be received by all individuals in a human population in order to double the incidence of genetic disease in that population. Thus genetic risk is usually not the dose-limiting factor for the gonads in radiation protection situations. The current annual maximum permissible dose to the ovaries is 0.5 Sv (NCRP 1987).

LENS OF THE EYE

Physiology and Histology

The eye is optically the direct equivalent of a camera. There is a retina, which represents the film of the camera. The iris is a variable aperture, which controls the amount of entering light. The lens is used for focusing. The lens is composed of viscous transparent fibers contained within a transparent capsule (Merriam et al. 1972). The adult human lens is about the size of a pea. It is situated immediately posterior to the iris and is held in place by muscles. These muscles also can change the shape of the lens and, thereby, focus an image on the retina in the back of the eye. In a relaxed state, with no tension on the capsule, the lens has a spherical shape. There is usually some tension on the lens, however, caused by the elastic pull of the ligaments that attach radially around the lens. This constant tension causes the lens to remain relatively flat under most normal conditions (Guyton 1981).

Cell Biology

The lens consists mostly of fiber cells and a layer of epithelial cells on the interior surface of the anterior capsule. Cell division occurs only at the periphery of the lens, and the daughter cells differentiate into fiber cells as they migrate toward the interior.

Unique Aspects

As a tissue, the lens is unusual in two respects. First, it has no blood supply. Second, there seems to be no mechanism for removal of dead cells. Both of these factors may have an impact on its response to ionizing radiation.

A unique aspect of radiation-induced cataracts is that they develop and progress in a specific fashion that is different from naturally occurring cataracts. This is distinct from most other radiation effects, which cannot usually be distinguished from the natural background.

Radiation Effects

The first studies of ionizing radiation effects on the eye were conducted as early as 1897, and the first case of radiation-induced cataract was reported in 1908 (Merriam et al. 1972). Originally, the lens was considered a relatively radioresistant eye structure. It is now recognized that the lens is the most sensitive part of the eye, and that cataract formation is the relevant biological effect (Hall 1988).

A cataract is a loss of transparency in the lens of the eye. Most cataracts are the result of aging, although environmental factors can also play a role. The primary mechanism is thought to be the killing of fiber cells, which become opaque and migrate to the posterior pole of the eye. A radiation-induced cataract generally forms in a characteristic way. In the early stages this can be seen through an ophthalmoscope as a dot of about millimeter size. As the cataract grows, granules and vacuoles will begin to surround it. Sometimes, with further growth, the opacity will develop a clear center, resulting in a ring or doughnut appearance. Eventually, granules and vacuoles appear in the anterior region, and the opacity may go on to encompass the entire lens. There is typically a latent period of 6 months to 35 years between radiation exposure and cataract formation.

The lower the radiation dose, the longer the latent period. This long latency period is due to the very slow rate at which lens cells are renewed. Among different species there is a correlation between the life span and the length of the latent period (Lett et al. 1991), so the development of cataracts relates in some way to the cell turnover rate or aging process.

Dose Considerations

As opposed to most other biological effects of ionizing radiation, cataractogenesis appears to be a nonstochastic effect. That is, it is a nonrandom process, has an intensity proportional to dose, and has a threshold. In humans, the threshold is about 2 Gy; however, fractionation of the dose will lessen the effect and increase the threshold by a factor of at least two-fold. For an acute dose of 7.5 Gy there is virtually a 100% incidence of cataracts, while fractionation of the dose will reduce the incidence to approximately 60% (Merriam et al. 1972). Doses in the range of 2.5–6.5 Gy have a latency period of about 8 years. Progressively higher doses result in both increased severity and decreased latency. The current annual maximum permissible occupational dose to the lens of the eye is 150 mSv (NCRP 1987).

Studies in rats suggest that high linear energy transfer (LET) radiations are more effective at producing cataracts. The relative biological effectiveness (RBE) for high-LET particles varies with dose. High-LET particles, such as neutrons, have RBEs of about 10 at high doses, but the RBE may be as high as 50 at low doses (Bateman et al. 1967, Merriam et al. 1984, Worgul et al. 1985). This is not due to the increased effectiveness of neutrons at low doses but rather to the decreased effectiveness of x-rays at low doses. This is important because it suggests that the threshold for high-LET particles may be quite low. Furthermore, in contrast to low-LET radiation, which produces fewer cataracts with dose fractionation, recent reports suggest that fractionation of high-LET radiation actually increases incidence of cataracts (Worgul et al. 1989a, 1989b).

Because the threshold for cataract formation is relatively low, it may be difficult or impossible to keep the radiation dose to the lens below the threshold when performing radiation therapy in the vicinity of the eye (Merriam et al. 1972). For this reason, special care needs to be taken to protect the lens during radiation therapy of the head (Dean et al. 1988, Haik et al. 1986, Harnett et al. 1987, Kirsner et al. 1987, McCormick et al. 1988, Parker et al. 1987, Rustgi 1986, Seddon et al. 1986, Taylor et al. 1988). In addition to dose reduction techniques, work is also being done with chemical radioprotectors, such as WR-77913, which have been shown to reduce cataract incidence in irradiated rat eyes when administered prior to irradiation (Clark et al. 1987, Menard et al. 1986). Also, compounds that modify lens protein metabolism have been shown to suppress cataract formation even when administered following radiation (Pandolfo et al. 1986, Taura et al. 1985). In any event, surgical treatment of cataracts has advanced greatly in recent years, and vision can usually be restored in patients who do not have other eye disease.

REFERENCES

Albert, R.E., and Shore, R.E. Carcinogenic effects of radiation on human skin. In: *Radiation Carcinogenesis,* edited by A.C. Upton, R.E. Albert, F.J. Burns, and R.E. Shore. New York, Elsevier, 1986, pp. 335–345.

Bateman, J.L., and Bond, V.P. Lens opacifications in mice exposed to fast neutrons. *Radiat. Res. 7(suppl.):*239–249, 1967.

Bianchi, M. Cytotoxic insult to germinal tissue. In: *Cytotoxic Insult to Tissue,* edited by C.S. Potten, J.H. Hendry. Edinburgh, Churchill Livingstone, 1983, pp. 258–328.

Bowden, G.T., Jaffe, D., and Andrews, K. Biological and molecular aspects of radiation carcinogenesis in mouse skin. *Radiat. Res. 121:*235–241, 1990.

Chabot, G.E., and Skrable, K.W. Beta-gamma point source on the skin problem–activity estimation and dose analysis. *Health Phys. 55:*729–739, 1988.

Charles, M.W., Williams, J.P., and Coggle, J.E. Skin carcinogenesis following uniform and nonuniform beta irradiation. *Health Phys. 55:*399–406, 1988.

Clark, J.I., Osgood, T.B., and Trask, S.J. Inhibition of phase separation by reagents that prevent X-irradiation cataract in vivo. *Exp. Eye Res. 45:*961–967, 1987.

Cole, S. Long-term effects of local ionizing radiation treatment on Langerhans cells in mouse footpad epidermis. *J. Invest. Dermatol. 87:*608–612, 1986.

Davis, M.M., Hanke, C.W., Zollinger, T.W., Montebello, J.F., Hornback, N.B., and Norins, A.L. Skin cancer in patients with chronic radiation dermatitis. *J. Am. Acad. Dermatol. 20:*608–616, 1989.

Dean, G., Alderson, M., and Maximilien, R. Increased risk of cataract in patients receiving radiotherapy to the eye: a pilot study. *Br. J. Radiol. 61:*309–311, 1988.

Fajardo, L.F. *Pathology of Radiation Injury.* New York, Masson Inc., 1982.

Fry, R.J. Radiation protection guidelines for the skin. *Int. J. Radiat. Biol.* 57:829–839, 1990.

Goldstein, L.S. X-ray-induced dominant lethal mutations in mouse oocytes detected by an in vitro assay. *Radiat. Res. 112:*374–380, 1987.

Guyton, A.C. *Textbook of Medical Physiology*, 6th Ed. Philadelphia, WB Saunders, 1981.

Haik, B.G., Jereb, B., Smith, M.E., Ellsworth, R.M., and McCormick, B. Radiation and chemotherapy of parameningeal rhabdomyosarcoma involving the orbit. *Ophthalmology 93:*1001–1009, 1986.

Hall, E.J. *Radiobiology for the Radiologist,* 3rd Ed. Philadelphia, JB Lippincott, 1988.

Harley, N., Kolber, A.B., Shore, R.E., Albert, R.E., Altman, S.M., and Pasternack, B. The skin dose and response for the head and neck in patients irradiated with x-ray for tinea capitis: implications for environmental radioactivity. In: *Epidemiology Related to Health Physics. Proceedings of the 16th Midyear Topical Meeting of the Health Physics Society*, CONF-83011. Springfield, VA, National Technical Information Service, 1983, pp. 125–142.

Harnett, A.N., Hungerford, J.L., Lambert, G.D., Hirst, A., Darlison, R., Hart, B.L., Trodd, T.C., and Plowman, P.N. Improved external beam radiotherapy for the treatment of retinoblastoma. *Br. J. Radiol. 60:*753–760, 1987.

Hennings, H., Michael, D., Cheng, C., Steinert, P., Holbrook, K., and Yuspa, S.H. Calcium regulation of growth and differentiation of mouse epidermal cells in culture. *Cell 19:*245–254, 1980.

Hopewell, J.W. The skin: its structure and response to ionizing radiation. *Int. J. Radiat. Biol. 57:*751–773, 1990.

Jaffe, D.R., and Bowden, G.T. Ionizing radiation as an initiator in the mouse two-stage model of skin tumor formation. *Radiat. Res. 106:*156–165, 1986.

Jaffe, D., and Bowden, G.T. Ionizing radiation as an initiator: effects of proliferation and promotion time on tumor incidence in mice. *Cancer Res. 47:*6692–6696, 1987.

Jaffe, N., Sullivan, M.P., Ried, H., Boren, H., Marshall, R., Meistrich, M., Maor, M., and da Cunha, M. Male reproductive function in long-term survivors of childhood cancer. *Med. Pediatr. Oncol. 16:*241–247, 1988.

Kawamura, H., Strickland, J.E., and Yuspa, S.H. Association of resistance to terminal differentiation with initiation of carcinogenesis in adult mouse epidermal cells. *Cancer Res. 45:*2748–2752, 1985.

Kilkenny, A.E., Morgan, D., Spangler, E.F., and Yuspa, S.H. Correlation of initiating potency of skin carcinogens with potency to induce resistance to terminal differentiation in cultured mouse keratinocytes. *Cancer Res. 45:*2219–2225, 1985.

Kirsner, S.M., Hogstrom, K.R., Kurup, R.G., and Moyers, M.F. Dosimetric evaluation in heterogeneous tissue of anterior electron beam irradiation for treatment of retinoblastoma. *Med. Phys. 14:*772–779, 1987.

Lett, J.T., Lee, A.C., and Cox, A.B. Late cataractogenesis in rhesus monkeys irradiated with protons and radiogenic cataract in other species. *Radiat. Res. 126:*147–156, 1991.

McCormick, B., Ellsworth, R., Abramson, D., Haik, B., Tome, M., Grabowski, E., and LoSasso, T. Radiation therapy for retinoblastoma: comparison of results with lens-sparing versus lateral beam techniques. *Int. J. Radiat. Oncol. Biol. Phys. 15:*567–574, 1988.

Meistrich, M.L. Critical components of testicular function and sensitivity to disruption. *Biol. Reprod. 34:*17–28, 1986a.

Meistrich, M.L. Relationship between spermatogonial stem cell survival and testis function after cytotoxic therapy. *Br. J. Cancer 7(suppl):*89–101, 1986b.

Menard, T.W., Osgood, T.B., Clark, J.I., Spence, A.M., Steele, J.E., Krohn, K.A., and Livesey, J.C. Radioprotection against cataract formation by WR-77913 in gamma-irradiated rats. Int. *J. Radiat. Oncol. Biol. Phys. 12:*1483–1486, 1986.

Merriam, G.R., Szechter, A., and Focht, E.F. The effects of ionizing radiations on the eye. *Front. Radiat. Ther. Oncol. 6:*346–385, 1972.

Merriam, G.R., Worgul, B.V., Medvedovsky, C., Zaider, M., and Rossi, H. Accelerated heavy particles and the lens. I. Cataractogenic potential. *Radiat. Res. 98:*129–140, 1984.

National Academy of Sciences. *Health Effects of Exposure to Low Levels of Ionizing Radiation: BEIR V.* Washington, DC, National Academy Press, 1990.

National Council on Radiation Protection and Measurements. *Recommendations on Limits for Exposure to Ionizing Radiation,* Report No. 91. Bethesda, MD, National Council on Radiation Protection and Measurements, 1987.

National Council on Radiation Protection and Measurements. *Limit for Exposure to "Hot Particles" on the Skin,* Report No. 106. Bethesda, MD, National Council of Radiation Protection and Measurements, 1989.

Pandolfo, L., Livrea, M.A., and Bono, A. Effects of bendazac L-lysine salt on X-ray-induced cataract in the rabbit lens. *Exp. Eye Res. 42:*167–175, 1986.

Parker, R.G., and Greenberg, P. Treatment planning for particle radiation therapy. *Front. Radiat. Ther. Oncol. 21:*84–93, 1987.

Piltingsrud, H.V., and Gels, G.L. An evaluation of the external radiation exposure dosimetry and calculation of maximum permissible concentration values for airborne materials containing 18F, 15O, 13N, 11C and 133Xe. *Health Phys. 49:*805–824, 1985.

Potten, C.S. *Radiation and Skin.* London, Taylor & Francis, 1985.

Rustgi, S.N. Dose distribution under external eye shields for high energy electrons. *Int. J. Radiat. Oncol. Biol. Phys. 12:*141–144, 1986.

Seddon, J.M., Gragoudas, E.S., Polivogianis, L., Hsieh, C.C., Egan, K.M., Goitein, M., Verhey, L., Munzenrider, J., Austin Seymour, M., and Urie, M. Visual outcome after proton beam irradiation of uveal melanoma. *Ophthalmology 93:*666–674, 1986.

Sevcoca, M.J., Sevc, J., and Thomas, J. Alpha irradiation of the skin and the possibility of late effects. *Health Phys. 35:*803–806, 1978.

Shore, R.E. Overview of radiation-induced skin cancer in humans. *Int. J. Radiat. Biol. 57:*809–827, 1990.

Shore, R.E., Albert, R.E., Reed, M., Harley, N., and Pasternack, B.S. Skin cancer incidence among children irradiated for ringworm of the scalp. *Radiat. Res. 100:*192–204, 1984.

Straume, T., Dobson, R.L., and Kwan, T.C. Neutron RBEs and the radiosensitive target for mouse immature oocyte killing. *Radiat. Res. 111:*47–57, 1987.

Taura, T., Giblin, F.J., and Reddy, V.N. Further observations on the effect of galactose on the development of x-ray-induced cataract in mice. *Exp. Eye Res. 41:*527–543, 1985.

Taylor, D.M., and Thorne, M.C. The potential for irradiation of the lens and cataract induction by incorporated alpha-emitting radionuclides. *Health Phys. 54:*171–179, 1988.

Tubiana, M., Dutreix, J., and Wambersie, A. *Introduction to Radiobiology.* New York, Taylor & Francis, 1990.

Turner, C.D., and Bagnara, J.T. *General Endocrinology*, 6th Ed. Philadelphia, WB Saunders, 1976.

Withers, H.R. The dose-survival relationship for irradiation of epithelial cells of mouse skin. *Br. J. Radiol. 40:*187–194, 1967.

Withers, R.H., Hunter, N., Barkely, H.T., and Reid, B.O. Radiation survival and regeneration characteristics of spermatogenic stem cells of the mouse testis. *Radiat. Res. 57:*88–103, 1974.

Worgul, B.V., Medvedovsky, C., Powers Risius, P., and Alpen, E. Accelerated heavy ions and the lens. IV. Biomicroscopic and cytopathological analyses of the lenses of mice irradiated with 600 MeV/amu 56Fe ions. *Radiat. Res. 120:*280–293, 1989a.

Worgul, B.V., Merriam, G.R., and Medvedovsky, C. Accelerated heavy particles and the lens. II. Cytopathological changes. *Invest. Ophthalmol. Vis. Sci. 27:*108–114, 1985.

Worgul, B.V., Merriam, G.R.,Jr., Medvedovsky, C., and Brenner, D.J. Accelerated heavy particles and the lens. III. Cataract enhancement by dose fractionation. *Radiat. Res. 118:*93–100, 1989b.

Chapter 12

Acute Radiation Injuries and Their Medical Management

NIEL WALD

ABSTRACT

The purpose of this chapter is to help health physicists to provide useful information and advice to physicians facing the need to diagnose and treat acute radiation injury, including the acute radiation syndromes in particular, and to participate effectively in emergency planning and training for such events. Some basic concepts of human radiobiology underlying the medical effects of radiation accidents are briefly reviewed. The medical management of the acute radiation syndromes, and, more briefly, some aspects of local tissue damage from external exposure and the medical problems associated with radionuclide contamination and incorporation are considered. Diagnostic and treatment considerations relevant to radiation emergency planning are discussed. Some of the current limitations in national medical resources for large-scale radiation emergency response planning and management are mentioned.

INTRODUCTION

In this presentation, I will try to focus on those features of human radiobiology that underlie potential health impairments from significant radiation overexposure. I will emphasize the aspects relevant to their early recognition and the information and resources required for their prompt management. This should provide a basis for health physicists to give useful information and advice to physicians confronted with the nonroutine health problems of the very occasional acute radiation injury patient. These concepts should also underlie realistic and effective emergency planning and training activities for radiation accidents, important facets of good radiation protection practice.

There are five potential major medical radiation problems associated with radiation accidents. The first two, those usually produced by external sources of exposure, are the acute radiation syndromes (Anno et al. 1989, Finch 1987, Gerstner 1958, Keller 1946, Thoma and Wald 1959, United Nations Scientific Committee on the Effects of Atomic Radiation [UNSCEAR] 1988, Young 1987) and local radiation injury (Saenger et al. 1980). The other three are produced by radionuclide contamination (National Council on Radiation Protection and Measurements [NCRP] 1980) that may be on the external surface of the body, in traumatic wounds, and/or incorporated by inhalation, ingestion, or even passage through the intact skin.

Unfortunately these radiation injuries are not mutually exclusive and, in addition, they may be accompanied by thermal and/or mechanical trauma as well. My assignment is to focus on the phenomena associated with the acute radiation syndrome, but I will, where pertinent, also mention the other forms of radiation injury.

SOME RELEVANT HUMAN RADIOBIOLOGY

In order to understand the basic pathophysiology underlying all of these diverse forms of radiation injury, we must consider the final common pathway leading to clinical significance, the basic unit of which is the living cell. The two most frequent major biological results of radiation interactions at the cell level are direct cell death, i.e., the killing of functional cells, and, generally at lower levels of exposure, reproductive death, i.e., damage to the mitotic capability of cells such that they are unable to divide despite their continued survival.

The relative intrinsic radiosensitivity of different cell lines in the body's organs, modified by their reproductive rate and functional life span, determines which organ systems have a particular vulnerability. The functions of those particular organ systems, then, are the ones in greatest jeopardy. Significant loss of function of key organ systems, in turn, defines the nature of the ensuing clinical response of the exposed individual.

Any listing of cells in the order of their radiosensitivity would place the blood cell lines at the top. Most of them have a relatively short life span and high turnover rate in order to maintain the functional integrity of the hematological tissue and to provide the requisite large numbers of cells in circulating blood. Thus they are most vulnerable to the detrimental effect of radiation on cell replication or mitosis. Another high-turnover system is the gastrointestinal tract lining, whose cells also have great radiosensitivity. Other organs have an intermediate life span and level of radiation sensitivity. The supporting connective tissues, muscle, and bone are less radiosensitive, and the postfetal nervous system is among the least sensitive to visible cell damage from ionizing radiation. In these latter tissues, note that the cell turnover is minimal.

Let us consider the blood-forming tissue as a good model system for human radiobiology. In this system the production of the various blood cell lines (erythrocyte, granulocyte, lymphocyte, monocyte, and platelet) goes on primarily in the red bone marrow throughout the body. The stem cells of the lymphocyte line, although initially part of this population in early fetal life, migrate to lymphoid tissues such as the thymus and the lymphoid areas in the gastrointestinal tract. The stem cells maintain their own essential population number and also generate the cells that mature into the functional, differentiated members of the various blood cell lines. When mature, they migrate into the circulating blood and tissues to carry out their specialized functions.

Radiation exposure at an LD_{50} level will deplete the hematological radiosensitive precursor cell population drastically. For example, Bond and colleagues (1965) found that only 2 or 3 colony-forming cells out of 1000 retain their proliferative capacity after an in vivo $LD_{50/30}$ exposure (about 6 Gy) in mice, and Fliedner (1990) calculated from the granulocyte response of total-body irradiated patients that in the dose range of about 320–420 cGy (around the LD_{50}), 95% to 97.6% of the stem cells are destroyed. The deficit is filled in by an apparent increase in fat cells as well as by some bleeding into the tissue from the small blood vessels that supply the marrow with oxygen and nutrients. As the stem cell compartment recovers from injury, it replenishes itself, and subsequently the maturing and differentiating populations, in the bone marrow. Eventually, fully mature cells are again released into circulation.

The same sequence can be seen in the postirradiation morphological changes in the lymph nodes as well, although depletion is faster since this line is vulnerable to direct cell killing as well as to reproductive arrest, and its recovery is slower.

In terms of the three-compartment model often used in radiation cytokinetics (Bond et al. 1965), a depletion wave begins with the more radiosensitive cells in the stem cell compartment, which then reduces the number of cells that are differentiating and dividing and finally reduces the number of cells that are released into the circulating cell compartment. It should be noted that by the time the depletion wave reaches the circulating cell compartment, regeneration of the stem cell compartment is well underway, with its consequent repopulation of the dividing and maturing cell compartment. This ultimately leads to renewed release of mature cells into circulation.

When the depletion wave reduces the numbers of circulating cells of the various blood cell lines significantly, it leads to clinical consequences. They are those one would expect when the cells that defend the body against infection and bleeding are temporarily reduced or absent.

The ordinary clinical blood count, done serially during this period, will show a characteristic sequence of changes, including a rapid depletion of lymphocytes within the first 48 hours postexposure; a gradual decrease in red cells, reaching a low point about 5–6 weeks postexposure; and a gradual fall in platelets to a nadir about 30 days postexposure (Wald 1971).

There is a transient increase in granulocytes in circulation within minutes to hours resulting from the mobilization of these cells from tissues as a nonspecific stress response. This is followed by a gradual fall as cells are used up and then by a plateau or even an increase in granulocyte population in the 2nd week in an only transiently effective attempt of injured precursor cells to continue replicating. This is followed by the more drastic fall in the granulocyte circulating numbers, which is rather abruptly reversed when the repopulation of the stem cell and the dividing and maturing populations in the bone marrow results in a wave of newly released mature cells entering circulation.

At the human $LD_{50/60}$, or midlethal exposure level, this maximum depletion occurs with surprising regularity at about the 30th postexposure day. At higher exposure levels in the supralethal range there is a similar response pattern, but it is more condensed in time and greater in magnitude.

Another important organ system with a high cell turnover rate is the gastrointestinal tract. A similar sequence of events takes place, but here the wave of cell depletion leaves a transiently destroyed lining of the gastrointestinal tract. Unfortunately this depletion, combined with the deficits in blood cells, has resulted in irreversible clinical effects leading to a fatal outcome in the accidental exposure cases in which it has occurred. The doses at which these phenomena have been seen are in the range of 5 Gy and higher.

At about 20 to 30 Gy and higher, the *dysfunctions* of the central nervous and cardiovascular systems are dramatic and irreversible. They are related to interruptions of the normal regulatory control systems, which ordinarily ensure the necessary responses in heart rate, respiration, blood pressure, body temperature, etc., to the stresses of life. On the other hand, the visible *morphological* changes at the cell level of the central nervous system are limited to a few tissues, including the granule cell layer of the cerebellum and the meningeal lining of the brain. Also, there is a breakdown of the blood-brain barrier that leads to cerebral edema.

Another effect on the respiratory tract, interstitial pneumonitis or inflammation of the lung, can be elicited by external exposure beginning at around 8 or 9 Gy depending on many variables, such as dose rate and age (U.S. Nuclear Regulatory Commission [U.S. NRC] 1989, Van Dyk et al. 1981). It usually appears in 1–3 months (UNSCEAR 1988).

Although fatal disruptions of neurovascular function occur only after very high radiation exposure, there is also a transient response of the nervous system to relatively low exposures beginning at about 1 Gy, and affecting almost all patients at about 3 Gy. Its manifestations include anorexia (loss of appetite), nausea, and vomiting, occurring within a few hours after such exposures and only lasting for about 48 hours. This transient effect is probably due to a neurohumoral substance acting on the chemoreceptor trigger zone of the medulla (Young 1986), the vomiting center in the brain, and on the innervation of the stomach by the vagus nerve from the brain. It can be mitigated to some extent by administering various neuropharmacological blocking agents (Luthra et al. 1981).

Another organ system responding to exposures as low as several Gy is the skin and its appendages, the hair and sweat glands. Erythema or reddening would be expected in 50% of those exposed to about 6 Gy, transepithelial injury with moist desquamation (blistering) over about 20 Gy, and dermal necrosis or skin loss over 30 Gy (U.S. NRC 1989).

THE ACUTE RADIATION SYNDROMES

The clinical effects of radiation are determined by biological responses depending on many variables, including dose, dose rate, and radiation energy in particular. At the cell level irradiation may lead to transient depletions of various cell lines that are important to the performance of such crucial systems as the blood-forming tissue and gastrointestinal tract. Inflammatory changes may affect the lungs, while neural transmission defects at higher levels of exposure may impair the nervous and cardiovascular systems. The clinical manifestations of the malfunctions of these organ systems can conveniently be considered as several subdivisions of the acute radiation syndrome.

Hematological Form

The hematological form is the least severe, occurring at around the LD_{50} exposure level (about 3–4 Gy without treatment). It is initiated by the prodromal symptoms of nausea and vomiting for approximately 48 hours, followed by a latent period during which there may not be any apparent clinical manifestations of radiation injury. This lasts approximately 2 weeks and is followed by increasing evidence of impairment of blood formation. The manifestations are those of infection and hemorrhage. Evidences of infections include fever, inflammations of the mucous membranes, enteritis, pneumonitis, and septicemia. Infections with viruses such as herpes and cytomegalovirus, as well as bacteria and fungi, may be life threatening in these granulocytopenic and immunosuppressed patients. The manifestations of bleeding include punctate hemorrhages in the skin, or petechiae; larger areas of bruising; nose bleeding; blood in the urine, feces, or vomitus; and internal bleeding in the abdomen, chest, or head. These are accompanied by the marked blood count abnormalities already described. Repair and repopulation of the hematopoietic stem cells leads to renewed blood cell production and release into circulation, abating the clinical signs and symptoms at about the 5th postexposure week if the individual has survived earlier complications.

Gastrointestinal Form

At somewhat higher levels of exposure, beginning at about 10 Gy, the combination of hematopoietic damage and impairment of the lining of the gastrointestinal tract leads to clinical sequelae known as the gastrointestinal form of the acute radiation syndrome. Here, the prodromal manifestations may be more severe and include diarrhea as well as the previously described phenomena. The latent period is shorter, and within a few days evidences of gastrointestinal malfunctions appear. These include bleeding in the gastrointestinal tract and signs of infection, malnutrition, and electrolyte imbalances. A number of accidentally exposed humans in whom this has occurred have had other complications and have not survived (Guskova 1988, Guskova and Baranov 1988), although vigorous supportive therapy with blood products and antibiotics has extended the lives of experimental dogs with gastrointestinal impairment (Perman et al. 1962). Autopsy findings include a severe enterocolitis with loss of the integrity of the gastrointestinal lining, resulting in denudation and ulceration of the intestinal wall.

Neurovascular Form

The most dramatic form of the acute radiation syndrome is the neurovascular form, which occurs after acute radiation exposure beginning at about 20–30 Gy. Here the prodromal symptoms are prompt and severe, accompanied by alterations in the state of consciousness ranging from apathy, lethargy, and somnolence to hyperexcitability, tremors, convulsions, and gait disturbances. The cardiovascular system is increasingly unresponsive to efforts to maintain blood pressure, and the inadequacies of circulation lead to further neurovascular malfunctions, resulting in coma and death within hours to a day or two.

Pulmonary Form

Acute radiation pneumonitis has been observed in therapeutically irradiated patients 1 to 3 months after receiving over 8 or 9 Gy to the chest (UNSCEAR 1988). It was diagnosed in seven Chernobyl accident patients. In some of the Chernobyl cases, interstitial edema and respiratory failure occurred within 3 days. Also, a slower pattern of interstitial pneumonitis was observed in some cases, developing several days before death 14–30 days after irradiation (Guskova 1988, Guskova and Baranov 1988). The definition of the syndrome from these patients' experiences is complicated by other factors such as chemotherapy for neoplastic diseases, smoke inhalation, multiple system injury, and immunosuppressive therapy.

DIAGNOSIS AND TRIAGE OF THE ACUTE RADIATION SYNDROMES

In order to deal with these radiation-induced alterations at the cell and organ level, efforts must be made to tide the individual over the period during which the organ damage results in the acute radiation syndrome. The first step in this clinical approach is the early recognition of the causative factor itself and anticipation of the severity of the resultant damage. There are a number of early evaluation procedures that will help to provide this information. They involve simple diagnostic techniques including the medical history taking and physical examination as well as a blood cell count and cytogenetic analysis. Table 12.1 presents a simple triage scheme based on clinical observations and blood count data. Table 12.2 provides a list of the appropriate procedures for early injury evaluation of potential radiation overexposure patients, with the time of onset of key findings and the minimum exposure level that would be expected to produce these changes (modified from Wald 1983). With the possible exception of the chromosome analysis, they can also be utilized in carrying out the triage of a

Table 12.1. Preliminary Evaluation of Acute Radiation Injury Following Overexposure

Evaluation Procedures

1. Observe and record time of onset of clinical signs and symptoms
2. Perform daily blood count

Findings	Potential Outcome
Nausea, vomiting, diarrhea within *minutes*. Ataxia, disorientation, shock, coma in *minutes to hours*	Neurovascular syndrome
Nausea and/or vomiting and some blood count derangement in *2 days*	Minor hematological syndrome
Marked leukocyte and lymphocyte count derangement in *3 days*	Major hematological syndrome
Diarrhea within *4 days* and marked platelet derangement within *6 to 9 days*	Gastrointestinal syndrome

TABLE 12.2. RADIATION INJURY EVALUATION PROCEDURES AND THEIR RESULTS IN RELATION TO TIME AND MAGNITUDE OF RADIATION EXPOSURE

Procedure	Finding	Time of Onset	Approx. Minimum Exposure Required
History	Nausea, vomiting	Within 48 hours	1.5 Gy
Physical exam	Erythema	Within hours-days	3.0 Gy
	Epilation	Within 2–3 weeks	3.0 Gu
Blood count and differential	Lymphocytes <1000/ mm^3	Within 24–48 hours	1.0 Gy
Chromosome analysis	Dicentrics, rings, and/or fragments	Within hours	0.10 Gy

large number of individuals in order to reserve such resources as medical personnel, blood components and other therapeutic agents for those individuals in whom treatment may enhance the likelihood of survival. The large scale use of the cytogenetic test for chromosome damage in circulating blood lymphocytes requires sufficient trained personnel as were apparently available for the triage of 154 Chernobyl workers (U.S.S.R. State Committee on the Utilization of Atomic Energy 1986) or an automated aberration scoring system (Bayley et al. 1991).

Having obtained an idea of the extent of injury and its related prognosis by the procedures described above, it is certainly helpful to have an independent evaluation of the magnitude of exposure by physical dosimetric techniques. Obviously, if the dose estimate agrees with the evidence of biological impairment one can have a greater degree of confidence in the evaluation. If not, the clinical findings should be given most reliance.

ACUTE RADIATION SYNDROMES: TREATMENT CONSIDERATIONS

Given an adequate diagnostic and prognostic evaluation by medical examination and confirmatory dosimetry, the next major consideration is the approach to medical management of the problem. I have selected information to give basis for providing useful health physics input into the management of radiation accident patients and into the process of planning and training for such contingencies.

Although the need for treatment is usually a planning assumption, it should be recognized that treatment efficacy in the acute radiation syndrome and its effect on the human $LD_{50/60}$ are not well defined. In preparing the section on acute radiation injury for the NRC's Reactor Safety Study in 1975 (U.S. NRC 1975), Watson and I used all the then-available radiation accident and medical therapy patient data to generate an estimate of the effect of treatment on the mid-lethal exposure level (Wald and Watson 1977). In our view the $LD_{50/60}$ for cases receiving minimal or no treatment was approximately 3.4 Gy, whereas supportive treatment was felt to increase that figure to approximately 5.1 Gy. Based on the medical patient population receiving 10 Gy or more of whole body exposure with subsequent bone marrow transplantation as treatment for leukemia, the estimated effect of such intensive therapy was considered to raise the

LD_{50} to some unknown level over that exposure. We used 1.05 Gy as a conservative figure since about 90% of these seriously ill patients survived at least 60 days.

In an updating (U.S. NRC 1989) of the consequence model used in the U.S. NRC's 1975 Reactor Safety Study, WASH-1400 (U.S. NRC 1975), relevant clinical data were reanalyzed and considered to support an estimate of 4.5 Gy as the human $LD_{50/60}$ *with* supportive therapy. Fliedner and colleagues (1984), suggested that supportive therapy could raise the LD_{50} by a factor of 1.5–2.0, and that the addition of bone marrow transplantation under favorable circumstances could well shift the LD_{50} value by another factor of 3.0, or in the range of 13.5–18 Gy. UNSCEAR (1988) estimated the $LD_{50/60}$ for the three treatment categories to be 2.5, 5, and probably around 9 Gy. Among the Chernobyl patients, the overall LD_{50} was about 5.2 Gy (U.S. Department of Energy 1987) but, given the large number of complicating exposure, injury, and treatment variables, this is difficult to relate to other experiences or predictions.

At the LD_{50} level of exposure, the damage to important organs including the hematopoietic and immune systems in particular leads to health impairments that, if untreated, produce fatalities from infection and hemorrhage. In the past, since there were no tested definitive means of reversing radiation-induced changes at the cell level, treatment was essentially supportive, i.e., using measures to support survival of the patient until spontaneous regeneration of the various cell lines took place, resulting in the restored functional capabilities of the various important organ systems.

Supportive Treatment

There are four general principles in the supportive therapy of the acute radiation syndrome. These are the maintenance of a "clean" environment; prophylactic oropharyngeal and gastrointestinal antibiotic therapy; use of intravenous fluids, nutrients, and blood fractions; and vigorous therapy of infections.

To maintain a "clean" environment requires attention to both the internal and external environment, particularly from a microbiological standpoint. To minimize the occurrence of infection, any ongoing evidence of infection at the outset should be treated vigorously prior to the period of hematological depression. In addition, to minimize the potential hazard of the normal microbiological population of the gastrointestinal tract, antibiotics can be used prophylactically. Various regimens utilizing oral antibacterial agents as well as antifungal agents have been recommended for this purpose (Browne et al. 1990). These are subject to change as improved pharmaceuticals replace older ones.

To control the external environment and reduce the likelihood of infections, a variety of possible procedures are available. One may subject the air in the patient's room to filtration through absolute filters, which will collect even bacteria the size of *E. coli.* The patient area may be placed under somewhat positive pressure relative to the general environment. The patient may be placed in a "life island" or other device that provides laminar flow of filtered air from the head of the patient in a horizontal direction past his feet in order to minimize the deposition of bacteria in the area of the patient. Food may be prepared in the room and can be limited to only canned or packaged materials. Minimizing perforations of the skin or mucous membranes is helpful, and unusual care should be given to maintain sterility of such perforations as are necessary.

In preparing for the hematological complications, intravenous fluids, nutrients, and blood fractions must be available. These include blood components for red cell, fresh platelet, and, possibly, white cell transfusions, as well as the laboratory techniques for the necessary typing of these materials and bone marrow. Also, laboratory procedures must be available for the analysis of any samples related to the search for possible infections with microbiological, viral, or fungal agents. The appropriate antibiotic sensitivity tests and the necessary antibiotics in sufficient dosages for vigorous bactericidal therapy must also be available. Antifungal and antiviral preparations also are needed.

Bone Marrow Transplantation

In the late 1950s, a new form of therapy, bone marrow transplantation, was developed for the treatment of hematological disorders (Thomas et al. 1975). It has now been used in almost 20,000 patients with various forms of leukemia (Gale et al. 1991), as well as in some with lymphomas and other neoplasms. The procedure now generally involves preconditioning with antineoplastic chemotherapy, total body irradiation of about 10–12 Gy in one brief exposure or about 12–15 Gy divided over 3 to 7 days, and then transfusion with bone marrow cells. In a review for the NRC's NUREG/CR-4214 report (U.S. NRC 1989) of relevant publications from 1981–1987 with sufficient pertinent data including 60-day follow-up, I found that the postirradiation 60-day survival rate was 85% to 90% in these seriously ill patients.

Bone marrow transplantation has been used in several radiation accidents for treatment of the hematological acute radiation syndrome. In the Vinca reactor accident (Jammet et al. 1959, Mathe et al. 1959) it was given on the 14th and 27th days to the one patient who died. Four survivors were transplanted between 27 and 36 days postexposure with a very prompt granulocyte increase but little lymphocyte change, suggesting spontaneous regeneration was already under way before the transplantations.

In the Pittsburgh accelerator accident (Gilberti and Wald 1991, Thomas et al. 1971, Wald et al. 1968) one of the three men involved was given marrow from his identical twin on the 8th day after the accident. An abrupt reversal of both granulocytopenia and lymphopenia began on the 19th postexposure day, culminating in complete hematological recovery. This was well before spontaneous regeneration could be expected. It occurred despite severe local radiation injuries that ultimately necessitated amputation of all four limbs (Schenk and Gilberti 1970).

In the care of about 200 of the Chernobyl reactor accident patients who were exposed to large doses of total body radiation, 13 received bone marrow transplants and 6 received hematopoietic cells from fetal liver (Baranov et al. 1989). The only two marrow recipients who survived had transient engraftments of up to 36 days. None of the fetal liver recipients survived. It must be remembered that these patients all had nonhematologic radiation injuries as well as thermal burns and trauma complicating the clinical response.

Thus although the record for therapeutic efficacy of bone marrow transplantation in radiation accident patients is not impressive, the benefit in patients with medical illnesses requiring total body irradiation and in the Pittsburgh accident case suggests that it should still be considered in accident patients with life-threatening exposures, probably in the range of 8 to 12 Gy (Browne et al. 1990). It is necessary to make an individual benefit-versus-risk judgment in each instance (Gale and Reisner 1988).

Hematopoietic Growth Factors

Fortunately, at about the same time as the bone marrow transplantation research was progressing, in the mid-1960s, it became apparent that a group of specific glycoproteins was involved in the control of growth of each of the blood cell lines (Metcalf 1985, Sachs 1987). More recently, with the advent of recombinant gene technology, sufficient amounts of various leukocyte growth factors have become available for hematological disease therapy (Nienhuis 1988). These include interleukin-3 (IL-3) (which stimulates the production of red and white blood cell and platelet precursors in the bone marrow), granulocyte colony-stimulating factor (G-CSF), and granulocyte-macrophage colony stimulating factor (GM-CSF).

Medvedev (1991) mentions the use in Chernobyl patients of "the latest 'growth factors' stimulating the multiplication of blood cells" and quotes testimony of a Chernobyl unit no. 4 shift foreman that "all those with serious or moderately serious injuries were given bone marrow transplants and 'growth factors'–medications that accelerated the growth of bone marrow cells." The first scientific report of use of a recombinant human hematopoietic growth factor, GM-CSF, involved eight patients in a large-scale cesium-137 contamination accident in Goiania, Brazil (Butturini et al. 1988). It was not available until 35 days after the onset of exposures. Increases in granulocytes were seen in the four who survived, suggesting a beneficial role in their recovery.

Recombinant GM-CSF was also used in two other accidents. One involved three patients in San Salvador, all of whom survived the hematological acute radiation syndrome (although one died after 197 days, probably from radiation pneumonitis and other complications) (International Atomic Energy Agency [IAEA] 1990). The other, in Israel, involved one patient who received a bone marrow transplant and interleukin-3 in addition, but who died in 36 days after a very high radiation exposure (IAEA 1992).

The hematopoietic growth factors are relatively new clinical tools, and there are uncertainties about their mechanisms of action, the timing of administration, and the radiation dose range over which they may be beneficial in radiation accident patients. However, their clinical toxicities are low, and the risk-benefit ratio may be favorable over a relatively broad range of radiation exposures (Gale and Butturini 1990). It has been suggested that this may be 4–8 Gy of uniform total body exposure.

The efficacy of treatment for severe gastrointestinal and pulmonary effects of external radiation exposure has not been demonstrated in accidentally irradiated humans. However, there have been many cases of medically prescribed whole body irradiation of leukemic patients at more than 10 Gy, a level at which some gastrointestinal impairment might be anticipated, in which intensive care, including bone marrow transplantation, has resulted in at least 60-day survival (U.S. NRC 1989). Also, there have been successful outcomes from careful management of experimentally induced damage of this degree in many laboratory animals (Perman et al. 1962, Vriesendorp and van Bekkum 1984). It is therefore expected that similar care, with the possible additional use of hematopoietic growth factors, should enhance the survival rate of normal people following accidental exposures at these or somewhat higher dose levels.

Acute Radiation Syndrome Treatment Resources

The very useful results of a consensus development conference on the treatment of acute radiation injuries and the resources needed have recently been published (Browne et al. 1990). The resources necessary for intensive care, including bone marrow transplantation, for such patients are presently available in a number of major medical centers in the United States. Unfortunately, no data on the number of such centers willing and able to admit radiation accident patients on an emergency basis have been found (U.S. NRC 1989). The clinical resources of several hundred hospitals whose internal medicine services are currently approved for internal medicine residency training by the American Board of Internal Medicine should be able to help with supportive therapy for this type of problem.

LOCAL TISSUE DAMAGE

Local tissue damage is the most frequent form of external radiation injury. Although the most common tissue involved is the hand, other areas include the abdomen, the male gonads, and the buttocks. The diagnosis of this type of injury rests primarily on the history of the exposure and the eventual appearance, up to several weeks later, of typical skin changes. These may be only mild erythema, although wet radiodermatitis with blister formation and radionecrosis with actual devitalization of tissue may occur depending on the source and magnitude of exposure (Saenger et al. 1980). Experimental attempts have been made to use thermography and isotopic blood flow studies to arrive at earlier diagnoses, but they are not as yet practical procedures. It must always be remembered in local tissue damage that the possibility of whole body radiation exposure or exposure of other locales, such as the eyes, must also be evaluated.

Although local radiation injury may superficially resemble a thermal burn, it must be recognized that the penetration of the radiation is generally deeper. This affects the skin's supporting tissues, resulting in poor healing and chronic breakdown of the "healed" area. The underlying pathophysiology is the over-response of the endothelial cells comprising the interior lining of the small arterioles that provide blood to the damaged area. After a period of inhibition, cell production may overcompensate and ultimately fill the lumen of the vessel, obliterating the open space through which blood must flow. This obliterative endarteritis then results in progressive impairment of the tissue deprived of its blood supply, with ultimate loss of the tissue.

The treatment of local tissue damage lies largely in the realm of the surgeon, in aiding healing by avoiding complications and in determining when the extent of impairment is such that removal of devitalized tissue becomes necessary (Schenk and Gilberti 1970). Other than surgical intervention, only palliative measures are currently available.

From the standpoint of emergency management, there are no unusual requirements other than the availability of surgical supervision, conventional wound dressing supplies, and antibiotics.

RADIONUCLIDE CONTAMINATION

In the event of external exposure, where the worst consequence may be death in a period of 60 days, there is a period of 1 week to 10 days in which one can bring together the necessary resources to prepare for its management. It is paradoxical that in radionuclide contamination incidents, the worst consequence is generally the increased risk of a delayed occurrence of leukemia or cancer in anywhere from 2–40 years, but the preparation time to deal with such patients is very short since the removal of contamination should be done as quickly as possible. Adequate preparation for such cases is particularly difficult since there are so many variables in radionuclide contamination situations, such as differing routes of entry, forms of radionuclides, metabolic behavior of associated compounds, and the like. NCRP Report No. 65 (NCRP 1980) is a useful compilation of data needed in dealing with the complexities of radionuclide contamination incidents.

Early Treatment of External Contamination

After any life-threatening problems involving cardiac arrest, impairment of respiration, or uncontrolled bleeding have been dealt with, the primary objective is the separation of the patient from the contamination to the greatest extent possible. The first step usually is physical removal of the contaminated individuals to a controlled area where decontamination procedures can be performed.

Since the physico-chemical nature of the compound involved is generally not immediately known, a series of agents that are generally helpful may be used. The sequence of their trial is based on the philosophy that the mildest intervention should be tried earliest. Progressively more vigorous procedures should be used if the previous ones are ineffective. A typical sequence of decontaminating agents in decreasing order of mildness would be soap and water, abrasive soap, detergents, oxidizing agents (e.g., Clorox), complexing agents (e.g., citric acid), and chelating agents (e.g., EDTA, DTPA).

Success in this effort lies in either the removal of all contamination or the removal of all unfixed contamination. Another indication for stopping decontamination is the emergence of marked redness and hyperemia or of abrasions of the skin, since these may enhance incorporation of the material from the surface into the body.

When removable activity has been separated from the skin of a contaminated individual, consideration should be given to determination of the presence of any internal radionuclide contamination. Bioassay measurements and whole-body counting may supplement the history and physical examination, including measures such as nasal swabs. If internal contamination is a possibility, its early treatment depends on whether its route of entry was inhalation, ingestion, or contamination deposited in a wound.

Early Treatment of Inhaled Contamination

If inhalation appears to be the most significant route of intake, prompt irrigation of the nose, mouth, and pharynx is indicated. Unfortunately, there is no simple method for enhancing the clearance of a contaminant deposited in the lung. One possibility is bronchopulmonary lavage with an appropriate antidote, which may be used if a sufficient quantity of a sufficiently hazardous and insoluble material has been inhaled to outweigh the attendant risks. The procedure requires general anesthesia as well as the services of a pulmonary specialist who is well versed in the performance of this type of lavage.

Early Treatment of Radionuclide Ingestion

The management of radionuclide ingestion also begins with irrigation of the nose, mouth, and pharynx. If a large amount of material was ingested shortly before treatment is instituted, it may be useful to remove the gastric contents. After a few hours this is no longer effective, but one may hasten the transit time of the material through the gastrointestinal tract by using a purgative. In addition, the chemical antidote for blocking, diluting, mobilizing, or chelating the contaminant may be utilized.

Early Treatment of Radionuclide-Contaminated Wounds

Extensive irrigation with water or saline as well as decontamination of the surrounding skin with the previously discussed agents are the major early treatments for radionuclide-contaminated wounds. When irrigation has resulted in the maximum reduction of the radiation level, treatment of the injury as if it were a noncontaminated one may be appropriate. In the special case where long-lived, very hazardous, insoluble material is embedded in a wound, excision may be considered. Radioactivity counting with wound probes that can be sterilized and counting of excised tissue or washwater removed are useful aids in facilitating the medical benefit-versus-risk judgments that are necessary during the performance of this type of treatment.

Internal Radionuclide Decorporation

When the local surface contamination has been removed or stabilized and internal deposition has been diagnosed and evaluated as to medical significance, more definitive removal treatment may be warranted. There are a number of chemical agents that are of particular efficacy for certain radionuclides under certain conditions. They can be classified by their functions as blocking, diluting, mobilizing, or chelating agents. These are described in detail in the previously cited NCRP Report No.65 (NCRP 1980). Table 12.3 (modified from Wald 1983) indicates some of the appropriate agents in each category used for particular radionuclides. Many of these agents are available to physicians for other purposes,

TABLE 12.3. DECORPORATION METHODS AND AGENTS FOR SOME IMPORTANT RADIONUCLIDES

Method	Isotope	Agent
Dilution	^{3}H	Water
	^{32}P	Phosphorus (Neutrophos)
Blocking	^{137}Cs	Ferric ferrocyanide (Prussian blue)
	^{131}I, ^{99m}Tc	KI (Lugol's solution)
	^{89}Sr, ^{85}Sr	Al-phosphate (Phosphojel)
		Al-hydroxide (Amphojel)
		Na-Alginate (Gaviscon)
Mobilization	^{86}Rb	Chlorthalidone (Hygroton)
Chelation	^{252}Cf, ^{242}Cm, ^{241}Am, ^{239}Pu, ^{144}Ce, rare earths, ^{143}Pm, ^{140}La, ^{90}Y, ^{65}Zn, ^{46}Sc	DTPA
	^{210}Pb	EDTA, penicillamine
	^{203}Hg, ^{60}Co	Penicillamine

but some must be freshly made.

Chelation, in particular, may accomplish the binding of many of the most hazardous transuranics and some other nuclides by converting them to a more soluble form that then is excreted in the urine, reducing the body burden. Although most effective when done very promptly after contamination, these agents have been used several years after an unsuspected internal contamination with continued success (Rosen et al. 1980). They are not available on the open market but are made available to qualified physicians by the Department of Energy as investigational drugs, requiring written informed patient consent for use.

TREATMENT CONSIDERATIONS CONCERNING PSYCHOLOGICAL STRESS

An important consideration in radiation emergencies, especially when the general public is involved, is the problem of psychological stress. The perception of the risk of death or injury from nuclear energy is disproportionately large among the general public or media personnel as compared with scientists (Rothman and Lichter 1982) or scientifically knowledgeable people (Slovic et al. 1979). Dealing with the resultant anxiety and even panic is a significant responsibility in emergency planning and management. Helpful antidotes include better education before any emergency occurs, the efficient dissemination of factual information during the emergency by a single credible source, and the provision of evidence prior to, during, and after the emergency that a practical plan for management of the problem is in place that can be implemented effectively to mitigate any detrimental impact of the occurrence. These procedures apply to individual patients as well as to the general public. The recent Oak Ridge Associated Universities (ORAU) International REAC/TS Conference on the medical basis for radiation accident preparedness in December 1990 (Ricks et al. 1991) was devoted entirely to the psychological perspective.

PHYSICIAN PREPAREDNESS FOR RADIATION ACCIDENTS

The general availability of medical personnel experienced and/or trained in the area of radiation injury management is quite limited, consisting mainly of physicians in a few academic institutions, national laboratories, private companies, and the medical employees of large nuclear industry organizations. While this meets the needs of nuclear industry workers, it may not be adequate for the general public. In addition, specialized resources such as the decontaminating and decorporating agents may not be readily available locally.

An effort is being made to prepare physicians and hospitals for radiation emergencies. Useful references include medical journal articles (Leonard and Ricks 1980; Wald 1983, 1985; Thomas 1990); conference proceedings on radiation accident preparedness (Hubner and Fry 1980, Ricks and Fry 1990); books (Browne et al. 1990, Conklin and Walker 1987, Mettler et al. 1990); and book chapters (Linnemann 1983, Saenger 1984, Wald 1982); monographs on management of radionuclide-contaminated patients (IAEA 1978, 1990, International Commission on Radiological Protection [ICRP] 1977, NCRP 1980); a self-instruction cassette program on preparedness and response in radiation accidents (Saenger et al. 1982); courses on the medical care of radiation accident patients by ORAU, Radiation Management Consultants (RMC), the University of New Mexico, and other organizations; guides to the initial hospital management of radiation accident patients distributed by the American Medical Association (AMA) (1984) and the American National Standards Institute (ANSI) (1985); and a manual on radiation accident emergency planning from the U.S. Public Health Service (Schleien 1983).

These have been available to physicians not specializing in radiation medicine in the past several years. They should help to make up for the fact that many of the nation's experienced radiation medicine specialists entered this field during the 1940s, when the military and civilian uses of nuclear energy were beginning. This cohort is now withdrawing from professional activities and may be replaced by physicians specializing in occupational medicine, emergency medicine, and family practice. The deficit in physicians with experience in radiation injuries places some responsibility on members of the health physics profession to be prepared to advise and assist their medical colleagues in the management of what hopefully will be very rare radiation emergencies.

REFERENCES

American Medical Association. *A Guide to the Hospital Management of Injuries Arising from Exposure to or Involving Ionizing Radiation.* Chicago, American Medical Association, 1984.

American National Standards Institute. *Guide for Hospital Emergency Departments on Handling Radiation Accident Patients*, Project N13.28. New York, American National Standards Institute, 1985.

Andrews, G. A. The medical management of accidental total-body irradiation. In: *The Medical Basis for Radiation Accident Preparedness*, edited by K.F. Hubner and S.A. Fry. New York, Elsevier/North Holland, 1980, pp. 297–310.

Anno, G. H., Baum, S. J., Withers, H. R., and Young, R. W. Symptomatology of acute radiation effects in humans after exposure to doses of 0.5–30 Gy. *Health Phys. 56*:821–838, 1989.

Baranov A., Gale, R. P., Guskova, A. Piatkin, E., Selidovkin, G., Muravyova, L., Champlin, R. E., Danilova, N., Yevseeva, L., Petrosyan, L., Pushkareva, S., Konchalovsky, M. Gordeeva, A., Protasova, T., Reisner, Y., Mickey, M.R., and Terasaki, P. I. Bone marrow transplantation after the Chernobyl accident. *N. Engl. J. Med. 321:*205–212, 1989.

Bayley, R., Carothers, A., Chen, X., Farrow, S., Gordon, J., Ji, L., Piper, J., Rutovitz, D. Stark, M., and Wald, N. Radiation dosimetry by automatic image analysis of dicentric chromosomes. *Mutat Res. 253:* 223–235, 1991.

Bond, V. P., Fleidner, T. M., and Archambeau, J. O.`*Mammalian Radiation Lethality.* New York, Academic Press, 1965.

Browne, D., Weiss, J. F., MacVittie, T. J., and Pillai, M. V. eds. *Treatment of Radiation Injuries.* New York, Plenum Press, 1990.

Butturini, A., De Souza, P. C., Gale, R. P., Cordiero, J. M., Lopes, D. M., Neto, C., Cunha, C. B., DeSouza, C. E. P., Ho, W. G., Tabak, D. G., Sanpai, J. M., and Burla, A. Use of recombinant granulocyte-macrophage colony stimulating factor in the Brazil radiation accident. *Lancet 2:*471–475, 1988.

Conklin, J. J., and Walker, R. I. *Military Radiobiology.* Orlando, FL, Academic Press, 1987.

Finch, S. C. Acute radiation syndrome. *J.A.M.A. 258:*664–667, 1987.

Fliedner, T. M. Hematological indicators to predict patient recovery after whole-body irradiation as a basis for clinical management. In: *The Medical Basis for Radiation Accident Preparedness. II. Clinical Experience and Follow-up Since 1979,* edited by R.C. Ricks and S.A. Fry. New York, Elsevier, 1990, pp. 445–460.

Fliedner, T. M., Nothdurf, W., and Heit, H. Biological factors affecting the occurrence of radiation syndromes. In: *Response of Different Species to Total Body Irradiation,* edited by J.J. Broerse and T.J. MacVittie. Boston, Martinus Nijhoff Publishers, 1984, pp. 209–219.

Gale, R. P., and Butturini, A. Use of hematopoietic growth factors in radiation accidents. *Int. J. Radiat. Oncol. Biol. Phys. 19:*1291–1295, 1990.

Gale, R. P., Butturini, A., and Bortin, M. M. What does total body irradiation do in bone marrow transplants for leukemia? *Int. J. Radiat. Oncol. Biol. Phys. 20:*631–634, 1991.

Gale, R. P., and Reisner, Y. The role of bone marrow transplants after nuclear accidents. *Lancet 1:*923–926, 1988.

Gerstner, H. B. Acute clinical effects of penetrating nuclear radiation. *J.A.M.A. 168:*381–388, 1958.

Gilberti, M. V., and Wald, N. The Pittsburgh radiation accident: twenty-three year follow-up of clinical and psychological aspects. In: *The Medical Basis for Radiation Accident Preparedness III: Psychological Perspectives,* edited by R.C. Ricks, M.E. Berger, and F.M. O'Hara. New York, Elsevier, 1991, pp. 199–206.

Guskova, A.K., ed. Appendix: Acute radiation effects in victims of the Chernobyl muclear power plant accident. In: United Nations Scientific Committee on the Effects of Atomic Radiation. *1988 Report to the General Assembly.* New York, United Nations, 1988, pp. 613–631.

Guskova, A. G., and Baranov, A. E. Treatment of acute radiation disease (ARD)–experience from Chernobyl. In: *Berzelius Symposium XV,* Umed, 1988, pp. 157–185.

Hubner, K. F., and Fry, S. A., eds. *The Medical Basis for Radiation Accident Preparedness.* New York, Elsevier/North Holland, 1980.

International Atomic Energy Agency. *Manual on Early Medical Treatment of Possible Radiation Injury.* Vienna, International Atomic Energy Agency Safety Series 47, 1978.

International Atomic Energy Agency. *The Radiological Accident in Sor Van, Israel.* Vienna, International Atomic Energy Agency, in preparation, 1992.

International Commission on Radiological Protection. *The Principles and General Procedures for Handling Emergency and Accidental Exposures of Workers,* Publication No. 28. New York, Pergamon, 1977.

Jammet, H., Mathe, G., Pendic, B., Duplan, J.-F., Maupin, B., Latarjet, R., Kalic, D., Schwartzenberg, L., Djukic, Z., and Vigne, F. Etude de six cas d'irradiation totale aigue accidentelle. *Rev. Franc. Etudes Clin. Biol. 4:*210–225, 1959.

Keller, P.D. A clinical syndrome following exposure to atomic bomb explosions. *J.A.M.A. 131:*504–506, 1946.

Leonard, R. B., Ricks, R. C. Emergency department radiation accident protocol. *Ann. Emerg. Med. 9:*462–470, 1980.

Linnemann, R. E. A systems approach to the initial management of radiation injuries. In: *Systems Approach to Emergency Medical Care,* edited by D. Boyd, R.F. Edlich, S. Micik. Norwalk, CT, Appleton-Century-Crofts, 1983, 341–369.

Luthra, Y. K., Mattson, J. L., and Yochmowitz, M. G. Inhibition of radioemesis by catecholamines in dogs. *Radiat. Res. 85:*583–591, 1981.

Mathe, G., Jammet, H., Pendic, B., Schwarzenberg, L., Duplan, J.-F., Maupin, B., Latarjet, R., Larrieu, M.-J., Kalic, D., and Djukic, Z. Transfusions et greffes do moelle osseuse homologue chez des humains irradies a haute dose accidentellment. *Rev. Franc. Etudes Clin. Biol. 4:*226–238, 1959.

Medvedev, G. *The Truth about Chernobyl.* New York, Basic Books, 1991.

Metcalf, D. The granulocyte-macrophage colony-stimulating factors. *Science 229:*16–22, 1985.

Mettler, F. A., Kelsey, C. A., and Ricks, R. C. *Medical Management of Radiation Accidents.* Boca Raton, FL, CRC Press, 1990.

National Council on Radiation Protection and Measurement. *Management of Persons Accidentally Contaminated with Radionuclides,* NCRP Report No. 65. Washington, DC, National Council on Radiation Protection and Measurements, 1980.

Nienhuis, A. W. Hematopoietic growth factors. *N. Engl. J. Med. 318:*916–918, 1988.

Perman, V., Cronkite, E. P., Bond, V. P., and Sorensen, D. K. The regenerative ability of hemopoietic tissue following lethal x-irradiation in dogs. *Blood 19:*724–737, 1962.

Ricks, R. C., and Fry, S. A., eds. *The Medical Basis for Radiation Accident Preparedness II. Clinical Experience and Follow-up Since 1979.* New York, Elsevier, 1990.

Ricks, R.C., Berger, M.E., and O'Hara, F., eds. *The Medical Basis for Radiation Accident Preparedness. III: Psychological Perspective.* New York, Elsevier, 1991.

Rosen, J. C., Gur, D., Pan, S. F., and Wald, N. Long-term removal of ^{241}Am using Ca-DTPA. *Health Phys. 39:*601–609, 1980.

Rothman, S., and Lichter, S. R. The nuclear energy debate: scientists, the media and the public. *Public Opinion Aug/Sept:*47–52, 1982.

Sachs, L. The molecular control of blood cell development. *Science 238:*1374–1379, 1987.

Saenger, E. L. The nuclear physician's role in planning for and handling radiation accidents. In: *Nuclear Medicine Annual 1984,* edited by L.M. Freeman and H.W. Weissman. New York, Raven Press, 1984, pp. 1–22.

Saenger, E. L., Andrew, G. A., Linnemann, R. E., and Wald, N. *Radiation Accident Preparedness–Medical and Managerial Aspects. An Audio-Visual Self-Instruction Course.* New York, Science-thru-Media, 1982.

Saenger, E. L., Keriakes, J. G., Wald, N., and Thoma, G. E. Clinical courses and dosimetry of acute hand injuries to industrial radiographer from multicurie sealed gamma sources. In: *The Medical Basis for Radiation Accident Preparedness,* edited by K.F. Hubner and S.A. Fry. New York, Elsevier/North Holland, 1980, pp. 327–333.

Schenk, R., and Gilberti, M. V. Four extremity radiation necrosis. *Arch. Surg. 100:*729–734, 1970.

Schleien, B. *Preparedness and Response in Radiation Accidents.* Rockville, MD, National Center for Devices and Radiological Health, U.S. Public Health Service, 1983.

Slovic, P., Fischoff, B., and Lichtenstein, S. Rating the risks. *Environment 21:*14–20, 36–39, 1979.

Thoma, G. E., and Wald, N. The diagnosis and management of accidental radiation injury. *J. Occup. Med. 1:*421–447, 1959.

Thomas, E. D. The role of bone marrow grafts in irradiation accidents. *Int. J. Radiat. Oncol. Biol. Phys. 19:*1289–1290, 1990.

Thomas, E. D., Rudolph, R. H., Fefer, A., Storb, R., Slichter, S., and Buckner, C. D. Isogenic marrow grafting in man. *Exper. Hematol. 21:*16–17, 1971.

Thomas, E. D., Storb, R., Clift, R. A., Fefer, A., Johnson, F. L., Neiman, P. E., Lerner, K.G., Glucksberg, H., and Buckner, C. D. *Bone-marrow transplantation. N. Engl. J. Med. 292:*832–843, 1975.

United Nations Scientific Committee on the Effects of Atomic Radiation. *1988 Report to the General*

Assembly. Annex G. Early Effects in Man of High Doses of Radiation. New York, United Nations, 1988, 545–647.

U.S. Nuclear Regulatory Commission. *Reactor Safety Study: An Assessment of Accident Risks in U.S. Commercial Nuclear Power Plants, Appendix VI,* WASH-1400 (NUREG-75/014). Springfield, VA, National Technical Information Service, 1975.

U.S. Nuclear Regulatory Commission. *Health Effects Models for Nuclear Power Plant Accident Consequence Analysis. Low LET Radiation. Part II: Scientific Bases for Health Effects Models,* NUREG/CR-4214 (SAND85–7185), Rev.1, Part II. Albuqurque, NM, Sandia National Laboratory, 1989.

U.S.S.R. State Committee on the Utilization of Atomic Energy. The accident at the Chernobyl nuclear power plant and its consequences. *Working Document for International Atomic Energy Agency Chernobyl Postaccident Review Meeting, 25–29 August 1986,* Part 1, Annes 7. Moscow, U.S.S.R. State Committee on the Utilization of Atomic Energy, 1986, pp. 1–49.

Van Dyk, J., Keane, T. J., Kan, S., Rider, W. O., and Fryer, C. J. Radiation pneumonitis following large single dose irradiation: a reevaluation based on absolute dose to lung. *Int. J. Radiat. Oncol. Biol. Phys. 7:*461–467, 1981.

Vriesendorp, H. M., and van Bekkum, D. W. Susceptibility to total body irradiation. In: *Response of Different Species to Total-Body Irradiation,* edited by J.J. Broerse and T.J. MacVittie. Boston, Martinus Nijhoff, 1984, pp. 43–57.

Wald, N. Haematological parameters after acute radiation injury In: *Manual on Radiation Haematology.* Vienna, International Atomic Energy Agency, 1971, pp. 253–264.

Wald, N. Radiation injury. In: *Cecil Textbook of Medicine,* 16th Ed., edited by J.B. Wyngaarden and L.H. Smith. Philadelphia, WB Saunders, 1982, pp. 2228–2234.

Wald, N. Diagnosis and therapy of radiation injuries. *Bull. N.Y. Acad. Med. 59:*1129–1138, 1983.

Wald, N. Radiation accidents and their emergency management. In: *Emergency and Disaster Medicine. Proceedings of the Third World Congress, Rome, 1983,* edited by C. Manni and S.I. Magalini. New York, Springer-Verlag, 1985, pp. 61–70.

Wald, N., Pan, S., and Thomas, E. D. Cytogenetic observations in accidental human radiation injury treated by bone marrow transplantation. In: *12th International Congress of the International Society of Hematology,* Abstract A-7:4. Boston, International Society of Hematology, 1968.

Wald, N., and Watson, J. A. Medical modification of human acute radiation injury. In: *Proceedings of IVth International Congress of International Radiation Protection Association,* 24–30 April 1977, Paris. Paris, International Radiation Protection Association, 1977, pp. 1183–1190.

Young, R. W. Mechanism and treatment of radiation-induced nausea and vomiting. In: Davis, C.J., Lake-Bakaar, G.V., and Grahame-Smith, D.G., eds. *Nausea and Vomiting: Mechanisms and Treatment.* New York, Springer-Verlag, 1986, pp. 94–109.

Chapter 13

Modulation of Mortality by Tissue Trauma and Sepsis in Mice after Radiation Injury

G. DAVID LEDNEY, THOMAS B. ELLIOTT, AND MARCUS M. MOORE

ABSTRACT

The nuclear disasters at Hiroshima, Nagasaki, and Chernobyl underscore the need for useful animal models to (*a*) evaluate the combined effects of radiation and tissue trauma and (*b*) develop successful therapeutic modalities for sepsis in irradiated individuals inflicted with tissue trauma. In mice, mortality subsequent to radiation and tissue trauma depends on the (*a*) timing of the trauma relative to radiation, (*b*) dose and quality of radiation, (*c*) nature of the inflicted trauma (burn or wound), (*d*) genetic makeup of the host, and (*e*) microbiologic agents associated with the host. Therapies for sepsis after wound trauma were developed in gamma ray- and neutron-irradiated mice. Single-agent therapy for infections with antimicrobials or immunomodulators is not as useful as combined modality therapy with antimicrobials and immunomodulators. Topical treatment of the injury with antimicrobials in addition to systemic therapy with antimicrobials or immunomodulators is necessary to effect survival. Sepsis in mice subsequent to neutron irradiation and wound trauma was more difficult to treat than sepsis after gamma ray exposure and wounding. The increased biological effectiveness of neutrons compared to gamma rays for radiosensitive tissues makes therapy for sepsis less successful in neutron-irradiated hosts.

INTRODUCTION

Approximately 100 years ago Roentgen discovered x-rays and Becquerel discovered radioactivity. Since that time man has attempted to harness the atom for civilian as well as military purposes. The promise of the expanded peaceful use of nuclear energies for diagnosis of disease, therapy for malignancies, and unlimited electrical power has been clouded by major accidents where radionuclides have been released into the environment from power reactors (Three Mile Island, United States, and Chernobyl, U.S.S.R.) and from abandoned medical radiation devices (Juarez, Mexico, and Goiania, Brazil). The increased risk to human health as well as the loss of life in Chernobyl and Goiania have had sobering influences upon the world.

Nearly 50 years ago nuclear weapons were developed in the United States and employed in Hiroshima and Nagasaki with devastating effects. Today many nations control nuclear weapons or reactors capable of producing weapon-grade material. In addition to these devices there are numerous medical centers and power plants using radiation for peaceful purposes. While these radiation resources are under apparent control, the possibility that accidents will occur or that terrorists will obtain and use these devices is of increasing concern.

The expanded probability of exposure to radiation environments requires the development of understanding of both the damage induced by radiation and the treatments available to counteract that damage. Further, the treatments provided must be tailored for the additional specific injuries expected or associated with radiation damage (Browne et al. 1990). For example, burn trauma in the individuals exposed to radiation at Chernobyl complicated therapeutic efforts (Baranov et al. 1989). Burns and mechanical traumas to the irradiated inhabitants of Hiroshima and Nagasaki contributed heavily to mortality (Fujita et al. 1990). In retrospective analyses of mortality after the Hiroshima event, burn and wound traumas in irradiated casualties may have resulted in underestimates of the LD_{50} (Fujita et al. 1989).

Irradiated individuals with mechanical or burn injuries are termed "combined injured." Combined injury is defined here as the infliction of tissue damage on an individual when (*a*) one of the injuries is radiation and (*b*) the recovery and repair from it or the other injury has not taken place before the occurrence of the second. In this chapter, we summarize our findings on factors contributing to mortality after combined injury in mouse models. In addition, we show that survival is possible in lethal models of combined injury in mice if hematopoietic recovery is augmented and if overwhelming sepsis is controlled.

MATERIALS AND METHODS

Animals

B6CBF1/CUM female mice were obtained from Cumberland View Farms in Clinton, Tennessee. B6D2F1/J female mice were obtained from Jackson Laboratory in Bar Harbor, Maine. C3H/HeN female mice were obtained from the National Cancer Institute Animal Breeding Facility in Frederick, Maryland. The mice were maintained as previously described (Madonna et al. 1991). Research was conducted in a facility accredited by the American Association for Accreditation of Laboratory Animal Care (AAALAC). All procedures involving animals were reviewed and approved by an institutional animal care and use committee.

Radiation

Detonation of a nuclear device will result in the release of neutrons and gamma rays. The proportion of each of these received by an exposed individual will depend on such factors as type and yield of the weapon and distance from the epicenter. To simulate nuclear detonation environments, irradiations with mixed fission neutron and gamma photons were performed using the Armed Forces Radiobiology Research Institute (AFRRI) Training, Research, and Isotope, General Atomics (TRIGA) Mark-F research reactor. This reactor is a movable-core pool-type facility with maximum operational steady-state power of 1 MW. All reactor irradiations were performed at a total dose rate of 38 cGy/min by altering either experimental animal placement in the radiation field, reactor power, or shielding design. Desired neutron to gamma ratios were obtained by varying shielding configurations of water, lead, borated polyethylene, and paraffin at selected reactor powers. The total dose rate varied less than 2% over the entire radiation field. Mice were irradiated in aerated aluminum tubes that rotated at 1.5 rpm.

Gamma ray exposures such as that possible at a reactor site accident or in a fallout zone were simulated by irradiation with cobalt-60. B6CBF1 and B6D2F1 mice were irradiated bilaterally at 40 cGy/min in the AFRRI ^{60}Co radiation facility. C3H/HeN mice were irradiated unilaterally at 40 cGy/min with a ^{60}Co Theratron unit. All irradiations of mice were done in aerated Plexiglas restrainers. The tissue/air ratio was 0.988 for bilateral ^{60}Co irradiation and 0.98 for unilateral ^{60}Co irradiation. Dosimetric techniques for measuring reactor- and ^{60}Co-produced radiations were described elsewhere (Zeman and Ferlic 1984).

Skin Injuries

Mice were anesthetized by inhalation of methoxyflurane before injury. Full-thickness, nonlethal skin wounds of varying sizes, from 7% to 15% of the total body surface area (TBSA), were inflicted by removing a section of dorsal skin fold and underlying panniculus carnosus muscle with a steel punch. Details for inflicting skin wounds have been previously published (Madonna et al. 1991). Full-thickness burns were inflicted on the shaved dorsal surface area by a 12-second ignition of 95% ethanol (Steritz et al. 1982). The burn site varied in size from 7% to 15% of the TBSA.

Bacteria

All endogenously acquired bacteria found in either normal, injured, irradiated, or combined-injured mice and isolated on phenylethanol agar or MacConkey's agar were identified by using combinations of Gram's stain, colony morphology, and specific biochemical tests (Lennette et al. 1985).

Treatment Agents

Immunomodulators are substances that, when used, alter nonspecific or specific immune functions within the host. Synthetic trehalose dicorynomycolate (S-TDCM) was a gift of Ribi ImmunoChem Research, Inc., in Hamilton, Montana. S-TDCM activates nonspecific host resistance against induced bacterial infections in gamma ray- (Madonna et al. 1989) and neutron-irradiated mice (McChesney et al. 1990). Therefore, we examined its usefulness in endogenously acquired infections in combined injured mice. The method of S-TDCM preparation was previously described (Madonna et al. 1991). A dose of 200 μg S-TDCM was given intraperitoneally (ip) in 0.5 ml of 0.2% Tween-80 saline.

Oxacillin sodium, gentamicin sulfate, ofloxacin (R. W. Johnson Pharmaceutical Research Institute, Raritan, NJ), and ceftriaxone sodium (Hoffman-LaRoche,

Inc., Nutley, NJ) were used systemically. All these antimicrobials were prepared in pyrogen-free water, and 0.1 ml was injected subcutaneously (sc) above the right or left gluteus medius of each mouse daily for 10 consecutive days. The daily dose of oxacillin was 150 mg/kg; gentamicin, 7.5 mg/kg; ofloxacin, 40 mg/kg; and ceftriaxone, 75 mg/kg. Garamycin cream (0.1% gentamicin sulfate) was obtained from Schering Corp. in Kenilworth, New Jersey; a generic 0.1% gentamicin sulfate was obtained from Goldline Labs in Fort Lauderdale, Florida. The creams were applied once daily for 10 consecutive days in 0.5 to 0.7-g amounts sufficient to cover the wounded sites.

Statistical Analyses

Survival data of mice in experimental groups were obtained for 30 days after irradiation. Comparisons were made by the generalized Savage (Mantel-Cox) procedure (Lee 1980). Probit analysis of numbers of mice surviving 30 days was made on log-transformed doses (Finney 1971, 1979).

RESULTS

Influence of Radiation Quality, Radiation Dose, and the Type of Skin Trauma on Survival of Combined-Injured Mice

In uncontrolled nuclear radiation environments, individuals may be exposed to either alpha, beta, gamma, or neutron particles. External contamination of the injury site with alpha and beta emitters does not normally constitute an immediate life-threatening situation for the individual. Internal and external contamination with radionuclides, while presenting a possibility of increased risk for life shortening and malignancy, does not usually preclude immediate lifesaving treatments for personnel with combined injury. Rather, burn or wound trauma in conjunction with exposure to neutron or gamma ray irradiation or combinations thereof pose the more serious life-threatening situation.

To simulate a variety of radiation environments, groups of B6D2F1 mice were irradiated with three different neutron/gamma dose ratios (n/τ) produced by the reactor. The n/τ's employed were 0.33, 1, and 19. To simulate exposure in a high gamma ray fallout field, groups of mice were irradiated with "pure" gamma radiation either from a ^{60}Co source or the reactor. Nonlethal 2.5-cm by 3.8-cm (15% TBSA) burns or wounds were inflicted 1 to 2 hours after each exposure, and 30-day survival responses were compared to control mice that were irradiated but uninjured. Thus complete dose-response survival curves were obtained at each radiation quality with each type of injury. Radiation doses lethal to 50% of mice in 30 days ($LD_{50/30}$) are presented in Figure 13.1. In irradiated animals, the $LD_{50/30}$ is one endpoint used to compare relative sensitivity or resistance to radiations of different quality. In all groups of mice, i.e., control, irradiated, irradiated and burned, and irradiated and wounded, the $LD_{50/30}$ decreased as the proportion of neutrons in the total dose increased. At each n/τ, postirradiation burn trauma and wound trauma reduced the $LD_{50/30}$ from the comparable radiation control group about 10% and 20%, respectively.

Injury subsequent to radiation not only increases the mortality incidence (i.e.,

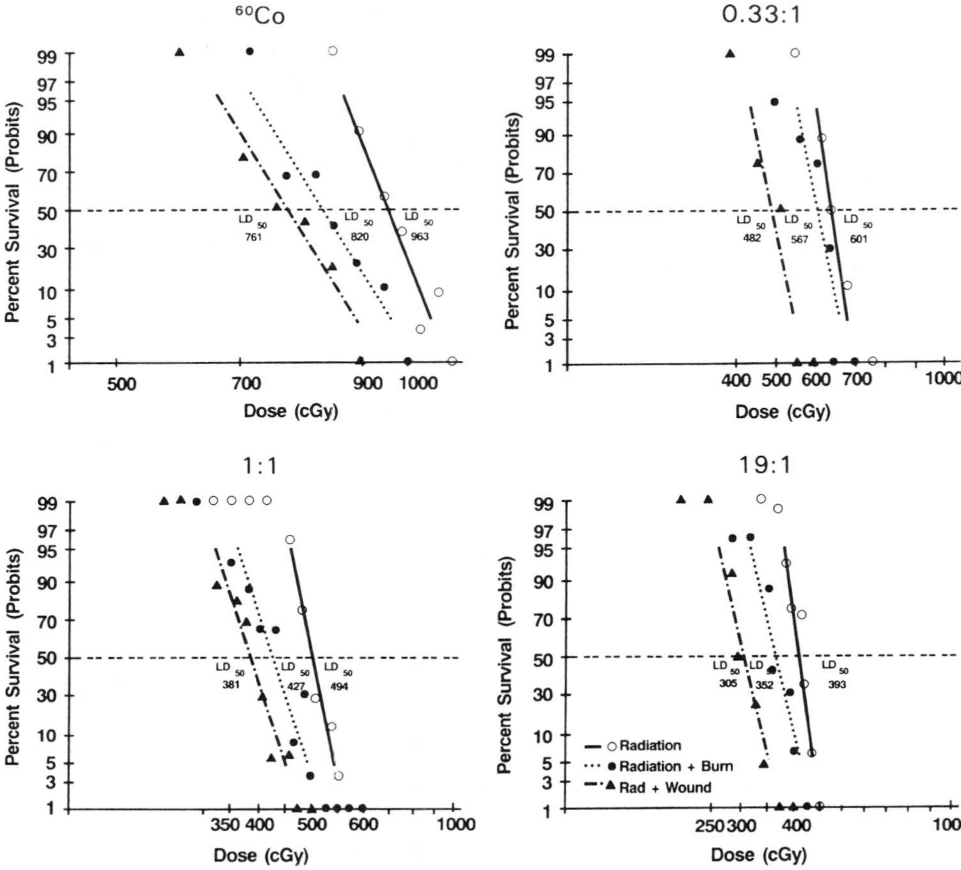

FIG. 13.1. Radiation doses required to produce 50% lethality in mice. All B6D2F1 mice were irradiated at 0.4 Gy/min, and $LD_{50/30}$ values were statistically determined from complete dose-response survival curves. The *open circles* indicate survival for irradiated mice. The *closed circles* indicate survival for mice given 15% TBSA burn trauma 1 to 2 hours after irradiation. The *triangles* indicate survival for mice given 15% TBSA wound trauma 1 to 2 hours after irradiation.

reduces the $LD_{50/30}$) but also reduces the survival time of mice that died. Plotted in Figure 13.2 are the mortality percentages of mice dying after irradiation with various proportions of neutron and gamma rays and or irradiated mice inflicted with burn or wound trauma. Except in isolated instances, the greatest mortality occurred during the 1st week after receiving a high proportion of neutrons or when wound trauma was inflicted after irradiation. When mice were irradiated with an $n/\tau = 1$ or less, or when burn trauma had been inflicted after irradiation, the majority of mortality occurred in the 2nd week postirradiation.

INFLUENCE OF TIME OF INJURY RELATIVE TO IRRADIATION ON SURVIVAL OF COMBINED-INJURED MICE

In a series of studies, three groups of B6CBF1 mice were irradiated with three (9, 10, or 11 Gy) lethal doses of ^{60}Co, respectively. At eight time points before

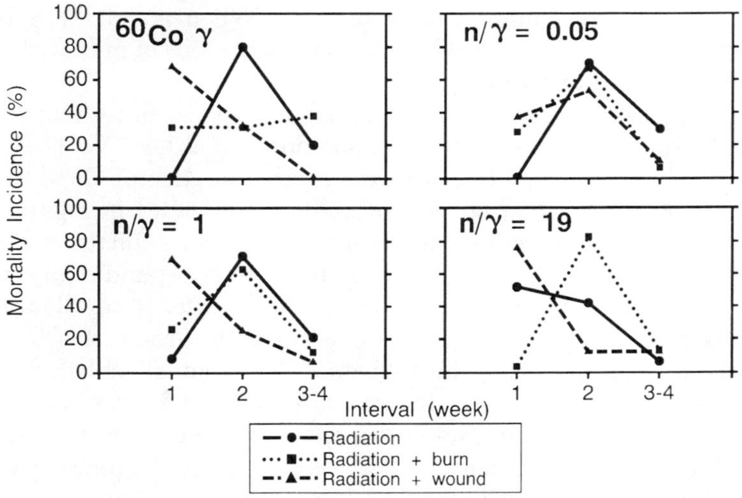

FIG. 13.2. Incidence of mortality of mice irradiated with ^{60}Co-produced gamma rays or reactor-produced neutrons and gamma rays. The mortality percentages are based on a total of 30 to 60 mice irradiated, irradiated and burned, or irradiated and wounded at their respective $LD_{50/30}$'s at the indicated gamma ray or neutron to gamma ratios.

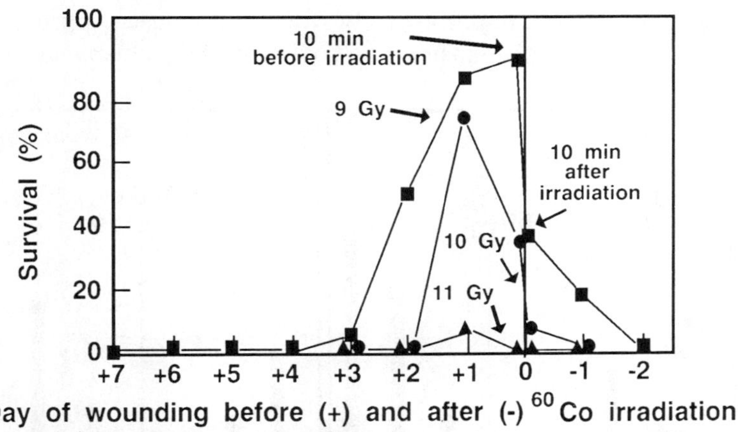

FIG. 13.3. Thirty-day survival of B6CBF1 female mice (n = 16/time point) after receiving 1.3-cm by 1.9-cm skin wounds and ^{60}Co irradiation. All control irradiated mice died; all control wounded mice lived. (From Ledney, G. D., Exum E. D., Jackson, W. E., III. Wound-induced alterations in survival of ^{60}Co irradiated mice: importance of wound timing. Experientia 1985;41:614–616.)

irradiation and three time points afterward, 1.3-cm by 1.9-cm skin wounds were inflicted. Survival data are shown in Figure 13.3. The 30-day survival of animals wounded before irradiation increased as the time interval between injury and irradiation shortened. Further, the number of 30-day survivors increased when mice were wounded within 10 minutes after a nominally lethal radiation dose (9 Gy). Survival times of all control-irradiated mice that died within the 30-day period were 11 to 14 days. However, survival was extended to 17 to 20 days for mice irradiated with nominally lethal doses (9 Gy) when wounds were inflicted from 2 days before to 10 minutes after exposure. When wounds were inflicted 1 or 2 days after irradiation, the survival times for all groups of mice with combined injury were decreased to about 7 days.

In a second series of experiments, a comparative mortality study was done with B6D2F1 mice irradiated with sublethal doses of either ^{60}Co (7 Gy) or with an n/τ = 19 (3 Gy). Groups of mice were injured at each of three time points either before or after irradiation. Mortality data are shown in Figure 13.4. The incidence of death from combined injury was greater for animals irradiated with the high n/τ (19) than with ^{60}Co. In all irradiated mice, wound injury resulted in more deaths than burn injury. Burn or wound injury after irradiation resulted in more deaths than injuries given before exposure. As seen in the first series of experiments, injuries occurring shortly before (10 minutes) sublethal irradiation resulted in fewer mortalities than injuries inflicted 1 or 2 days before radiation exposure. In the second group of studies, a relatively similar decrease in mortality incidence was noted for animals injured soon after (10 minutes) irradiation compared to injuries inflicted 1 or 2 days later.

In a third set of experiments, we determined if there was a positive correlation between wound-induced survival and the number of endogenous colony-forming units found on the spleen (E-CFU-S). The E-CFU-S assay is based on the survival and proliferation of hematopoietic cells forming discrete nodules on the surface of the spleen in an appropriately irradiated mouse. Approximately 8 to 14 days after irradiation the spleen is removed and histochemically stained and

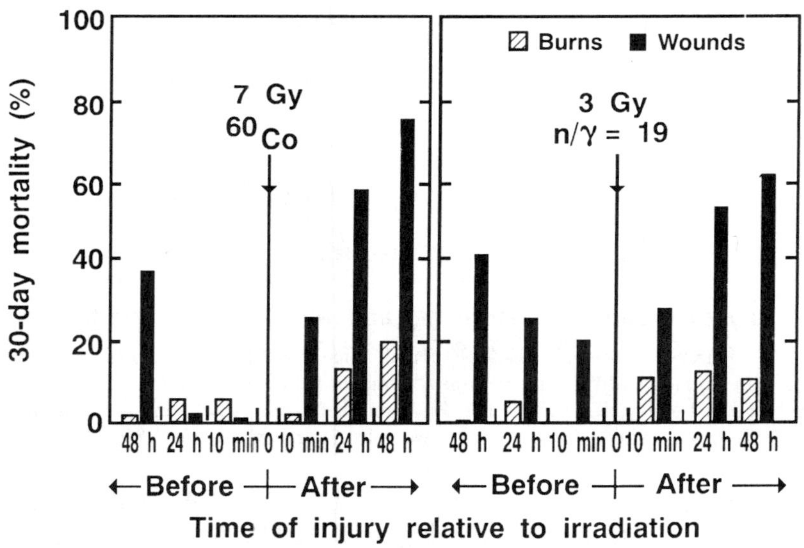

FIG. 13.4. Thirty-day mortality of B6D2F1 female mice after either 7 Gy ^{60}Co (20 mice/time point) or 3 Gy mixed field (neutron/gamma ratio of 19; 24 mice/time point) irradiation and administration of 2.5-cm by 3.8-cm skin wounds or skin burns. In these experiments, less than 4% of mice died in burned, wounded, gamma-irradiated, or neutron-irradiated control groups. (From Ledney, G. D., Madonna, G. S., McChesney, D. G., Elliott, T. B., and Brook, I. Complications of combined injury. Radiation damage and skin wound trauma in mouse models. In: Treatment of radiation injuries. New York: Plenum Press, 1990)

preserved. The number of nodules are then counted. The E-CFU-S assay has been used as an indicator for chemical agents useful for protecting against radiation-induced hematopoietic failure in mice (Kinnamon et al. 1980). Along these lines S-TDCM increased E-CFU-S in both gamma ray and neutron-irradiated mice (Stewart et al. 1991). However, no quantitative relationship between colony number and survival is suggested. Thus to determine if survival from combined injury positively correlated with increases in E-CFU-S, groups of B6CBF1 mice were irradiated with either 9, 10, or 11 Gy. Skin wounds (1.3 cm by 1.9 cm) were inflicted either 2 days, 1 day, or 10 minutes before irradiation, or 10 minutes or 1 day after irradiation. The mean values of 10-day E-CFU-S for 12 to 16 mice per treatment group are reported in Figure 13.5. A positive correlation was noted between the survival from combined injury (Fig. 13.3) and the number of spleen colonies. Thus in combined-injury situations employing nominally lethal radiation doses (9 Gy), trauma increased the hematopoietic proliferative compartments of irradiated mice.

Influence of Genetic Strain on Survival of Combined Injured Mice

The survival-mortality responses of mice to x-irradiation is controlled by their gene makeup (Kohn and Kallman 1956). We irradiated inbred C3H/HeN mice and hybrid B6D2F1 mice with mixed field ($n/\tau = 1$) or ^{60}Co gamma radiation to

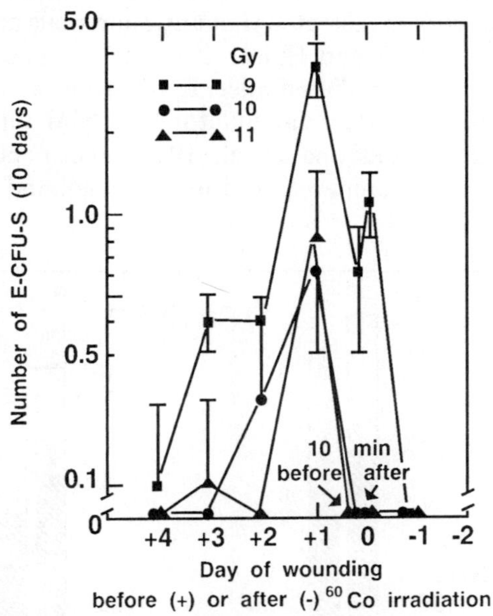

FIG. 13.5. Number of endogenous E-CFU-S in wounded irradiated mice. The time points tested were based on the survival studies in Figure 13.3. No E-CFU-S were found in control irradiated mice or in control wounded mice. Data presented in Figures 13.3 and 13.5 were obtained from all mice exposed to radiation at the same time. (From Ledney, G. D., Exum E. D., Jackson, W. E., III. Wound-induced alterations in survival of ^{60}Co irradiated mice: importance of wound timing. Experientia 1985;41:614–616.)

determine if survival responses to subsequent wound trauma were different in these two mouse strains.

Groups of B6D2F1 and C3H/HeN mice were given doses ranging from 250 cGy to 600 cGy (n/τ = 1). A comparative study was done with these mouse strains given ^{60}Co gamma photons in doses ranging from 550 cGy to 1100 cGy. In both sets of experiments, additional groups of mice received skin wounds 1 hour after irradiation. Figure 13.6 presents the $LD_{50/30}$ and relative biological effectiveness (RBE) for all of these experiments. $LD_{50/30}$ values for B6D2F1 mice were higher than for C3H/HeN mice, indicating resistance to the lethal effects of radiation or radiation wounding. Wound trauma significantly (p<.009) decreased slope values for a n/τ = 1 and gamma-irradiated B6D2F1 mice and for gamma-irradiated C3H/HeN mice. The slope for n/τ = 1 irradiated C3H/HeN mice (29.1) was not changed (p<.22) by inflicting wound trauma. Slope values for B6D2F1 and C3H/HeN mice were similar after gamma irradiation (35.8 and 37.6, respectively) and gamma irradiation with subsequent wound trauma (22.3 and 20.8, respectively). Slope values for these strains of mice were significantly different after n/τ = 1 irradiation (40.3 versus 23.4, p = .0002). The p value was 0.096 between the strains given n/τ = 1 radiation and wound trauma.

Sepsis in Irradiated Mice after Wound Trauma

Antimicrobial defenses are compromised by radiation, and death from sepsis (i.e., spread of bacteria or their products from a focus of infection) may occur if injury is severe. We have demonstrated that antimicrobials are useful in managing endogenously derived (Brook and Elliott 1991) or exogenously induced (Brook and Ledney 1990) sepsis in irradiated mice. In addition, we recently reported our findings on the use of the immunomodulator S-TDCM for managing sepsis in gamma-irradiated mice (Madonna et al. 1989) and neutron-irradiated mice (McChesney et al. 1990). Compared to animals irradiated only, little research

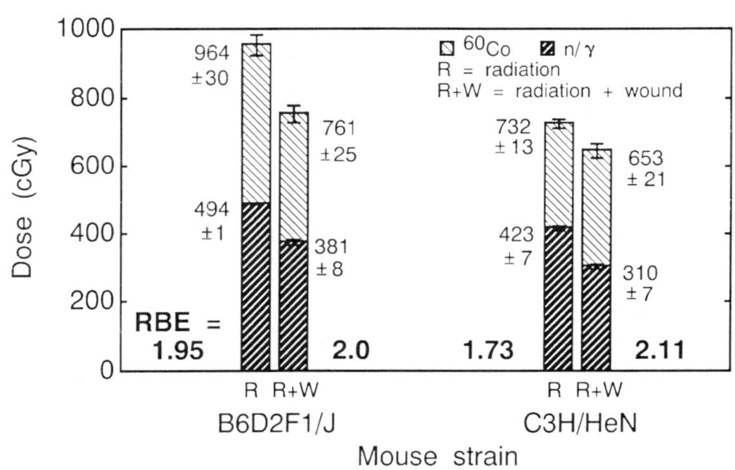

FIG. 13.6. $LD_{50/30}$'s and relative biological effectiveness (RBE) values of normal and skin-wounded mice after mixed-field (n/τ = 1) or ^{60}Co irradiation. The RBE values were calculated at $LD_{50/30}$.

has been done on the management of sepsis in combined-injured animal models (Elliott et al. 1990).

In the first series of studies, we determined the species and relative incidence of bacteria on the wound and in the liver of irradiated-wounded, wounded, and irradiated mice (Table 13.1). S-TDCM alone and combined with the antimicrobial ofloxacin, a new fluoroquinolone, were evaluated as therapies for infections. We previously demonstrated that ofloxacin (40 mg/kg/day per os (po) for 7 consecutive days starting 1 day after po challenge with *Klebsiella pneumoniae* was effective against exogenously induced infection in lethally irradiated (8.0 Gy) B6D2F1 female mice (Brook and Ledney 1991).

Gram-negative and gram-positive bacteria were found on the wound site of irradiated-wounded mice treated with S-TDCM or treated with saline and were also found on the wound site of irradiated mice, while only gram-positive bacteria were found on the wound site of nonirradiated animals. No gram-negative bacteria were found either on the wound site or translocated to the liver in mice treated with ofloxacin or ofloxacin and S-TDCM–early (day 4 to 5) mortality

TABLE 13.1. ISOLATION OF BACTERIA FROM WOUNDS AND LIVERS OF GAMMA-IRRADIATED-WOUNDED MICE AFTER S-TDCM AND OFLOXACIN THERAPY[a]

Experimental Group	4–5 Days of Culture		6–11 Days of Culture	
	Wound	Liver	Wound	Liver
Treatment				
Saline or S-TDCM	*S. aureus* *S. faecium* *E. coli* *P. mirabilis*	*P. mirabilis*	[b]	[b]
Ofloxacin	*S. aureus* *S. faecium*	[c]	*S. aureus* *S. xylosus*	*S. aureus* *A. viridans* *S. faecium* *S. faecalis*
S-TDCM/ofloxacin	*S. aureus* *S. faecium*	*S. faecium* *S. aureus*	*S. aureus* *S. xylosus* *S. faecium*	*S. aureus* *S. faecium* *A. viridans* *S. faecalis*
Control				
Wounded	*S. aureus* *S. xylosus* *Streptococcus spp.* *S. epidermidis*	[c] *Streptococcus spp.*	*S. aureus* *S. xylosus*	[c]
Irradiated	[d]	[c]	[d]	*S. aureus* *E. coli* *K. oxytoca*

[a] C3H/HeN mice were wounded 1 h after 8.0 Gy gamma-irradiation, and antimicrobial therapy began 4 h later. S-TDCM (200 μg) was given i.p. 1 h after irradiation, immediately after wounding. Mice in each group were euthanized either on day 4, 5, 6, 8, or 11 after irradiation and injury, and the wound site and liver were cultured to identify the bacteria. Bacteria are listed in order of frequency of isolation in each group/time. Control mice received 9.0 Gy irradiation.

[b] No mice available for testing because of mortality.

[c] No bacteria isolated.

[d] Mice not wounded in this group.

from gram-negative sepsis was prevented by ofloxacin. However, gram-positive bacterial species were found on the wound site and in the liver of all mice treated with ofloxacin. From these findings we concluded that irradiated-wounded mice were dying with gram-positive bacterial sepsis, and that the source of the infections was the wound, because the bacteria found in the liver were similar to those bacteria colonizing the wound site.

Survival from Sepsis in Irradiated Mice Inflicted with Wound Trauma and Given Systemic Combined Modality Therapy with Antimicrobials, S-TDCM, and Topical Antimicrobials

Bacteria colonizing the wound site were systemically disseminated in irradiated mice, as noted in Table 13.1. Because topical and systemic antimicrobials applied together are more effective in treating wound infections (Stringel 1989), we evaluated several common topical antimicrobial preparations for their efficacy in treating wound infections in mice injured after irradiation. Common disinfectants were also evaluated. In normal, nonirradiated mice inflicted with wounds, treatment with several disinfectants did not alter bacterial colonization of the wound site. The agents tested were 10% povidone-iodine (Pharmadine ointment, Sherwood), 0.5% povidone-iodine (Operand aerosol, Redi-Products), and 0.25% sodium hypochlorite (diluted Dakin's solution, Clorox). Because wound colonization by bacteria was not altered by these disinfectants in normal mice, they were not evaluated in irradiated mice. The effective agents tested and the survival data obtained are presented in Figure 13.7. In this experimental series, as well as in other tests, gentamicin cream increased survival time most ($p<.05$) and hence was used in later experiments with systemic antimicrobials and S-TDCM alone or in combination.

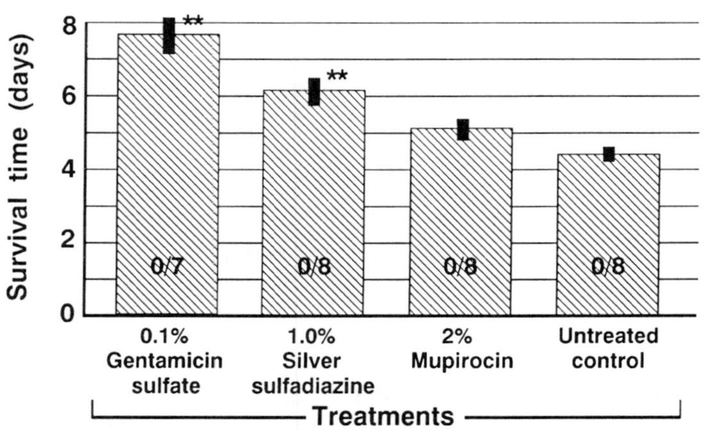

FIG. 13.7. Survival of irradiated wounded mice after topical application of antimicrobials. Mice were wounded 1 hour after 8 Gy gamma irradiation. Treatment with either gentamicin sulfate, silver sulfadiazine, or mupirocin commenced 4 hours after wounding and was applied for 5 consecutive days. Topical application of gentamicin sulfate and silver sulfadiazine significantly increased ($p<.05$) the survival of treated mice compared to untreated control mice. The mean survival time for gentamicin-treated mice differed significantly from silver sulfadiazine-treated mice ($p = .0213$) .

Systemic antimicrobial therapy included sc treatments with oxacillin, ofloxacin, and gentamicin in both gamma-irradiated and n/τ = 1 irradiated mice. Systemic ceftriaxone was evaluated only in n/τ = 1 irradiated mice. The survival data for topical and systemic antimicrobial treatments are presented in Figure 13.8.

In mice wounded after 8.0 Gy gamma irradiation, topical gentamicin sulfate treatments with or without systemic gentamicin resulted in about 50% survival. When topical gentamicin was used with oxacillin, all mice survived. The increased survival was statistically significant at $p<.05$ when compared to the survival obtained for topical gentamicin treatments and $p<.01$ for all other comparisons. In mice wounded after 3.8 Gy n/τ = 1 irradiation, topical gentamicin sulfate in combination with all the antimicrobials tested systemically increased survival significantly ($p<.01$). There were no significant differences in survival between the antimicrobial treatment groups ($p>.05$). The enhancement of survival with oxacillin and 0.1% gentamicin sulfate cream may be due to the synergistic action between the semisynthetic penicillin and the aminoglycoside gentamicin in *Staphylococcus aureus* infections (Rahal 1978). *S. aureus* was frequently found colonizing the wound and disseminating to the liver in irradiated-wounded mice (Table 13.1).

To demonstrate the efficacy of S-TDCM treatments with combination antimicrobial therapies (topical plus systemic), we used 0.1% topical gentamicin sulfate in combination with systemic gentamicin. In models of lethal n/τ = 1 and gamma irradiations followed by skin wound trauma, topical gentamicin and S-TDCM therapy with or without systemic gentamicin treatment resulted in

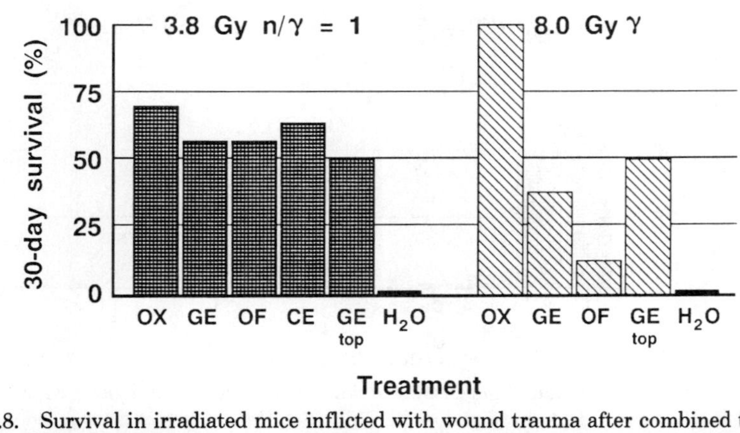

FIG. 13.8. Survival in irradiated mice inflicted with wound trauma after combined therapy with topical 0.1% gentamicin sulfate and systemic antimicrobials. C3H/HeN mice were wounded 1 hour after irradiation. Antimicrobial therapies commenced 4 hours after injury and were provided daily for 10 days. Groups of 20 mice (n/τ = 1 irradiated) and groups of 16 mice (gamma irradiated) were treated topically with 0.1% gentamicin sulfate (GE, top) and systemically (sc) with either oxacillin (OX), GE, ofloxacin (OF), or ceftriaxone (CE). Control groups were treated with 0.1 ml sterile water plus gentamicin cream or were given no antimicrobial therapy. In n/τ = 1 irradiated mice, all antimicrobial treatments were equally effective ($p>.05$) and increased survival ($p<.05$) over water-treated controls. In gamma-irradiated mice, topical gentamicin alone or with systemic oxacillin and gentamicin significantly increased survival ($p<.05$) compared to all other treatments.

approximately 60% survival (Fig. 13.9). S-TDCM effectively increased survival (p<.05) when combined with antimicrobial treatments, compared to antimicrobial treatments given without S-TDCM. Only S-TDCM combined with gentamicin produced 30-day survival approximating 50% in both $n/\tau = 1$ irradiated and gamma-irradiated mice. In this combined modality treatment series for sepsis in combined-injured mice, 0/20 survived after oxacillin treatment, 3/20 survived after ceftriaxone injection, and 1/20 survived the 30-day observation period after ofloxacin application. We did not test systemic antimicrobials without topical gentamicin sulfate in S-TDCM-treated, irradiated-wounded mice nor the combination gentamicin treatments in $n/\tau = 1$ irradiated-wounded mice.

DISCUSSION

Our data confirm the complex issues involved in developing appropriate animal models for variables associated with combined injury. In radiation casualties where there are associated tissue injuries, the primary concern is medical stabilization of the patient followed by repair of life-threatening tissue injuries. Schemes for treating victims of radiation accidents have been published elsewhere (Browne et al. 1990). Emergency care to the irradiated-traumatized patient is of greater concern than decontamination of non-life-threatening internally or externally deposited radioactive isotopes or activation products (sodium-24 $T_{1/2}$ = 15 hours; phosphorus = 32 $T_{1/2}$ = 14.3 days). Eventually, decontamination must be done to reduce the body burden of contaminating or induced isotopes. The removal of radioactive material reduces the radiation dose absorbed by the individual.

FIG. 13.9. Survival in irradiated mice inflicted with wound trauma and given combined therapy with topical 0.1% gentamicin sulfate and S-TDCM. C3H/HeN mice were wounded 1 hour after irradiation. S-TDCM (200 μg ip) was given immediately after wound trauma. Gentamicin therapy was applied topically and/or given sc 4 hours after injury and daily thereafter for 9 days. Groups of 20 mice ($n/\tau = 1$ irradiated) and groups of 16 mice (gamma irradiated) were treated as indicated. T = topical 0.1% gentamicin sulfate; S = systemic 7.5 mg/kg gentamicin sulfate; C = control. Topical gentamicin with systemic gentamicin was not tested in $n/\tau = 1$ irradiated mice. Systemic gentamicin sulfate with S-TDCM but without topical gentamicin was not evaluated in gamma-irradiated or $n/\tau = 1$ irradiated mice. In both $n/\tau = 1$ irradiated and gamma-irradiated mice, gentamicin therapy with S-TDCM and with or without systemic gentamicin therapy significantly increased (* = p<.05) survival, compared to all other treatment groups.

Once a radiation accident victim is stabilized, the second major concern is treatment for infection. Infection is a leading cause of death in otherwise survivable trauma incidents or radiation exposures. Severe immunological suppression induced by the combination of radiation and tissue injury makes the treatment for infection more difficult, as is noted in the present work in mice and in man (Baranov et al. 1989, Champlin 1990). Successful therapy for infections with topical antimicrobials, as reported here, suggests that bacterial colonization of the wound site must be controlled if further advances in therapy for combined injury are to be made (Kaplan 1985). Effective antibacterial therapies will allow sufficient time for induced hematopoietic proliferation or endogenous hematopoiesis to restore the aplastic condition induced by radiation. Possible increases in hematopoiesis by S-TDCM (Stewart et al. 1991) in consort with topical gentamicin sulfate increased survival from combined injury. The advantage of using specific cell growth factors, i.e., G-CSF and GM-CSF for treating sepsis and hematopoietic aplasia in combined-injured hosts remains to be determined. GM-CSF was used in several patients exposed to radiation in the accident at Goiania (Brandao-Mello et al. 1991, Butturini et al. 1988). Several patients responded to GM-CSF with increased peripheral granulocyte numbers; survival may have been increased by this cell growth factor. The animal models of combined injury discussed in this chapter could be used to evaluate the efficacy of cell growth factors prior to their use in radiation accidents.

In previous work, we observed that wound closure takes 1 to 2 weeks in unirradiated mice. Sublethal doses of 7.0 Gy ^{60}Co gamma rays further delayed wound closure by 1 week, while 2.5 Gy of reactor-produced neutrons ($n/\tau = 19$) further delayed wound closure by 2 weeks. Thus not only does the delay in wound closure promote continued contact with bacteria but also the unhealed site may come under continued exposure to nuclear fallout.

In radiation accidents, significant time and energy may be expended in determining the absorbed dose. The immediate concern is proper triage; a later concern is effective long-term therapy. Biological dosimeters, as well as physical dosimeters, have been used to reconstruct absorbed doses. Biological dosimeters, as adjuncts to measuring radiation doses by other means, may be more difficult in the combined-injured patient than in the individual receiving radiation only. Trauma is well known to change general cell populations (white blood cells) as well as specific cell populations (macrophages and suppressor T lymphocytes). Trauma in irradiated mice significantly altered the white blood cell and platelet patterns from that observed in irradiated animals (Madonna et al. 1991). Likewise, trauma in irradiated mice reduced the possibility of using diamine oxidase (DAO) as an indicator of radiation damage (DeBell et al. 1987). DAO is an enzyme found in high concentrations in intestinal epithelial cells, a target nearly as radiosensitive as bone marrow cells. Perturbations induced by trauma in other cell systems used as biological dosimeters is a possibility and should be examined in combined injury models. In managing dose estimations of irradiated patients, staff charged with reviewing biological and physical dosimetry may need to take these concerns into account.

In conclusion, the combined-injured individual represents a greater challenge to medical staff than the irradiated individual not compromised by additional tissue trauma. The issues of constructing radiation dose, decontamination of internal and external radioactive isotopes, activation products, and providing emergency and definitive care will all be made more difficult by the complex interactions of tissue trauma, radiation injury, and bacterial infection.

This work was supported by the Armed Forces Radiobiology Research Institute, Defense Nuclear Agency. The views presented in this paper are those of the authors; no endorsement by the Defense Nuclear Agency or the Department of Defense has been given or should be inferred. Research was conducted according to the *Guide for the Care and Use of Laboratory Animals* prepared by the Institute of Laboratory Animal Resources, National Research Council.

REFERENCES

Baranov, A., Gale, R. P., Guskova, A., Piatkin, E., Selidovkin, G., Muravyova, L., Champlin, R. E., Danilova, N., Yevseeva, L., Petrosyan, L., Pushkareva, S., Konchalovsky, M., Gordeeva, A., Protasova, T., Reisner, Y., Mickey, M. R., and Terasaki, P. I. Bone marrow transplantation after the Chernobyl nuclear accident. *N. Engl. J. Med. 321:*205–212, 1989.Brandao-Mello, C. E., Oliveira, A. R., Valverde, N. J., Farina, R., and Cordeiro, J. M. Clinical and hematological aspects of Cs^{137}: the Goiania radiation accident. *Health Phys. 60:*31–39, 1991.

Brook, I., and Elliott, T. B. Quinolone therapy in the prevention of mortality after irradiation. *Radiat. Res.* 128:100–103, 1991.

Brook, I., and Ledney G. D. Quinolone therapy of *Klebsiella pneumoniae* sepsis following irradiation: comparison of pefloxacin, ciprofloxacin, and ofloxacin. *Radiat. Res. 122:*215–217, 1990.

Brook, I., and Ledney, G. D. Short and long courses of ofloxacin therapy of *Klebsiella pneumoniae* sepsis following irradiation. *Radiat. Res. 129:*250–253, 1992.

Browne, D., Weiss, J. F., MacVittie, T. J., and Pillai, M. V. Conference report: the first consensus development conference on the treatment of radiation injuries. *Int. J. Radiat. Biol. 57:*437–442, 1990.

Butturini, A., Gale, R. P, Lopes, D. M., Cunha, C. B., Ho, W. G., Sanpai, J. M., DeSouza, P. C., Cordiero, J. M., Neto, C., DeSouza, C. E. P., Tabak, D. G., and Burla, A. Use of recombinant granulocyte-macrophage colony stimulating factor in the Brazil radiation accident. *Lancet 2:*471–475, 1988.

Champlin, R. Medical assessment and therapy in bone marrow failure due to radiation accidents: role of bone marrow transplantation and hemopoietic growth factors. In: *Treatment of Radiation Injuries.* Washington, DC, Plenum Press, 1990, pp. 3–10.

DeBell, R. M., Ledney, G. D., and Snyder, S. L. Quantification of gut injury with diamine oxidase activity: Development of a fission neutron RBE and measurements with combined injury models. *Radiat. Res. 112:*508–516, 1987.

Elliott, T. B., Brook, I., and Stiefel, S. M. Quantitative study of wound infection in irradiated mice. *Int. J. Radiat. Biol. 58:*341–350, 1990.

Finney, D. J. *Probit Analysis*, 3rd Ed. New York, Cambridge University Press, 1971, pp. 1–99.

Finney, D. J. *Statistical Method in Biological Assay*, 3rd Ed. New York, MacMillan Publishing Co., 1979, pp. 370–379.

Fujita, S., Kato, H., and Schull, W. J. The LD_{50} associated with exposure to the atomic bombing of Hiroshima. *J. Radiat. Res. 30:*359–381, 1989.

Fujita, S, Kato, H., and Schull, W. J. *The LD50 Associated with Exposure to the Atomic Bombing of Hiroshima and Nagasaki: A Review and Reassessment*, RERF TR 17–87. Tokyo, Radiation Effects Research Foundation Technical Report, 1990, pp. 1–51.

Kaplan, J. Z. Care of thermally injured victims of a thermonuclear explosion. In: *The Pathophysiology of Combined Injury and Trauma: Radiation, Burn and Trauma, Proceedings of the First Inter-*

national Symposium on the Pathophysiology of Combined Injury and Trauma, Bethesda, MD. Baltimore, University Park Press, 1985, pp. 41–46.

Kinnamon, K. E., Ketterling, L. L., Stampfli, H. and F., Grenan, M. M. Mouse endogenous spleen counts as a means to screening for antiradiation drugs. *Proc. Soc. Exp. Biol. Med. 164:*370–373, 1980.

Kohn, H. I., and Kallman, R. F. The influence of strain on acute x-ray lethality in the mouse. 1. LD_{50} and death rate studies. *Radiat. Res. 5:*309–317, 1956.

Lee, E. T. *Statistical Methods for Survival Data Analysis.* New York, Lifetime Learning Publications, 1980, pp. 122–129.

Lennette, E. H., Balows, A., Hausler, W. J., and Shadomy, H. J., eds. *Manual of Clinical Microbiology,* 4th Ed. Washington, DC, American Society of Microbiology, 1985.

Madonna, G. S., Ledney, G. D., Elliott, T. B., Brook, I., Ulrich, J. T., Myers, K. R., Patchen, M. L., and Walker, R. I. Trehalose dimycolate enhances resistance to infection in neutropenic animals. *Infect. Immun. 57:*2495–2501, 1989.

Madonna, G. S., Ledney, G. D., Moore, M. M., Elliott, T. B., and Brook, I. Treatment of mice with sepsis following irradiation and trauma with antibiotics and synthetic trehalose dicorynomycolate (S-TDCM). *J. Trauma 31:*316–325, 1991.

McChesney, D. G., Ledney, G. D., and Madonna, G. S. Trehalose dimycolate enhances survival of fission neutron-irradiated mice and *Klebsiella pneumoniae* challenged irradiated mice. *Radiat. Res. 121:*71–75, 1990.

Rahal, J. J. Antibiotic combinations: the clinical relevance of synergy and antagonism. *Medicine 54:*179–195, 1978.

Steritz, D. D., Bondi, A., and McDermott, D. A burned mouse model to evaluate anti-pseudomonas activity of topical agents. *J. Antimicrob. Chemother. 9:*133–140, 1982.

Stewart, D. S., Ledney, G. D., Madonna, G. S., Stiefel, S. M., and Moore, M. M. Synthetic trehalose dicorynomycolate (S-TDCM) increases hematopoietic cell proliferation in fission neutron ($n/\tau = 1$) irradiated mice. *J. Mil. Med. Lab. Sci. 19:*200–206, 1991.

Stringel, G., Bawdon, R., Savrich, M., Guertin, L., and Horton, J. Topical systemic antibiotics in the prevention of wound infection. *J. Pediatr. Surg. 24:*1003–1006, 1989.

Zeman, G. H., and Ferlic, K. P. *Paired Ion Chamber Constants for Fission Gamma-Neutron Fields,* Technical Report TR 84-8. Bethesda, MD, Armed Forces Radiobiology Research Institute, 1984.

SUGGESTED READINGS

Browne, D., Weiss, J. F., MacVittie, T. J., and Pillai, M. V. *Treatment of Radiation Injuries.* New York, Plenum Press, 1990.

Conklin, J. J., and Walker, R. I. *Military Radiobiology.* New York, Academic Press, 1987.

Gruber, D., Walker, R. I., MacVittie, T. J., and Conklin, J. J. *The Pathophysiology of Combined Injury and Trauma. Management of Infectious Complications in Mass Casualty Situations.* New York, Academic Press, 1983.

Mettler, F. A., Jr., Kelsey, C. A., and Ricks, R. C. *Medical Management of Radiation Accidents.* Boca Raton, FL, CRC Press, 1990.

Scherer, E., Streffer, C., and Trott, K. R. *Radiation Exposure and Occupational Risks.* New York, Springer-Verlag, 1990.

Walker, R. I., Gruber, D. F., MacVittie, T. J., and Conklin, J. J. *The Pathophysiology of Combined Injury and Trauma: Radiation, Burn and Trauma.* Baltimore, MD, University Park Press, 1985.

Chapter 14

History of Radiation Accidents

ROBERT C. RICKS, SHIRLEY A. FRY, ANN H. SIPE, MARY ELLEN BERGER, FUN H. FONG, JR. AND CLARENCE C. LUSHBAUGH

ABSTRACT

Serious radiation accidents worldwide are documented in the Radiation Accident Registry as part of the program of the Radiation Emergency Assistance Center/Training Site (REAC/TS)[1]. This data base of medically important information forms a basis for follow-up of accident survivors, development of new therapeutic protocols, and generation of training materials for physicians, nurses, and health physicists. This chapter presents an overview of the Registry data base, accident classification criteria, and the importance of dosimetry.

INTRODUCTION

In 1945 the first fatal radiation accident of the "atomic age" occurred in Los Alamos, New Mexico. It was subsequently reported in the medical literature along with another accident that occurred shortly afterward (Hempelmann et al., 1952). These accidents resulted in two radiation deaths from deterministic (nonstochastic) effects and underscored the need for investigating all accidents by teams of physicians and health physicists. These accidents were "revisited" in a 1979 symposium in Oak Ridge, Tennessee (Hubner and Fry 1980), which explored how radiation accident investigations with good health physics support form the basis for prevention of radiation accidents. This symposium built upon the previous human experience in the radiation therapy of cancer, which was summarized in Lacassagne's and Gricouroff's review of the radiation injury literature from 1902–1957 (Lacassagne and Gricouroff 1958). Studies of accidents since then and the medical attempts to alleviate the radiation damage that ensued have been used by U. S. governmental agencies (Department of Energy [DOE], Nuclear Regulatory Commission [NRC], Environmental Protection Agency [EPA]) to form a basis for radiation accident prevention, medical response, and a better understanding of how to help the unfortunate radiation accident victim. Health physicists have played an important role in this response and must learn

[1]Work in the REAC/TS program is carried out in the Medical Sciences Division of Oak Ridge Associated Universities (ORAU) under Contract Number DE-AC05-76OR00033 between the U.S. Department of Energy, Office of Energy Research, and ORAU.

how to meet the demands produced by the average primary care physician's unfamiliarity with the health effects of radiation.

Much is to be learned from accurate reports of the clinical events and medical responses directed at minimizing them. Did the medical treatment succeed? What was done? What more is needed? Is an accurate appraisal by a health physicist essential to good medical response? Can it make the difference between life and death?

Only our grandfathers recall that some of the first radiation injuries were self-inflicted by Curie and Becquerel, as was reported as long ago as 1901 (Becquerel and Curie 1901)! These skin injuries consisted of radiation-induced ulcerations and were quickly forgotten when Professor Curie was killed in a traffic accident involving a horsedrawn lorry loaded with kegs of beer. The study of radiation accidents has become more sophisticated since then, and the role of the dosimeter and/or health physicist has become proportionally more important.

THE NEED FOR DOSE ESTIMATION AS SOON AS POSSIBLE?

We used to believe that just knowing what the magnitude of exposure had been was *not* important to the immediate medical care of the irradiated victim, and, in fact, we taught that such knowledge often biased the therapy that resulted. This belief is still true in the majority of cases of accidentally irradiated persons, but, on the other hand, in some instances in which large numbers of people were irradiated but were neither badged nor monitored (for example, Chernobyl, 1986), knowledge of the "dose" actually could have helped to sort out immediately the higher dose groups from those of little medical concern. At present, highly technical, labor-intensive laboratory analyses are used to determine dose when physical dosimetry is not available; these include determining the number and distribution of chromosome breaks in the nuclei of peripheral (or circulating) lymphocytes and analyses of temporal changes in numbers of circulating lymphocytes and granulocytes. The development of clinical signs of radiation injury such as vomiting, skin redness and blistering, fever, and low blood pressure also can be used to estimate radiation dose, based on past clinical experience and experimental studies (Ricks and Fry 1990). These techniques, developed using experimental animals, did ultimately produce dose assessments that presaged the necessity for transplantation of bone marrow (BMT) or fetal liver cells in 19 of the 237 hospitalized persons from Chernobyl. However, only two of the BMT patients survived, giving rise to arguments that are yet to be resolved as to whether poorly cross-matched (or typed) BMT's can tide the irradiated victim over until the irradiated marrow can recover (Baranov and Guskova 1990).

The Chernobyl experience also showed that platelet (thrombocyte) separators to supply platelets and prevent death by hemorrhage, the capabilities in local hospitals for preventing cross-infection between patients, and the use of highly specific antibiotic drugs were questionably effective. Otherwise, the irradiated patient can die from overwhelming infection, hemorrhage and toxemia. In an undetermined (still!) number of cases, the question of BMT efficacy is made unanswerable by the onset of radiation-induced pneumonitis, which can cause hypoxia and death from cardiac failure. Here the postmortem analysis by experts becomes paramount in evaluating treatment efficacy and the role of multiorgan failure from radiation damage.

Another facet of the analysis for therapeutic efficacy is the interaction of radiation injury with injury from other forms of energy deposition (for example, the additive effect of thermal burns of the skin with radiation damage of the bone marrow). The Chernobyl experience serves at this time (5 years later) to suggest that the responses to ionizing radiation in different systems of the body can interact and augment the total systemic toxic effect so that the severity of the clinical course and the total therapeutic needs are multiplied rather than simply added.

RADIATION EMERGENCY ASSISTANCE CENTER/TRAINING SITE REGISTRIES

About 1974 the NRC and DOE formulated radiation dose criteria to be used in the Radiation Accident Registry at Oak Ridge Associated Universities (ORAU), Oak Ridge, Tennessee. These criteria (Table 14.1) have been used at ORAU's Radiation Emergency Assistance Center and Training Site (REAC/TS) since then as the basis for deciding whether an alleged radiation accident victim did or did not receive a "medically significant" radiation exposure. Inclusion in the Registry as a bona fide radiation accident victim is warranted then only if these criteria are met or exceeded. These criteria represent a compromise between the need of NRC to control all unplanned radiation exposures and of DOE to set a limit that would exclude accidental radiation exposures with little or no probability of causing a "deterministic" biological response, even though doses below such a threshold are assumed to carry a risk of late development of stochastic effects such as various types of cancer (National Research Council 1990). The agreed upon dose criteria are admittedly low by a factor of about 2. They were established in order to include persons who may be hypersensitive to the effects of radiation exposure because of preexisting diseases or conditions.

In the early (1940–1960) accidents, the exposure and dosage units were expressed according to the concepts (R, rad, rem) in vogue at the time of the accident. Since then the dosimetry in these early accidents has been upgraded or modernized, included in the files, and reported in several symposia (Hubner and

TABLE 14.1. CRITERIA FOR SELECTION OF CASES FOR LONG-TERM MEDICAL FOLLOW-UP

Condition	Criteria
1. Dose to whole body; active blood-forming organs or gonads	250 mSv
2. Dose to skin of whole body or extremities	6000 mSv
3. Dose to other tissues or organs from external source	750 mSv
4. Internal burdens	1/2 NCRP maximum organ burden
5. Medical misadministration	All misadministrations, provided they also result in a dose (if a radiation source) or a burden (if a radiopharmaceutical) equal to or greater than the criteria for conditions 1, 2, 3, or 4 above.

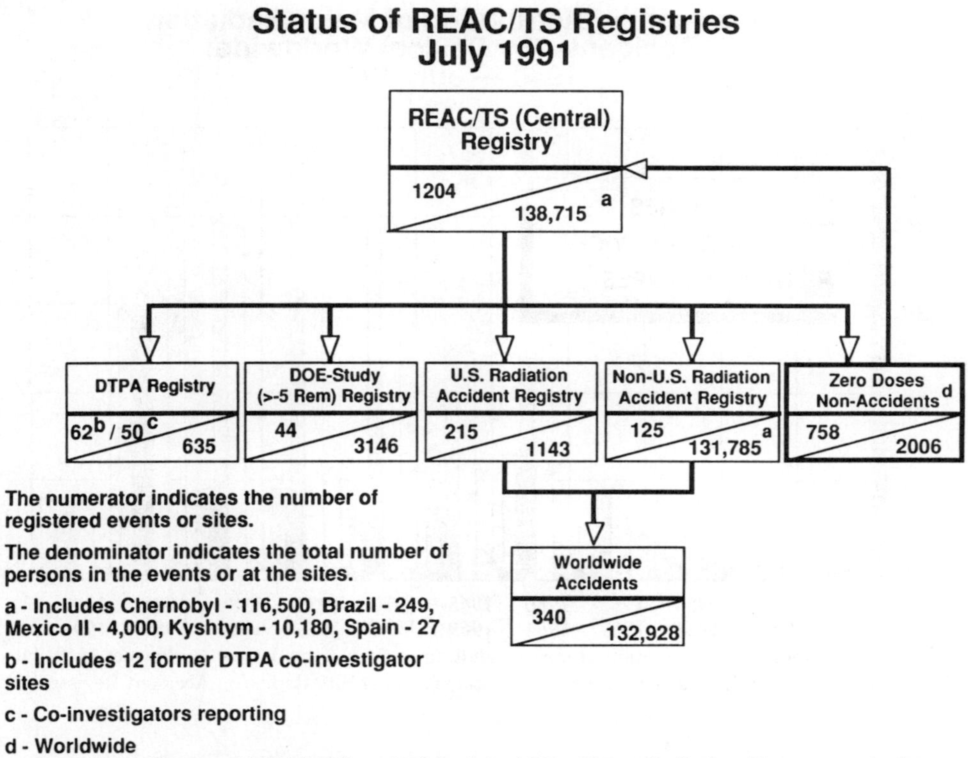

FIG. 14.1. Status of REAC/TS registries, July 1991. (From Radiation Emergency Assistance Center/Training Site Radiation Accident Registries.)

1. Nuclear Installations (Reactors)
2. Isotope Production
3. Materials Testings (Industrial Radiography)
4. Industrial Irradiation
5. X-ray and Radiotherapy Units (Medicine, Research)
6. Unsealed Radionuclides (Medicine, Research)
7. Transportation
8. ??

FIG. 14.2. Where radiation accidents occur.

Fry 1980, Ricks and Fry 1990). In addition, the illustrative block diagram (Fig. 14.1) shows that the scope of the REAC/TS radiation accident registries has been broadened to include a registry that documents the usage of the drug diethylenetriaminepentaacetic acid (DTPA), the agent of choice for chelating most internally deposited transuranic contaminants (National Council on Radiation Protection and Measurements [NCRP] 1980).

Medical and dosimetry data are maintained in the REAC/TS Central Registry. Information in the Registry can be used as a medical data base for formulating new or revised therapeutic approaches and for planning in the event of future accidents. This is especially true when one considers the recent development of

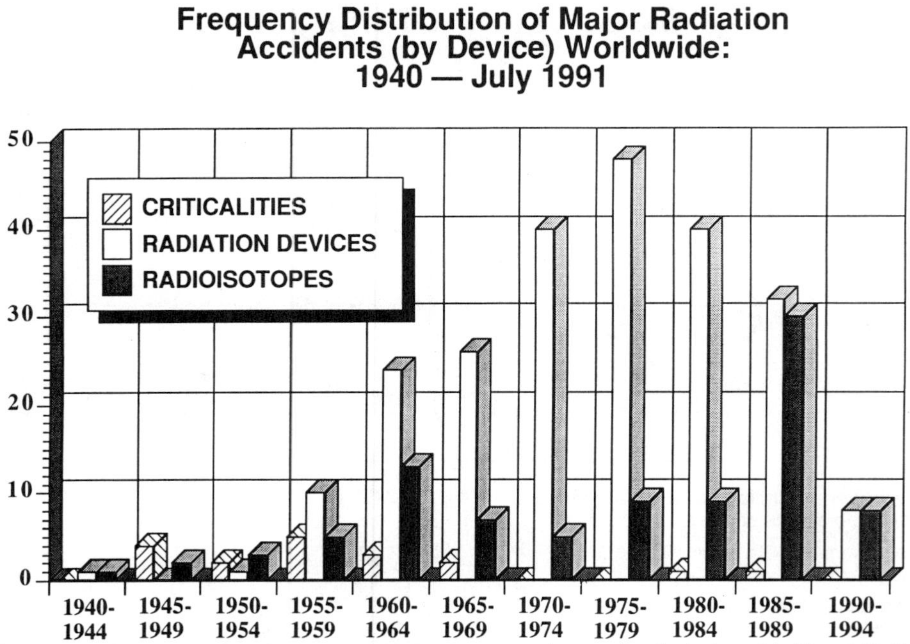

FIG. 14.3. Frequency distribution of major radiation accidents (by device) worldwide; 1940–July 1991. (From Radiation Emergency Assistance Center/Training Site Radiation Accident Registries.)

TABLE 14.2. MAJOR RADIATION ACCIDENTS WORLDWIDE (1944 THROUGH JULY 1991): CLASSIFICATION OF RADIATION ACCIDENTS BY "DEVICE"[a]

"Criticalities"		18
Critical assemblies	5	
Reactors	8	
Chemical operations	5	
Radiation devices		230
Sealed sources	151	
X-ray devices	62	
Accelerators	16	
Radar generators	1	
Radioisotopes		92
Transuranics	27	
Tritium	1	
Fission products	10	
Radium spills	2	
Diagnosis and therapy	46	
Other	6	
Total		340

[a] From Radiation Emergency Assistance Center/Training Site Radiation Accident Registries.

TABLE 14.3. MAJOR RADIATION ACCIDENTS WORLDWIDE HUMAN EXPERIENCE 1944–JULY 1991[a]

Number of Accidents	Persons Involved	Significant Exposures	Total Fatalities[b]
337	6,221	1,456	71
(1	116,500	500[c]	32)
		237[d]	
{1	27	27	13}
•1	10,180	1,054	0•
340	132,928	3,037	116

[a] From DOE-REAC/TS Radiation Accident Registries.
[b] Includes all deaths whether due to acute radiation effects or physical trauma.
[c] USDOE/NRC accident dose criteria.
[d] $\geq -$ 80 cGy.
( ) Chernobyl, U.S.S.R., data. { }, Zaragoza, Spain, data.
• Previously unrecorded accident, 1957, Kyshtym, U.S.S.R.

TABLE 14.4. MAJOR RADIATION ACCIDENTS WORLDWIDE (1944 THROUGH JULY 1991) ACUTE, DELAYED OR ASSOCIATED, AND NON-RADIATION DEATHS[a]

United States		Other	
New Mexico	3	Algeria	2
Ohio	10	Argentina	1
Oklahoma	1	Brazil	4
Rhode Island	1	Bulgaria	1
Texas	9	Canada	1
Washington	1	China (PR)	6
Wisconsin	1	El Salvador	1
	26	Germany	1
		Israel	1
		Italy	1
		Marshall Island	1
		Mexico	5
Non-Radiation Deaths			
Idaho	3	Morocco	8
Oklahoma	1	Norway	1
Pennsylvania	2	USSR	28
USSR	5	Spain	13
Total	11	Switzerland	1
		United Kingdom	2
		Yugoslavia	1
			79

[a] From Radiation Emergency Assistance Center/Training Site Radiation Accident Registries.

cytokines for regulation of hematopoiesis. Many cytokines (growth factors) have now been produced in high concentration through recombinant DNA technology. Of particular importance are granulocyte macrophage colony-stimulating factor (GM-CSF), granulocyte colony-stimulating factor (G-CSF), and interleukin-3 (IL-3). GM-CSF has now been used in selected patients following radiation accidents in the U.S.S.R. (1986), Brazil (1987), El Salvador (1989), and Israel (1990), but elucidation of an optimal treatment protocol requires further work.

The Registry also can be used to summarize radiation accidents based upon

TABLE 14.5. WORLDWIDE FATALITIES–1944 TO JULY 1991[a]

Country	Number of Accidents with Fatalities	Number of Fatalities
Algeria	1	2
Argentina	1	1
Brazil	1	4
Bulgaria	1	1
Canada	1	1
China	3	6
El Salvador	1	1
Germany	1	1
Israel	1	1
Italy	1	1
Marshall Islands	1	1
Mexico	1	5
Morocco	1	8
Norway	1	1
Russia	2	33
Spain	1	13
Switzerland	1	1
United Kingdom	1	2
United States	14	32
Yugoslavia	1	1
	36	116

[a] From Radiation Emergency Assistance Center/Training Site Radiation Accident Registries.

their cause, location, frequency by type, and the number of persons involved. These data are updated quarterly to include accidents worldwide as information becomes available. Frequent contact with international organizations such as the World Health Organization (WHO) and International Atomic Energy Agency (IAEA) facilitates the gathering of information from foreign sources. These data are presented in Figures 14.2 and 14.3 and Tables 14.2 through 14.5.

SUMMARY

The Registry data base provides valuable information that facilitates better diagnosis and therapeutic approaches for medical management of radiation accidents. Follow-up of previous accidents and information gained from future ones is invaluable to both the physicians and patients involved in these rare events. In addition, information contained in the REAC/TS Registry reflects the good record of the nuclear industry when the frequency of radiation accidents having serious clinical consequences for the individuals involved is considered. Such a good record is the result of good engineering and health physics supervision and control, facts not recorded in the REAC/TS Registry but certainly substantiated by its records.

REFERENCES

Baranov, A.E., and Guskova, A.K. Acute radiation disease in Chernobyl accident victims. In: *The Medical Basis for Radiation Accident Preparedness II: Clinical Experience and Follow-up Since*

1979, Proceedings of a Conference, Oak Ridge, TN, October 1988, edited by R.C. Ricks and S.A. Fry. New York, Elsevier Science Publishing Company, 1990.

Becquerel, H., and Curie, P. Action physiologique des rayons du radium. *Compt. Rend. Acad. Sciences 132:*1289–1291, 1901.

Hempelmann, L.H., Lisco, H., and Hoffman, J.G. The acute radiation syndrome: a study of nine cases and a review of the problem. *Ann. Intern. Med. 36:*279–510, 1952.

Hubner, K.F., and Fry, S.A., eds. *The Medical Basis for Radiation Accident Preparedness. Proceedings of the REAC/TS International Conference, 1979, Oak Ridge, TN, USA.* New York, Elsevier, 1980.

Lacassagne, A., and Gricouroff, G. *Action of Radiation on Tissues: An Introduction to Radiotherapy,* 2nd Ed., translated by C.C. Lushbaugh and G.R. Reiese. New York, Grune & Stratton, 1958.

National Council on Radiation Protection and Measurements. *Management of Persons Accidentally Contaminated with Radionuclides,* NCRP Report No. 65. Bethesda, MD, National Council on Radiation Protection and Measurements, 1980.

National Research Council. *Health Effects of Exposure to Low Levels of Ionizing Radiation, BEIR V.* Washington, DC, National Academy Press, 1990.

Ricks, R.C., and Fry, S.A., eds. *The Medical Basis for Radiation Accidents Preparedness II: Clinical Experience and Follow-up Since 1979, Proceedings of a Conference. Oak Ridge, TN, October 1988.* New York, Elsevier Science Publishing Company, 1990.

Section IV

Special Topics

Chapter 15

Contemporary Issues in Radiation Health

ERIC KEARSLEY

ABSTRACT

A number of changes in the philosophical and technical foundations of radiation protection practices have taken place over the last few years. Recently the International Commission on Radiological Protection (ICRP) has modified the fundamental radiological protection guidance in ways that will likely influence future regulatory actions. This new guidance provides the framework for a discussion of contemporary issues such as exposure limits, quality factor for neutrons, exposures below regulatory concern, and special natural background radiation sources such as radon and cosmic radiation.

INTRODUCTION

In recent years, the philosophical and technical foundations of radiation protection practices have changed profoundly. The numerical values for the risk coefficient for exposure to ionizing radiation has increased by at least a factor of three; estimates for the average exposure in the United States to background radiation have increased by a factor of three; authoritative bodies have recommended an increase in the quality factor for neutrons by a factor of two; and recommendations have been made to change exposure limits for the public and for those exposed in the workplace. The pace of recommendations from international bodies, such as the International Commission on Radiological Protection (ICRP) and the International Commission on Radiation Units and Measurement (ICRU), and national bodies, such as the National Council on Radiation Protection and Measurements (NCRP), are out of step with the mechanism of regulatory revision. The Nuclear Regulatory Commission (U.S. NRC 1991) implemented many of the 1977 recommendations of the ICRP (ICRP 1977) within weeks of publication of a subsequent major revision of the ICRP recommendations (ICRP 1991).

The new ICRP recommendations provide a comprehensive review of many of the current radiation health issues. The report introduces new radiation protection quantities, modifies the formulation of the quality factor, endorses the concept of exempting certain practices from regulatory control, discusses special natural background radiation problems such as radon or exposures expected in high-altitude aircraft or in space activities, and recommends new exposure limits. Since these recommendations will likely influence our thinking for some time to come, it may be instructive to use these recommendations as a framework to consider these issues in greater depth.

RADIATION QUANTITIES

Historically, the problem of relating the relative risks of exposure for different types of ionizing radiation has been dealt with by the quantity "dose equivalent" defined most recently in ICRU Report 33 (ICRU 1980) as

$$H = DQN \tag{1}$$

where D is the absorbed dose, Q is the quality factor, and N is a factor to account for other factors such as the dose rate, nonuniformity of exposure, etc. The value for Q was determined from the spectrum of linear energy transfer (LET) values for the charged particles in the volume of interest.

In ICRP Publication 60, the commission recommended that a slightly different quantity designated as the "equivalent dose" be adopted. The relationship between the equivalent dose and the absorbed dose for a specific tissue can be written

$$H_T = W_R H_{R,T} \tag{2}$$

The "radiation weighting factor", w_R, takes on the role of the old quality factor. The commission provides a table of recommended w_R values and recommends a method for calculating this parameter for other fields using the LET spectrum at a 10-mm depth in the ICRU sphere. As with Q, w_R has a value of unity for all low-LET sources (photons, electrons, etc.) irrespective of their energy or dose rate. The commission chose to maintain this simplification in the system although it is recognized that the variation in effectiveness of these radiations at low dose rates is at least a factor of 2 (ICRU 1986, NCRP 1990). The new quantity "effective dose" is defined by the commission as

$$E = \textstyle\sum_T w_T \sum_R w_R H_{R,T} \tag{3}$$

This is, of course, the analog of the old quantity, effective dose equivalent.

The commission also endorsed quantities introduced by the ICRU (ICRU 1985, 1988) that were intended to help distinguish between the quantities to be used for environmental monitoring (ambient dose equivalent and directional dose equivalent) and quantities to be used for personnel monitoring (individual dose equivalent, penetrating and individual dose equivalent, superficial).

NEUTRON QUALITY FACTOR

Following up on earlier statements by the ICRP regarding interim changes in the value for Q for neutrons (ICRP 1985), the commission has formally made a comprehensive change in the Q-LET relationship, resulting in significant increases in Q. As discussed in the previous section, the quality factor, Q, has been replaced by the radiation weighting factor, w_R.

Prior to reevaluation of the dosimetry data for the Hiroshima/Nagasaki

A-bomb survivors, the differences between the health effects observed at the two sites were attributed to differences in the neutron component of the exposures. That is, both the risk per unit dose and the relative contribution from neutrons were consistently lower in Nagasaki than in Hiroshima. This analysis provided what was widely accepted to be a sound basis for performing both neutron and photon risk estimates and from which, in turn, a human carcinogenesis relative biological effectiveness (RBE: the ratio of the dose of a reference radiation source such as cobalt-60 or 250 kVp x-rays to the dose of some radiation that gives the same biological effect), and therefore the neutron quality factor, could be determined. Under the new dosimetry system (Roesch 1987), the risk per unit dose is still lower in Hiroshima than in Nagasaki, but the difference is not considered to be statistically significant (National Research Council 1990).

Due to the lack of a robust human data base, human neutron risk assessments must rely on analyses of RBE data from in vivo and in vitro experimental data as well as insight from theoretical modeling of relevant radiation damage effects. Over the years a vast literature has been generated for a wide variety of radiation types and energies for biological effects as diverse as chromosome aberrations in tradescantia to loss of testis weight in rodents. Several broad conclusions may be drawn from these data. First, for low-LET radiation such as ^{60}Co gamma rays, the effects of the radiation for most endpoints follow what has come to be known as a linear-quadratic relationship with the absorbed dose. Second, for high-LET radiations such as fission spectrum neutrons, one most often observes a strictly linear relationship. One can immediately see that the RBE in such cases obviously is a function of the dose. The two relationships can be described mathematically as

$$U_g = aD_g + bD_g^2 \tag{4}$$

$$U_n = cD_n \tag{5}$$

where the subscripts refer to either gamma rays or neutrons. U is the observed effect, D is the dose, and the small-case letters are constants. Setting these two equations equal, one can derive a relationship between D_g and D_n under isoeffect conditions.

$$D_g = -\frac{a}{2b} + \frac{a}{2b}\sqrt{1 + \frac{4bcD_n}{a^2}} \tag{6}$$

Dividing by D_n, one obtains an expression for the RBE:

$$RBE = \frac{D_g}{D_n} = \frac{a}{2bD_n}\left\{-1 + \sqrt{1 + \frac{4bcD_n}{a^2}}\right\} \tag{7}$$

From equation 7, the limit of the RBE for small doses, sometimes called the RBE_M, is

$$RBE_M = \frac{c}{a} \tag{8}$$

At low doses, RBE_M is simply the ratio of the linear terms of the dose-response equations for the two types of radiation. RBE_M is generally considered to be the RBE of concern for radiation protection at occupational levels of exposure. For either large D_n *or* small a, equation 7 may be written

$$\lim_{D_n \to \infty} RBE = \lim_{a \to 0} RBE = \sqrt{\frac{c}{bD_n}} \tag{9}$$

Note that in cases in which the value of a is small (i.e., when the dose-response relationship shows little or no contribution from a linear term), there is no upper limit for the RBE.

Both the ICRP (1985) and the NCRP (NCRP 1987c) have recommended increases in the value of the quality factor for neutrons by a factor of two. A joint ICRU-ICRP task group analyzed RBE data and made recommendations concerning a new formulation of the quality factor, which was published as ICRU Report 40 (ICRU 1986). Figure 15.1. is a plot of the quality factor values as a function of neutron energy recommended by ICRU Report 40. Figure 15.2. is a plot of the quality factor for photons as a function of photon energy. The novel feature in the curve is the low value for Q for energetic photons. These recommendations take account of the observation that at low exposure levels, the effectiveness of energetic photons is substantially less than for low-energy photons.

It is widely recognized (NCRP 1990) that the extremely large values of RBE at low dose levels are primarily due to the shape of the response curve for the reference radiation. That is, in cases in which the linear term for the low-LET reference radiation is very small, there is no upper limit to RBE. From a radiation protection point of view, it is important that the results of such analyses be

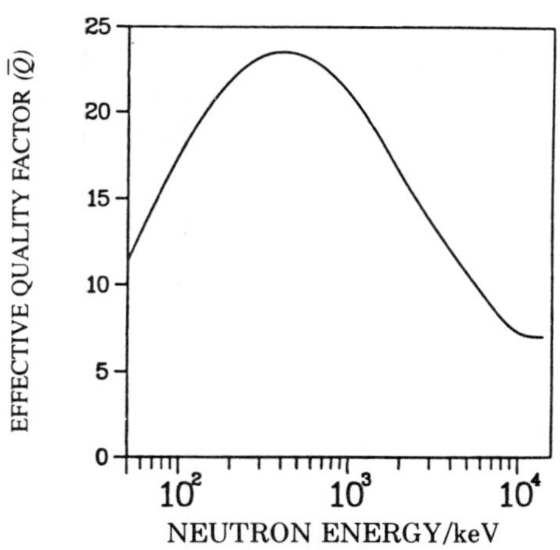

FIG. 15.1. Effective quality factor as a function of neutron energy. (From International Commission on Radiation Units and Measurements. The quality factor in radiation protection. Report of a joint task group of the ICRP and ICRU. ICRU Report 40. Bethesda, MD: International Commission on Radiation Units and Measurements, 1986.) Reprinted with the permission of the International Commission on Radiation Units and Measurements, Bethesda, MD.

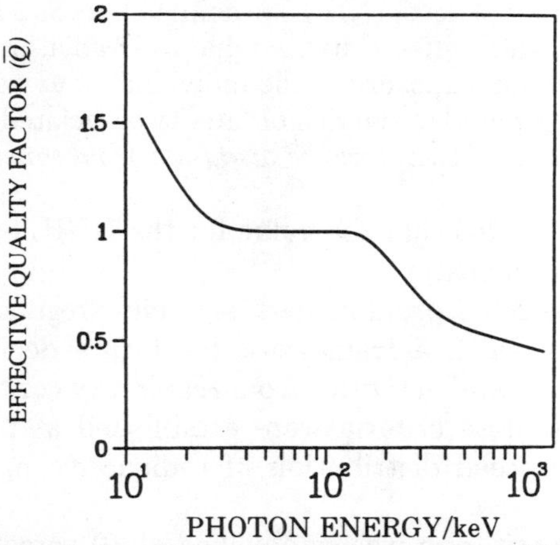

FIG. 15.2. Effective quality factor as a function of photon energy. (From International Commission on Radiation Units and Measurements. The quality factor in radiation protection. Report of a joint task group of the ICRP and ICRU. ICRU Report 40. Bethesda, MD: International Commission on Radiation Units and Measurements, 1986.) Reprinted with the permission of the International Commission on Radiation Units and Measurements, Bethesda, MD.

consistent with what is known about humans and radiation exposure. The BEIR V Committee concluded that the dose-response curve for A-bomb survivors followed a linear relationship for solid tumors and a linear-quadratic relationship for leukemia (NRC 1990). Since approximately 80% of the total number of tumors are of the solid type, the bulk of the data from which risk estimates are derived follow this relationship. Since most dose-response curves generated in the laboratory generally follow a function with a distinct quadratic component, there is some question as to the applicability of such data for performing risk estimates for high-LET exposures in the human. Regardless of whether the observed differences between the dose-response curve for humans is due to fundamental differences in the radiobiology of carcinogenesis in humans and animals or whether these differences are due to uncertainties in the human data or modeling techniques or some other reason, it is important that laboratory systems used to model human response follow similar functional relationships to avoid inconsistencies. Therefore, RBE_M values generated in the laboratory using an endpoint for which the dose-response curve exhibits a strong quadratic component may not be appropriate for making estimates for Q.

EXEMPTION FROM REGULATORY CONTROL

In ICRP Publication 60, the commission endorsed the concept that levels should be established for individual and collective doses at which no reasonable actions can be expected to achieve meaningful reductions in these exposures (ICRP 1991). Although the commission stopped short of making a specific quantitative recommendation, several sources such as cosmic radiation and potassium-40 in the body were listed as being outside consideration for specific regulatory oversight. The NCRP endorsed the concept of negligible individual risk level (NCRP 1987c):

> A negligible individual risk level (NIRL) is defined here as a level of average annual excess risk of fatal health effects attributable to irradiation, below which further effort to reduce radiation exposure to the individual is unwarranted. The NIRL is regarded as *trivial* compared to the risk of fatality associated with ordinary, normal societal activities and can, therefore, *be dismissed from consideration.*

The NCRP recommended that the value for the NIRL be chosen as 0.01 mSv from any one source in a year.

In July of 1990, the NRC promulgated its "below regulatory concern" (BRC) policy designed to establish a framework for future decisions on whether to exempt certain products and activities from regulatory control (U.S. NRC 1990). Specifically, the NRC dose criteria were established at 0.1 mSv/year and .01 mSv/year when widespread distribution of radioactive materials such as consumer products is involved.

The collective dose criterion was established at 10 person-Sv/year, with individual doses below 0.001 mSv/year excluded.

Shortly after the NRC announced this policy, the Health Physics Society issued a formal position statement, strongly endorsing this policy and hailing it as being "helpful in focusing the attention of health physicists, radiation users, regulators and the public alike away from insignificant problems and towards those that may involve significant risks."

The action of the NRC has generated a good deal of controversy. Legislation has been introduced in both houses of Congress to allow states to ban the disposal of BRC wastes in local landfills. In February 1991 the NRC announced that it would not accept petitions to exempt low-level radioactive waste from regulatory control until a consensus on the policy could be established. The NRC is considering a plan at this time to generate a consensus on this issue.

SPECIAL NATURAL BACKGROUND RADIATION SOURCES

In recent years there has been an increased awareness of the magnitude of the exposure associated with the concentration of radon in public and private buildings. The NCRP has estimated that approximately 55% of the total average annual exposure to the public is due to radon (NCRP 1987b). This is approximately two-thirds of the total exposure from all natural sources. The Environmental Protection Agency has taken actions in recent years to alert the public of the potential hazards of this source and to establish action levels for radon mitigation. There has been a good deal of controversy surrounding these actions, much of which has been thoroughly covered in *The Health Physics Society Newsletter.*

Historically, exposure from sources that originate as a part of "natural background" have not been considered a component of occupational exposure. In ICRP Publication 60, the commission refined the definition of occupational exposure to "exposures incurred at work as the result of situations that can reasonably be regarded as the responsibility of the operating management." The commission specifically placed radon exposure in the workplace within this category but did not specify a specific action level for mitigation efforts.

Besides addressing radon directly, the commission provided other examples of special sources of background radiation: the operation of jet aircraft and exposures in space.

Exposures from galactic and solar radiation sources have been a concern for a number of years. At the surface of the earth, we are normally protected by the earth's magnetic field as well as the thick protective blanket of our atmosphere. The exposure level from these sources is approximately 0.28 mSv/year, or only 8% of the average annual effective dose equivalent to the general public (NCRP 1987b). Our propensity to propel ourselves through the atmosphere in commercial aircraft contributes approximately .01 mSv to this figure; however, the absolute dose rate from cosmic radiation increases by about a factor of 100 at normal commercial jet altitudes (NCRP 1987a). Exposure rates from these sources depend very dramatically on the altitude, latitude, and levels of solar activity, but estimates higher than 0.05 mSv/h are not uncommon.

The exposure rates at altitude can be predicted to an acceptable degree of accuracy except during periods of intense solar activity. A high-flying aircraft such as the Concorde, which operates at a cruising altitude of 60,000 feet, carries simple radiation-monitoring instrumentation that alerts the crew to take evasive action (i.e., reduce altitude) during these periods. It is not clear whether the hazards of such maneuvers are indeed less than the hazards associated with the exposure, particularly if these maneuvers are performed in congested air lanes.

The collective dose equivalent for commercial flight crews was estimated by the NCRP to be 170 person-Sv (NCRP 1989a). This number is significant, particularly when compared with the average collective dose equivalent for other sectors of the economy (see Table 15.1).

The Federal Aviation Administration (FAA) has taken steps to require commercial airlines to provide instructional materials similar to the kinds of materials that NRC licensees are familiar with. Proposals to provide regulatory guidance in this area have been made, and a number of options are being considered.

Since the earliest days of space exploration, there has been concern about the radiation hazards associated with placing human beings outside the atmosphere. However, the primary focus of health effects research has been on the effects of

TABLE 15.1. OCCUPATIONAL EXPOSURE IN THE UNITED STATES

Occupational Category	Annual Collective Effective Dose Equivalent[a] (person-Sv)
Nuclear power plant personnel	551
Medical staff	420
Industrial personnel (other than nuclear fuel cycle)	390
Department of Energy	224
Flight crews	170
Uranium miners	112
U.S. Navy	51
Education and transportation	50
Well loggers	30

[a] Data derived from Table 4.3 of NCRP Report 101 (NCRP 1989a).

prolonged weightlessness on human physiology. Consideration of an extended flight to Mars with a projected duration of approximately 3 years (1 year to Mars; 1 year on Mars; and 1 year returning) has refocused attention on the radiation hazards involved. Dose estimates on the order of 1 Sv for such a mission have been made (NCRP 1989b). These estimates do not include the contribution from a possible solar particle event that could significantly increase this projected dose.

Outside the earth's magnetosphere the principal components to exposure are solar particle events and galactic cosmic radiation. Solar particle events, as the name implies, originate from activity on the sun and are associated with the formation of sunspots. The radiation is predominantly very high energy protons, helium ions, and sometimes heavier ions. The contribution to the dose from such sources is highly variable. Although the frequency of such events increases with increases in solar activity, some of the largest solar particle events have occurred during periods of minima in solar activity. Galactic cosmic radiation originates outside the solar system and is modulated only by variations in the solar wind that follows the solar cycle. The radiation is isotropic and is dominated by protons, helium ions, and high atomic number, energetic charged (HZE) particles.

The National Aeronautics and Space Administration (NASA) has embarked on a comprehensive research program to reduce the uncertainties associated with space radiation hazards. NASA's approach is to first develop more accurate models to describe the space radiation environment. This effort includes the problem of predicting the occurrence and intensity of solar particle events so that protective actions may be taken by or on behalf of astronauts. Second, NASA is developing models that provide shielding information for spacecraft materials and biological tissues. Third, NASA is interested in the biological effects of HZE particles.

This comprehensive research effort may very well have an important impact on the way risks for high-LET sources are determined.

EXPOSURE LIMITS

Perhaps the most widely publicized recommendation contained in ICRP Publication 60 is the new set of exposure limits. Very briefly, these new limits are summarized in Table 15.2.

There has been a good deal of discussion concerning the recommendation of limiting occupational exposure to 20 mSv/year, averaged over 5 years. In the light of the magnitude of current average annual occupational exposure levels in the United States, this limit would not appear to have a significant impact in the workplace, particularly in light of the fact that ICRP Publication 60 recommends an additional limit of 50 mSv in any 1 year. However, concern has been expressed that limits expressed as averages over 5 years may introduce administrative problems in the management of large radiation health programs. Another argument concerns the "ratcheting down" of exposure limits, leaving little room for establishing local administrative controls that are set well below limits and are used to take prompt action to minimize the likelihood that a limit might be exceeded.

TABLE 15.2. DOSE LIMITS RECOMMENDED BY ICRP PUBLICATION 60[a]

Application	Occupational	Public
Effective dose	20 mSv/year, averaged over defined periods of 5 years[b]	1 mSv in a year[c]
Annual equivalent dose		
Lens of the eye	150 mSv	15 mSv
Skin[d]	500 mSv	50 mSv
Hands and feet	500 mSv	50 mSv

[a] The limits apply to the sum of the relevant doses from external exposure in the specified period and the 50-year committed dose (to age 70 for children) from intakes in the same period.

[b] With the further provision that the effective dose should not exceed 50 mSv in any single year. Additional restrictions apply to the occupational exposure of pregnant women.

[c] In special circumstances, a higher value of effective dose could be allowed in a single year, provided that the average over 5 years does not exceed 1 mSv/year.

[d] The limitation on the effective dose provides sufficient protection for the skin against stochastic effects. An additional limit is needed for localized exposures in order to prevent deterministic effects.

TABLE 15.3. ATTRIBUTES OF DETRIMENT DUE TO EXPOSURE OF THE WORKING POPULATION[a]

Annual Effective Dose (mSv)	10	20	30	50
Approximate lifetime dose (Sv)	0.5	1.0	1.4	2.4
Probability of attributable death (%)	1.8	3.6	5.3	8.6
Weighted contribution from nonfatal cancer (%)	0.4	0.7	1.1	1.7
Weighted contribution from hereditary effects (%)	0.4	0.7	1.1	1.7
Aggregated detriment (%)	2.5	5	7.5	12
Time lost due to a death (y)	13	13	13	13

[a] Information extracted from Table 5 of ICRP Publication 60 (ICRP 1991).

In ICRP Publication 26, the commission established the basis for limits on the assessment of acceptable risks from rates of accidental death in industries not associated with ionizing radiation. In ICRP Publication 60, the commission abandoned this approach on the grounds that there is little evidence to suggest that society expects the same level of safety over a wide range of industries. Furthermore, in addition to loss of life or severe hereditary conditions, the commission has included other factors in the assessment of the total detriment, including length of life lost due to an attributable death and the incidence of nonfatal conditions.

The commission considered the consequences of continued uniform exposures over a 47-year working lifetime at a range of doses that could be adopted as a dose limit (10–50 mSv). The results of this analysis are summarized in Table 15.3. On the basis of this analysis, the commission concluded that (*a*) an aggregated detriment of 12%, which corresponds to a lifetime exposure at limits specified in ICRP Publication 26, was unacceptable, and (*b*) an aggregated detriment of 5%, which corresponds to a lifetime dose of 1 Sv or 20 mSv per year, could be considered "tolerable" and thus be a reasonable limit.

SUMMARY

Radiation health professionals are in a rather difficult position. Their primary objective is to enable the community to enjoy the benefits of an inherently hazardous commodity. They perform their work under the constant pressures of regulatory and political forces. While these pressures consume the bulk of their efforts, national and international authoritative advisory bodies continue to refine the philosophical foundations for radiation protection practices. Some health physicists dismiss these efforts as superfluous in light of the nature of the daily demands of managing radiation health programs. Some dismiss these efforts as inappropriate refinements in light of the substantial uncertainties involved with the assessment of actual risks. Almost all health physicists believe that the resources being committed to regulate, monitor, and control radiation exposures far outweigh the risks associated with the exposure levels actually encountered. This is ironic when one considers that the conservatism with which our society deals with these risks is largely responsible for their livelihood.

Perhaps the most important issue that radiation health professionals must face in the future is how to manage the societal forces that drive this profession. There is a tremendous educational challenge ahead that is directed at both the public at large as well as the health physics profession itself. The product of our efforts can be an improved ability to make informed judgments about how to allocate increasingly scarce risk management resources.

REFERENCES

International Commission on Radiation Units and Measurements. *Radiation Quantities and Units,* ICRU Report 33; Bethesda, MD, International Commission on Radiation Units and Measurements, 1980.

International Commission on Radiation Units and Measurements. *Determination of Dose Equivalents from External Radiation Sources,* ICRU Report 39. Bethesda, MD, International Commission on Radiation Units and Measurements, 1985.

International Commission on Radiation Units and Measurements. *The Quality Factor in Radiation Protection. Report of a Joint Task Group of the ICRP and ICRU,* ICRU Report 40. Bethesda, MD, International Commission on Radiation Units and Measurements, 1986.

International Commission on Radiation Units and Measurements. *Determination of Dose Equivalents from External Radiation Sources–Part 2,* ICRU Report 43. Bethesda, MD, International Commission on Radiation Units and Measurements, 1988.

International Commission on Radiological Protection. Recommendations of the International Commission on Radiological Protection, ICRP Publication 26. *Ann. ICRP 1(3):*1977.

International Commission on Radiological Protection. Statement from the 1985 Paris meeting of the International Commission on Radiological Protection. *Phys. Med. Biol. 30:*863–864, 1985.

International Commission on Radiological Protection. *1990 Recommendations of the International Commission on Radiological Protection,* ICRP Publication 60. Oxford, Pergamon Press, 1991.

National Council on Radiation Protection and Measurements. *Exposure of the Population in the United States and Canada from Natural Background Radiation,* NCRP Report No. 94. Bethesda, MD, National Council on Radiation Protection and Measurements, 1987a.

National Council on Radiation Protection and Measurements. *Ionizing Radiation Exposure of the Population of the United States,* NCRP Report No. 93. Bethesda, MD, National Council on Radiation Protection and Measurements, 1987b.

National Council on Radiation Protection and Measurements. *Recommendations on Limits for Exposure to Ionizing Radiation,* NCRP Report No. 91. Bethesda, MD, National Council on Radiation Protection and Measurements, 1987c.

National Council on Radiation Protection and Measurements. *Exposure of the U.S. Population from Occupational Radiation,* NCRP Report No. 101. Bethesda, MD, National Council on Radiation Protection and Measurements, 1989a.

National Council on Radiation Protection and Measurements. *Guidance on Radiation Received in*

Space Activities, NCRP Report No. 98. Bethesda, MD, National Council on Radiation Protection and Measurements, 1989b.

National Council on Radiation Protection and Measurements. *The Relative Biological Effectiveness of Radiations of Different Quality,* NCRP Report No. 104. Bethesda, MD, National Council on Radiation Protection and Measurements, 1990.

National Research Council. *Health Effects of Exposure to Low Levels of Ionizing Radiation: BEIR V.* Washington, DC, National Academy Press, 1990.

Roesch, W. C. *US-Japan Joint Reassessment of Atomic Bomb Radiation Dosimetry in Hiroshima and Nagasaki,* 1st Ed. Hiroshima, Japan, Radiation Effects Research Foundation, 1987.

United States Nuclear Regulatory Commission. *Below Regulatory Concern, Policy Statement.* Rockville, MD, United States Nuclear Regulatory Commission, 1990.

United States Nuclear Regulatory Commission. *Standards for Protection against Radiation, Final Rule,* Part VI, 10 CFR 20. Washington, DC, Government Printing Office, 1991.

Chapter 16

Radiation Biology and Operational Health Physics

ROGER J. CLOUTIER

ABSTRACT

The operational health physicist is required to have an understanding of radiobiology because of regulations, but most health physicists also desire the knowledge in order to better understand their profession. This increased knowledge allows them to make sound decisions and to describe more fully the possible effects of radiation to the radiation worker and the public. This chapter discusses the general regulations and recommendations related to radiobiology, the educational programs available, and the impact of several radiobiological topics on the practice of operational health physics.

INTRODUCTION

The purpose of this chapter is to discuss how much and what kind of radiobiology should be a part of the operational health physicist's education. I begin by reminding you of the health physicist's goal, which is to protect man and his environment from the harmful effects of radiation. We should then ask, "Will a knowledge of radiation biology actually improve radiation protection, or is it just nice to know?" Before tackling these questions, I want to discuss what was done in the past and what is done today.

The profession of radiation protection was called health physics because in the 1940s professionals from the health field and the "hard" sciences, mostly physics, were given the assignment of protecting the health of the Manhattan Project workers. These professionals had backgrounds in medicine, physics, electronics, etc., and were responsible for developing radiation detection instruments, measuring the biological effects of radiation, and establishing standards. To accomplish all these tasks, a multidisciplinary group was required.

As time passed, two groups of specialists evolved: research health physicists and operational health physicists. The research group continued to study both physical and biological problems related to radiation protection. The operational health physicist became responsible for the day-to-day health physics programs. The research groups became more knowledgeable about a single topic, and the operational group became familiar with many topics. In a joking manner, it is sometimes said that the operational health physicist knows a little about a lot of things, but not much about anything.

In 1950 the Atomic Energy Commission (AEC) initiated the first graduate-level fellowships in radiological protection. The academic programs were established at colleges and universities that had strong programs in one or another scientific discipline. The Vanderbilt University program was strong in nuclear physics, and the Rochester University program was strong in radiobiology. These schools, and the ones that were added later, made an attempt to provide the student with a good basic education in one of the sciences, while also attempting to provide a multidisciplinary education in radiation protection.

Today there is almost universal agreement that the research health physicist needs a solid education in one of the basic sciences followed by a specialized education in a subcategory of the chosen science.

However, a fundamental argument continues regarding the educational requirements for the operational health physicist. Some believe the operational health physicist should first receive a strong background in a basic science and then acquire the additional information needed to perform the health physics task. Others believe the student should be taught health physics as an undergraduate and receive a much broader, but not very deep, education.

RADIOBIOLOGY KNOWLEDGE REQUIREMENTS

The operational health physicist is governed almost totally by regulations, recommendations, and operating procedures. Regulations are generally believed to be based on well-established scientific facts, for "in creating regulatory agencies to protect and promote health and safety, Congress has repeatedly emphasized scientific data and analysis" (Crandall and Lave 1981). In most cases, a complete review of available scientific evidence does not provide clear, definitive answers to questions about actual risks. Uncertainties inevitably remain, inducing regulators to resort to intuitive judgment.

The following provides a short description of agencies and groups that provide regulations or guidance regarding the operational health physicist's need for knowledge about radiobiology. Because the health physicist is often the person selected to train the radiation worker, the health physicist must understand the fundamentals of radiobiology, or he or she will not be able to communicate the information very well. A more complete list of regulatory bodies can be found in *A Compendium of Major U.S. Radiation Protection Standards and Guides: Legal and Technical Facts* (Mills et al. 1989).

United States Nuclear Regulatory Commission

Section 19.12 of 10 CFR Part 19,[1] *Notices, Instructions and Reports to Workers: Inspections and Investigations*, requires that all persons working in or frequenting any portion of a restricted area be instructed in the health protection problems associated with exposure to radioactive materials or radiation. Regulatory Guide

[1]The Nuclear Regulatory Commission, Department of Energy, etc., regulations and guides are listed as suggested readings.

8.29, *Instructions Concerning Risks from Occupational Radiation Exposure,* describes the information that should be provided to the worker concerning biological risks from occupational radiation exposure. The Nuclear Regulatory Commission (NRC) states that the Guide is intended to "help develop an attitude of healthy respect for the risks associated with radiation, rather than unnecessary fear or lack of concern." To accomplish this goal, the Guide includes 32 questions about the effects of radiation along with the answers that the NRC feels are appropriate. The questions include "If I receive a radiation dose, does that mean I am going to get cancer?" The somewhat folksy response goes into the role of risk models and how they affect the risk estimate. Regulatory Guide 8.27 provides the training topics for personnel at light-water-cooled nuclear power plants. Appendix A includes the training requirements related to the biological effects of radiation.

United States Department of Energy

Department of Energy (DOE) Order 5480.11 describes the training requirements for radiation workers under the jurisdiction of the U.S. DOE. In addition to information about radiation and how to control an exposure, the Order requires the radiation worker and the radiation protection technician to be familiar with the following radiobiology topics:

1. Acute effects of exposure to radiation
2. Risks associated with occupational radiation exposure
3. Special considerations in the exposure of women of reproductive age
4. Dose equivalent limits

United States Occupational Safety and Health Administration

The *OSHA Hazard Communication Standard* (29 CFR 1910.120), called the "Right to Know" regulation, gives employees the right to information that relates to their health and safety on the job. It requires the employer to train employees to recognize hazards and to be able to protect themselves from the hazard.

National Council on Radiation Protection and Measurements

The National Council on Radiation Protection and Measurements (NCRP) is well known for its recommendations regarding acceptable radiation protection limits. *Recommendations on Limits for Exposure to Ionizing Radiation,* Report No. 91 (NCRP 1987), provides the most recent radiation protection guidelines. The NCRP has also issued several reports that recommend appropriate training requirements.

Radiation Safety Training Criteria for Industrial Radiography, Report No. 61 (NCRP 1978b) presents an outline for the initial training requirements, reproduced in Appendix B.

Report No. 71 (NCRP 1983), *Operational Radiation Safety–Training,* was issued to supplement Report No. 59 (NCRP 1978a), *Operational Radiation Safety Program.* Report 71 states that all good training programs begin with a job evaluation and needs assessment. From this information training objectives are set, as well as testing criteria. These naturally lead to course content and course structure. This list includes:

1. Acute effects of exposure to radiation
2. Risks associated with occupational exposures
3. Special considerations in the exposure of women of reproductive age
4. Modes of exposure–internal, external

Appendix A of Report No. 71 provides topics for radiation safety training for specific work groups.

Institute of Nuclear Power Operations

The Institute of Nuclear Power Operations (INPO) was created after the Three Mile Island nuclear power plant accident in order to improve the quality of nuclear power operations. The development and certification of training programs has been of high priority. Today most nuclear power plants have adopted the INPO training recommendations. The INPO programs were prepared with the assistance of radiological protection trainers, health physicists, and instructional technologists.

INPO 83–012 (INPO 1983) addresses the radiological protection training of general employees. The program involves the use of 50 questions, of which only 2 are on biological effects. INPO 82–006 (INPO 1982) relates to *Radiological Protection Technician Qualifications*. Section 5.2.5 is entitled "Biological Effects and Risks Associated with Exposure to Ionizing Radiation" and includes three parts: "Simple Mechanisms of Radiation Interactions with Cells," "Acute Radiation Syndrome," and "Somatic Effects of Radiation Exposure."

Section 11.4 of INPO 87–008, *Guidelines for Training and Qualifications of Radiological Protection Technicians* (INPO 1987), lists the topics that should be included in the section on biological effects (see Appendix C).

American Society for Testing and Materials

ASTM's *Standard Guide for Radiological Protection Training for Nuclear Facility Workers* (ASTM 1987) defines the elements of a training program for radiation workers consistent with the philosophy that occupational radiation exposures should be kept as low as reasonably achievable (ALARA). Section 6 describes the course content. The section "Possible Health Effects of Exposure to Ionizing Radiation" includes the same topics as those listed in NCRP Report No. 71; however, ASTM covers the subject in much more detail.

Universities

When the AEC initiated the original Health Physics Fellowship program, they did not wish to establish a rigidly uniform curriculum (Claus 1962). The AEC recognized that each center had its own specialities, strengths, and weaknesses; however, its Fellowship Board for graduate training did develop suggested core courses (Table 16.1). It is worth noting that 6 of the 28 semester hours were devoted to basic biology and radiation biology.

The U.S. DOE, successor to the AEC, now supports training at several

TABLE 16.1. EARLY RADIATION TRAINING AT UNIVERSITIES; CORE COURSES SUGGESTED BY AEC FELLOWSHIP BOARD[a]

Topic	First Semester (Sm-hr)[b]	Second Semester (Sm-hr)
1. Basic Nuclear Physics	3	3
2. Differential Equations	3	—
3. Applied Statistics	2	—
4. Basic Biology	2	—
5. Radiation Biology	—	4
6. Toxicology/Aerosols	—	2
7. Theoretical Radiation Dosimetry	—	4
8. Electronics and Instrumentation	3	—
9. Seminar (Organization, Management, Legal, Personnel, Safety, etc.)	1	1
Totals 14	14	

[a] Clause WD. Symposium on education and training in health physics, training programs in health physics. Health Phys 1962;8:93–95.
[b] semester hours

TABLE 16.2. PRESENT (1991) DOE FELLOWSHIP UNIVERSITIES

Operational Health Physics	Health Physics
Colorado State University	Colorado State University
Georgia Institute of Technology	Georgia Institute of Technology
Purdue University	Purdue University
Rutgers University	Rutgers University
Texas A&M University	Texas A&M University
University of Cincinnati	University of Florida
University of Florida	University of Lowell
University of Lowell	University of Michigan
University of Michigan	University of Missouri-Columbia
University of Missouri-Columbia	University of North Carolina at Chapel-Hill
University of North Carolina at Chapel-Hill	University of Pittsburgh
University of Pittsburgh	

universities (Table 16.2). Recently DOE published the seventh edition of *University Programs and Facilities in Nuclear Science and Engineering* (U.S. DOE 1991). About 50 higher education institutions are listed as providing courses in the area of health physics, radiobiology, and/or radiation protection. Only about half list radiobiology as a course or an area of study.

Christensen et al. published *Guidelines for Health Physics Program Content: Master's Level* (Christensen et al. 1986). The authors divide the training into 13 subject areas. Three instructional levels are suggested: descriptive, standard mathematically based, and theoretical/specialized treatment. Different instructional levels are recommended for general health physics and reactor health physics. The section dealing with chemical and biological effects of ionizing radiation has the same instructional level for both groups. Table 16.3 provides the entire guidelines.

TABLE 16.3. GUIDELINES FOR HEALTH PHYSICS PROGRAM CONTENT: MASTER'S LEVEL[a,b]

Subject Area	Instructional Level Used	
	General Health Physics	Reactor Health Physics
A. Physics of Ionizing Radiation		
1. Atomic and Nuclear Structure	S	S
2. Basic Quantum Theory	S	S
3. Radioactivity: decay kinetics	S	S
4. Charged Particle Interactions	S	S
5. Photon Interactions	S	S
6. Neutron Interactions	S	T
7. Sources of Ionizing Radiations	S	S
B. Radiation Detection and Dosimetry		
1. Radiation Quantities and Units	S	S
2. Detection Electronics	S	S
3. Quantification and Confidence Tests	S	S
4. Gas Ionization Detectors and Dosimeters	S	S
5. Scintillation Detectors	S	S
6. Semiconductor Detectors	S	S
7. Chemical Detectors and Dosimeters	S	S
8. TLD	S	S
9. Film	S	S
10. Miscellaneous Detectors and Dosimeters	S	S
11. Survey Instrumentation: selection criteria	S	S
12. Fixed Instrumentation: selection criteria	S	S
13. Radiochemical Separations and Analysis	S	S
C. Chemical and Biological Effects of Ionizing Radiation		
1. Radiation Biochemistry	D	D
2. Effects of Cellular Structure and Function	D	D
3. Effects on Organs and Systems	S	S
4. Effects on Organisms	S	S
5. Effects on Populations	S	S
6. Therapeutic Applications	S	S
D. External Radiation Protection		
1. Radiation Protection Guidelines: regulations	S	S
2. General External Dosimetry	S	S
3. Medical External Beam Calibrations and Dose Distributions	D	D
4. Source and Facility Design	S	S
5. Criticality Control	S	T
6. Accident Management and Emergency Planning	S	S
7. Administrative Control Measures	S	S
E. Internal Radiation Dosimetry and Protection		
1. Derivation of Internal Standards and Regulations	S	S
2. Pathways and Bioassays	S	S
3. Calculation of Dose	S	S
4. Air Sampling/Monitoring	S	S
5. Contamination Controls	S	S
6. Survey Methods	S	S
7. Decontamination Methodology	S	S
8. Administrative Control Measures	S	S

TABLE 16.3. *Continued*

Subject Area	Instructional Level Used	
	General Health Physics	Reactor Health Physics
F. Environmental Radiation		
1. Sources	S	S
2. Dosimetry; media sampling and processing	S	S
3. Distribution and Transport Dynamics	T	T
G. Radiation Protection Standards and Regulations		
1. Organizations	D	D
2. Regulations	S	S
3. Regulatory Guides and NUREGs	D	D
4. Standards (ICRP, NCRP, ANSI)	D	D
H. Waste Management and Disposal		
1. Sources	S	S
2. Categories	D	D
3. Techniques	S	S
4. Waste Characteristics	S	S
5. Packaging Technology	S	S
6. Sites	D	D
7. Legal Aspects	D	D
I. Accelerator Health Physics (unique issues)		
1. Radiation Environment	S	D
2. Accelerator Types	S	D
3. Shielding Problems	T	S
J. Reactor/Fuel Cycle Health Physics (unique issues)		
1. Organizational and Personnel Issues	D	D
2. Facility and Equipment Design	S	T
3. Operational Issues	S	S
4. Decommissioning Issues	S	S
5. Reprocessing Issues	S	S
6. Mining, Milling, and Fabrication Issues	S	S
K. Special Problems/Issues		
1. Property Decontamination/Decommissioning	S	S
2. Transportation of Radioactive Materials	S	S
3. Analysis of Radiation Accidents	D	D
4. Industrial Users and Users of Radiation	D	D
5. Risk Assessment	S	S
L. Non-Ionizing Radiation and Its Regulation		
1. Microwaves	S	S
2. Lasers and Masers	S	S
3. UV	S	S
4. Ultrasound	S	S
5. Magnetic Resonance Imaging	D	D
6. Other	D	D
M. Peripheral Topics		
1. Concepts and Practices of Industrial Hygiene	D	D
2. Principles of Management	D	D
3. Computer Applications in Health Physics	S	S
4. Professional Ethics	D	D

[a] From Christensen RC, et al. Guidelines for health physics program content: master's level. Health Phys 1986;51:11–15.

[b] Abbreviations: D, descriptive; S, standard mathematically based; T, theoretical/specialized treatment.

THE TRAINING RESOURCE AND DATA EXCHANGE GROUP

Most DOE contractors are members of the Training Resource and Data Exchange group (TRADE). This group has several special interest groups, including one for radiation protection training. This group identified a need for training material for radiation workers on the biological effects of ionizing radiation. The material was developed and produced to meet the specific needs of DOE contractors as they implement the training requirements of DOE Order 5480.11. The package contains a videotape and supporting materials (Oak Ridge Associated Universities, 1989). The topics required in radiation worker training programs include characteristics of ionizing radiation, man-made radiation sources, acute effects of exposure to radiation, risks associated with occupational radiation exposures, special considerations in the exposure of women of reproductive age, mode of exposure (including internal and external), and basic protective measures such as time, distance, and shielding.

TRADE also produced a *Guide to Good Practice in Radiation Protection Training* (Schenley 1988). The guide is designed to provide a useful resource for the development and self-evaluation of training programs at DOE facilities. This document suggests the course content for each of the following training programs (see Appendix D):

1. General employee radiation protection
2. Radiation worker training
3. Radiation protection technician

Although the document lists the many topics a health physicist must know, radiobiology is but a small part of the total knowledge the health physicist requires.

RADIOBIOLOGY

A review of the NRC, DOE, etc., training outlines clearly demonstrate that radiobiology must be a part of the training. However, each agency or group places a different emphasis on its importance. Let's look at a few of the topics and evaluate how they should be part of the occupational health physicist's education.

Acute Exposure

Almost everyone agrees that an understanding of the acute effects of whole-body irradiation is valuable to the operational health physicist. The most important aspect of learning about the acute effects of whole-body radiation may be that it leaves a lasting impression that radiation can cause serious biological damage.

However, we should pause for a moment and ask, "Does knowledge about the outcome of a serious radiation exposure cause people to behave differently?"

In the book entitled *Radiation: What It Is and How It Affects You* (Schubert and Lapp 1957), a chapter is devoted to the misuse and overuse of radiation. In this chapter, the authors lament that an earlier book titled *X-Ray and Radium*

Injuries–Prevention and Treatment "seems to have had no more effect upon the medical overuse and misuse of radiation than Webster's dictionary."

I believe that times have changed and that the radiation worker and the operational health physicist who have seen pictures of the biological damage caused by large exposures will act differently. However, an overemphasis on the acute effects of whole-body radiation may distract from the real need of the occupational health physicist to understand the effects of acute partial-body exposure and chronic effects. Very few acute whole-body exposures have occurred or are likely to occur, and only a few specialized health physicists are needed to assist medical personnel in evaluating the possible outcome of large exposures. The health physicist's role is to evaluate the physical conditions at the time of the accident and relate them to the observed biological damage and to assist physicians in the evaluation of the dose to the exposed person (Cloutier and Simpson 1990).

More training emphasis needs to be placed on the effects of partial-body exposures. Although these exposures are generally not life threatening, they occur more frequently. The occupational health physicists needs to be able to recognize such exposures and evaluate when medical assistance should be requested.

Chronic Low-Level Exposure

Over 99% of all radiation exposures that the operational health physicist deals with are of the chronic low-level type and are well below today's acceptable limit. Based on our present knowledge of radiobiology, we believe such exposures only result in delayed effects similar to those normally observed in an unexposed population; however, the risk of a particular effect is increased because of the exposure. We can not relate an individual's dose directly to any observable biological effect.

By examining large populations of exposed individuals, epidemiologists can sometimes correlate the population's collective dose to an increased risk (e.g., cancer). Thus the operational health physicist must know something about epidemiology and the possible delayed effects of radiation. Operational health physicists should be familiar with the risk estimates made by BEIR, the United Nations Scientific Committee on the Effects of Atomic Radiation, the International Commission on Radiological Protection, and the National Council on Radiation Protection and Measurements. They must also be familiar with the uncertainties in the estimates, for nothing is known with certainty. The data are confounded by the healthy worker effect, clusters, and the possibility of hormesis. The lifetime risk is dependent on the age at exposure, sex, etc. Superimposed on the risk estimate is the magnitude of the dose-rate effect factor, which is estimated to range from 2–10.

At the present time radiation protection is a public health issue based on epidemiological analysis of populations; however, information already exists that some segments of the population are especially radiosensitive. For example, individuals with a hereditary defect in the DNA repair mechanism, causing ataxia telangiectasia, can be identified by means of a special test. Several other disorders are also known to result in enhanced sensitivity to radiation. The knowledge that unique disorders result in greater radiosensitivity leads to the question about the value of radiation susceptibility testing.

Fred Li (1991) has analyzed this question in the following hypothetical situation. Ten thousand individuals are sampled for a condition that exists in only 1 of 2000 individuals. If a test has a *sensitivity* of 80% (i.e., among susceptible persons the proportion positive for the test is 80%) and a *specificity* of 90% (i.e., among nonsusceptible persons, the proportion who test negative for the test is 90%), then the *predictive value* of the test would only be 0.4%. This means that of the 10,000 tested, 1003 would test positive, yet only 4 would actually have the condition. Would such a test merit use when it could lead to the possible loss of a job and health insurance, and introduce anxiety and depression in 999 individuals who do not have the condition? The operational health physicist must be taught to understand the difficulties involved in using data that have uncertainties.

Operational health physicists are sometimes content to estimate the occupational exposures of their workers by using the presently approved quality factor (Q) or its surrogate and to give the estimated dose equivalent no further thought except to compare the dose equivalent to the accepted standard. However, the operational health physicist should also have an appreciation for the variability in the effectiveness of various types of radiation as described by the relative biological effectiveness (RBE) factor. This topic is not generally covered in much detail in training and educational programs because of its complexity. For example, the RBE values for carcinogenesis range from 1–80 and depend upon the type of radiation, dose, and dose rate. There is great difficulty in understanding the data found in NCRP Report No. 104 (NCRP 1990), *The Relative Biological Effectiveness of Radiations of Different Quality*, and applying it to the "real world." In the real world the operational health physicist has limited knowledge about the types of radiation, the energy spectrum, and the dose rate.

The operational health physicist needs to have an appreciation for the impact the RBE factor may play with regard to the observable delayed effects of radiation.

Other Topics

I have jumped over a host of other topics, such as the probability of causation, transfer of radionuclides through the food chain, internal dosimetry, methods of decontamination, and radioactive "fleas." These and many more topics are related to the broad topic of radiobiology and are of importance to the operational health physicist. An understanding of most of these topics can only be acquired by staying current with the literature in *Health Physics* and other journals and by attending continuing education programs presented by the Health Physics Society and the American Board of Health Physics.

SUMMARY

The health physicist's job is to protect radiation workers and the general public from the harmful effects of radiation. A basic understanding of radiobiology will provide health physicists with a basis for making sound decisions and permit them to more fully appreciate their responsibilities and the consequences of a faulty operational program. More important than a specific education in radiobiology is the need for health physicists to develop an interest in all topics related to radiation protection. They must know where information may be obtained, be it from books or from individuals. After obtaining the information, they must put forth effort to understand the information and not just take the word of "experts" before passing the information on to others.

APPENDIX A

Nuclear Regulatory Commission

Training Topics for Personnel at Light-Water-Cooled Nuclear Power Plants

Regulatory Guide 8.27 includes the list of training topics for personnel at light-water-cooled nuclear power plants. The "Biological Effects of Radiation" section includes the following:

a. Carcinogenesis
b. Genetic Effects
c. Acute Effects
d. Latent Effects
e. Collective Dose Concept
  (1) Group Total Man-Rem Risk
  (2) Individual Dose Risk
f. Dose-Effect Relationship
  (1) External Radiation
  (2) Internal Radiation

APPENDIX B

National Council on Radiation Protection and Measurements

Training Requirements Related to Radiobiology

NCRP Report No. 61 (1978), *Radiation Safety Training for Industrial Radiography*, presents an outline for the initial training requirements. Parts I, C and D of the training outline address topics related to radiobiology.

C. Nature and Consequences of Radiation Exposures
  1. Biological effect of radiation
    a. Types of effects
      (1) Somatic
      (2) Genetic
    b. Radiosensitivity
  2. Dose-effect relationships
    a. Classification of doses
    b. Effects of acute irradiation dose
    c. Chronic doses and late effects

3. Radiation quantities and units
    a. Exposure–roentgen, C/kg
    b. Absorbed dose–rad, gray
    c. Dose equivalent–rem, sievert
    d. Quality factor
4. Nature of radiation health problems
    a. External radiation
    b. Internal radiation contamination
5. Physical examinations
    a. Preoperational baseline data
    b. Exposure evaluation

D. Radiation Hazard in Proper Perspective
1. Philosophy of radiation benefits and risks
2. Personnel exposure
    a. Background
    b. Man-made sources
        (1) Occupational exposure
        (2) Medical exposure
3. Radiation risk to radiographers
4. Maximum permissible doses for occupational workers
    a. Quarterly and annual dose limits
    b. Lifetime dose limits

APPENDIX C

Institute of Nuclear Power Operations

Training Requirements Regarding Biological Effects and Risks Associated with Exposure to Ionizing Radiation

The Institute of Nuclear Power Operations produced the document *Guidelines for Training and Qualifications of Radiological Protection Technicians* (INPO 87–008). The suggested training related to the biological effects and risks associated with exposure to ionizing radiation, including the following:

a. describe the mechanisms of radiation interactions with cells
b. identify cell characteristics that affect radiosensitivity
c. define stochastic and non-stochastic effects
d. explain the difference between acute and chronic radiation exposure and for each describe the following:
    —somatic effects
    —genetic effects
    —teratogenic effects
e. for acute exposures, describe the following:
    —dose-response relationship
    —acute radiation syndrome
    —LD 50/30

f. for chronic exposures, describe the following:
   —long term health effects
   —quantitative risk assessments
   —comparison of radiation exposure risks to other health risks
g. describe the potential health effects and risks due to exposure to internally deposited radioactivity
h. describe the studies used to determine quantitative health risks
i. discuss the basis for and implications of the linear zero-threshold, dose-response curve
j. discuss why radiation exposures to both individuals and groups of workers should be kept as low as reasonably achievable (ALARA)
k. discuss the risk to the general public from operation of a nuclear plant and compare it to other risks accepted in everyday life

APPENDIX D

Training Resource and Data Exchange

Training Requirements Regarding Biological Effects

Training Resource and Data Exchange (TRADE) produced a *Guide to Good Practice in Radiation Protection Training* (Schenley 1988). This document suggests that the training should include the following, dealing with the biological effects of radiation.

General Employee Radiation Protection Training
Acute and Chronic Exposure
Effects of Different Dose Levels
Chronic Whole Body Exposure
Somatic and Genetic Effects
Effects of Prenatal Exposure
Radiation Worker Training
Acute Effects of Exposure
Risks Associated with Occupational Radiation Exposure
Special Consideration in the Exposure of Women of Reproductive Age
Mode of Exposure–Internal and External
Radiation Protection Technician
General
Acute Effects of Exposure to Radiation
Risks Associated with Occupational Radiation Exposure
Special Consideration in the Exposure of Women of Reproductive Age
Mode of Exposure–Internal and External

This report is based on work performed under contract number DE-AC05–76OR00033 between the U. S. Department of Energy and Oak Ridge Associated Universities.

REFERENCES

American Society for Testing and Materials. *Standard Guide for Radiological Protection Training for Nuclear Facility Workers,* ASTME 1168–87. Philadelphia, American Society for Testing and Materials, 1987, pp. 1017–1025.

Christensen, R. C., et al. Guidelines for health physics program content: master's level. *Health Phys. 51*:11–15, 1986.

Claus, W. D. Symposium on education and training in health physics, training programs in health physics. *Health Phys. 8*:93–95, 1962.

Cloutier, R. J., and Simpson, D. R. Radiation accident dose reconstruction. In: *The Medical Basis for Radiation Accident Preparedness II: Clinical Experience and Follow-up Since 1979, Proceedings of the Second International REAC/TS Conference.* New York, Elsevier Scientific Publishing Co., 1990, pp. 415–426.

Crandall, R. W., and Lave, L. B. *The Scientific Basis of Health and Safety Regulation.* Washington, DC, The Brookings Institution, 1981.

Institute of Nuclear Power Operations. *Radiological Protection Technician Qualifications*, INPO-82–006. Atlanta, Institute of Nuclear Power Operations, 1982.

Institute of Nuclear Power Operations. *General Employee Training in Radiological Protection*, INPO 83–012. Atlanta, Institute of Nuclear Power Operations,1983.

Institute of Nuclear Power Operations. *Guidelines for Training and Qualifications of Radiological Protection Technicians*, INPO 87–008. 1987.

Li, Fred. Genes, cancer, and radiation protection. Presentation at the 1991 Annual Meeting of the National Council on Radiation Protection and Measurements. Washington, DC, 1991.

Mills, W. A., et al. *A Compendium of Major U.S. Radiation Protection Standards and Guides: Legal and Technical Facts.* Oak Ridge, TN, Oak Ridge Associated Universities, 1989.

National Council on Radiation Protection and Measurements. *Operational Radiation Safety Program*, NCRP Report No. 59. Bethesda, MD, National Council on Radiation Protection and Measurements, 1978a.

National Council on Radiation Protection and Measurements. *Radiation Safety Training Criteria for Industrial Radiography,* NCRP Report No. 61. Bethesda, MD, National Council on Radiation Protection and Measurements, 1978b. National Council on Radiation Protection and Measurements. *Operational Radiation Safety–Training*, NCRP Report No. 71. Bethesda, MD, National Council on Radiation Protection and Measurements, 1983.

National Council on Radiation Protection and Measurements. *Recommendations on Limits for Exposure to Ionizing Radiation*, NCRP Report No. 91. Bethesda, MD, National Council on Radiation Protection and Measurements, 1987.

National Council on Radiation Protection and Measurements. *The Relative Biological Effectiveness of Radiations of Different Quality*, NCRP Report No. 104. Bethesda, MD, National Council on Radiation Protection and Measurements, 1990.

Oak Ridge Associated Universities. *Ionizing Radiation and Its Biological Effects*, ORAU 1989/E-175. Oak Ridge, TN, Oak Ridge Associated Universities, 1989.

Schenley, C., et al. *Guide to Good Practice in Radiation Protection Training*, ORAU 88/H-99. Oak Ridge, TN, Oak Ridge Associated Universities, 1988.

Schubert, J., and Lapp, R. E. *Radiation: What It Is and How It Affects You.* New York, Viking Press, 1957.

U.S. Department of Energy. *University Programs and Facilities in Nuclear Science and Engineering*, 7th Ed. Washington, DC, U.S. Department of Energy, 1991.

SUGGESTED READINGS

U.S. Department of Energy. *Radiation Protection for Occupational Workers*, DOE Order 5480. Washington, DC, U.S. Department of Energy, 1988, p. 11.

U.S. Department of Labor. *Code of Federal Regulations*, 29CFR 1910. Washington, DC, U.S. Government Printing Office, 1990, p. 120.

U.S. Nuclear Regulatory Commission. *Instructions Concerning Risks from Occupational Radiation Exposure*, Regulatory Guide 8.29. Washington, DC, U.S. Nuclear Regulatory Commission, 1981.

U.S. Nuclear Regulatory Commission. *Radiation Protection Training for Personnel at Light-Water-Cooled Nuclear Power Plants*, Regulatory Guide 8.27. Washington, DC, U.S. Nuclear Regulatory Commission, 1981.

U.S. Nuclear Regulatory Commission. *Code of Federal Regulations*, 10 CFR Part 19. Washington, DC, U.S. Government Printing Office, 1990.

Chapter 17

Radiation Protection of Human Subjects in Research Protocols

KENNETH L. MOSSMAN

ABSTRACT

This paper discusses radiation protection of human subjects participating in research studies. The Institutional Review Board, or Human Subjects Committee, is required to review human research studies to ensure that volunteers are not subject to undue risk. The Radiation Safety Committee and the Radioactive Drug Research Committee may also be required to review and approve the study. Four areas of concern in radiation protection are addressed: (*a*) informed consent, (*b*) interaction of the Institutional Review Board with institutional radiation committees, (*c*) radiation dose limits, and (*d*) special populations.

INTRODUCTION

Human subjects participating in research studies must be properly informed of the risks and benefits of their participation. Federal regulations require institutional review and approval of protocols involving human subjects. In its review of research protocols, an Institutional Review Board (IRB), or Human Subjects Committee, must evaluate the risks to the subjects and determine that the sum of the benefits and the knowledge to be gained from the study outweigh the risks. The IRB must also ensure that informed consent will be obtained from all subjects. The IRB shall have at least five members with sufficient qualifications to review specific research activities. One member of the IRB shall be a nonscientist (e.g., lawyer or clergyman), and at least one member must not be associated with the institution. The IRB, at its discretion, may consult individuals with competence in special areas (e.g., radiation protection and radiation dosimetry) to assist in the review of specific protocols (Department of Health and Human Services 45 CFR 46 1990).

IRB review and approval of protocols must be established for each university or other institution (e.g., hospital) that conducts human subjects research and that applies for federal funding. Institutions not in compliance with the law jeopardize federal funding of their programs.

In evaluating protocols involving use of radionuclides or radiation-producing equipment, the IRB may have to assess the radiation risks to subjects and put these risks into understandable lay terms in the informed consent form. To assist the IRB, the institution's Radiation Safety Committee, or health physicist or other radiation expert, may be asked to review the protocol.

This chapter briefly discusses radiation protection of human subjects in research studies. Four areas are explored: (*a*) informed consent, (*b*) interaction of the IRB with institutional radiation committees, (*c*) radiation dose limits, and (*d*) special populations.

INFORMED CONSENT

No research investigator may involve a human subject in a research study unless the investigator has obtained the legally effective informed consent of the subject or the subject's legally authorized representative. The purpose of informed consent is to protect the subject from undue research risks by apprising him or her of the potential risks and benefits of participation in the study, in language understandable by the subject. A properly executed informed consent form contains the following elements: (*a*) a statement describing the purpose of the research and expected length of time of the subject's participation as well as a description of the procedures to be used, (*b*) a description of direct benefits or lack thereof to the subject or to others as a result of the research and possible risks or discomforts to the subject, (*c*) disclosure of appropriate alternative procedures or treatments that may be advantageous to the subject, (*d*) a statement pertaining to confidentiality of records identifying the subject, (*e*) a statement describing any compensation to subjects and an explanation as to whether medical treatments are available should injury occur, (*f*) an explanation of whom to contact for answers to questions about the research, and (*g*) a statement that participation is voluntary and that withdrawal from the study will not penalize the subject. Depending on the nature of the study, additional elements of informed consent may be required. For instance, a statement may be required indicating that a particular procedure may involve risks to the subject (or to the embryo/fetus if the subject is or may become pregnant) that are presently unforeseeable (Department of Health and Human Services 45 CFR 46 1990).

As part of the protocol approval process, the IRB carefully reviews the informed consent form. The IRB must determine that all appropriate elements of informed consent are included in the form and that the document is in language understandable by the subject or representative.

In research protocols involving radiation exposure, the IRB may consult a health physicist or other radiation expert to assist in the evaluation of radiation risks and in placing such risks in perspective. For instance, in radiopharmaceutical studies, the health physicist may be required to calculate radiation doses to critical organs and to assess the associated risks. In providing this information to the IRB, the fact that dose and risk figures are estimates needs to be clearly articulated. Ideally, some range of values should be given to reflect the degree of precision in the estimates. Furthermore, the stochastic nature of radiogenic effects (e.g., neoplasia) needs to be clearly spelled out. No matter what the radiation dose, the occurrence, or lack of occurrence, of radiogenic effects cannot be predicted with certainty.

The health physicist may also be asked to review the informed consent form to be sure that risks of radiation exposure are stated accurately and in language understandable by the subject. Using risk comparisons may be of value in putting the radiation risk in perspective. For example, assume that a subject is given a whole body dose of penetrating radiation of 0.1 mGy. The following are examples of risk comparison statements that may be included in the informed consent form to help the subject better understand the significance of the dose:

> The amount of radiation received in this procedure is about ten times lower than the allowable dose to individual members of the public in a year. The risk from an exposure of this magnitude is generally considered to be negligible.
>
> The amount of radiation received in this procedure is equivalent to the amount of naturally occurring natural background radiation that all individuals in the United States receive in each one-month period.

The first statement is preferable because it involves comparison with a standard. The second statement is also useful but may be less effective in providing useful risk information because it involves comparison with another source of radiation exposure. Statements involving comparisons with risks unrelated to the research study should be avoided whenever possible. For example, comparing the risk of a diagnostic x-ray procedure with the risks of smoking a given number of cigarettes is irrelevant and not legitimate. Such comparisons are ineffective in providing meaningful risk information and may result in misinformation about the nature of the radiation risk.

INTERACTION OF THE INSTITUTIONAL REVIEW BOARD WITH INSTITUTIONAL RADIATION COMMITTEES

Research protocols involving radiation exposure of subjects may require review and approval by other committees in addition to the IRB. A Radiation Safety Committee (RSC) must be established in medical institutions licensed by the U.S. Nuclear Regulatory Commission (NRC) to use byproduct materials (U.S. NRC 10 CFR 35 1991a). Membership must consist of at least three individuals and must include an authorized user of each type of use permitted by the license, the Radiation Safety Officer, a representative from management, and a representative of the nursing service. The primary role of the RSC is to oversee the use of byproduct material in laboratory and clinical settings in the context of radiation safety practices and to maintain personnel radiation doses as low as reasonably achievable (ALARA). Accordingly, the RSC may be required to review human research protocols to ensure that appropriate radiation safety procedures are in place and that personnel doses are maintained ALARA. In the case of protocols not involving byproduct materials (e.g., use of machine-generated radiation), approval of the RSC in NRC-regulated states is not mandated by law. Nevertheless, it may be desirable for the RSC to review protocols and make recommendations to the IRB. In fact, many institutions may have such an internal requirement. In Agreement States, RSC review of protocols not involving byproduct materials may be a requirement.

When the research protocol involves use of radioactive drugs authorized by the institution's federal or state radioactive materials license, approval by the institution's Radioactive Drug Research Committee (RDRC) is required, if the radioactive drug is not otherwise approved by the Food and Drug Administration (FDA). In other words, if an investigator has an investigational new drug (IND) for the study of a particular radioactive drug, approval by the RDRC is not required. The RDRC regulations are specific for radioactive drugs for which a new drug application (NDA) has not been filed or an IND issued. In cases where an NDA or IND does not exist, the RDRC may approve the protocol under the following specific provisions: (*a*) The level of active ingredient in the drug is known not to cause additional detectable pharmacological effects, and (*b*) the radiation exposure is justified by the quality of the study being undertaken and the importance of the information the study seeks to obtain. At a minimum the RDRC shall consist of five members. The committee shall include the following three individuals: a physician specializing in nuclear medicine, a person with special competence in radiation dosimetry and radiation protection, and a person with specialized training to formulate radioactive drugs. The other members shall be qualified in various disciplines pertinent to the field of nuclear medicine (FDA 21 CFR 361 1990). In many institutions, the FDA-approved RDRC is a subcommittee of the RSC. An institution may not arbitrarily establish an RDRC and appoint members. Both the NRC (or Agreement State) and the FDA must approve the RDRC.

When multicommittee review of a research protocol is necessary, the ultimate responsibilities for determining research risks and protocol approval resides with the IRB because of its broader mandate to review *all* research risks. As a member of the RSC and the RDRC, the health physicist plays a key role in information exchange between these committees and the IRB.

RADIATION DOSE LIMITS

The FDA regulations assume that participation in a research protocol by a volunteer will result in no direct benefit to the volunteer. Proof of benefit rests with the investigator. Research protocols involving radiation exposure of volunteers for which no direct benefit is expected are subject to specific restraints as set forth by the FDA (21 CFR 361 1990). Protocols in this category include basic science studies regarding the metabolism, kinetics, distribution, and localization of radioactively labeled drugs. The RDRC must approve such protocols based on the specific provisions described above and, in addition, ensure that the radiation dose to an adult subject is within the limits set forth in Table 17.1. For research subjects under 18 years of age, the radiation dose limits are 10% of the limits listed in Table 17.1 (FDA 21 CFR 361 1990). These RDRC requirements apply only to protocols that utilize radioactive drugs not approved through the NDA or IND mechanism. They do not apply to protocols where these radioactive drugs are not used (FDA 21 CFR 361 1990).

For research protocols in which participation may provide direct diagnostic or therapeutic benefit, the above restrictions are not applicable. In such instances, the radiation dose to the subject will be based on clinical requirements of the individual case. Even when participation may provide no direct diagnostic or

TABLE 17.1. LIMITS ON RADIATION DOSE TO ADULT RESEARCH SUBJECTS[a]

Target Organ	Single Dose (mSv)	Annual and Total Dose Commitment (mSv)
Whole body	30	50
Active blood forming organs	30	50
Lens of the eye	30	50
Gonads	30	50
Other organs	50	150

[a] From Food and Drug Administration. Prescription drugs for human use generally recognized as safe and effective and not misbranded: drugs used in research. Washington, DC: U.S. Government Printing Office, Title 21, Code of Federal Regulations, Part 361 (21 CFR 361), revised as of 1 April 1990.

therapeutic benefit, the above restrictions do not apply unless the protocol involves the use of a radioactive drug for which there is no NDA or IND.

In accordance with federal regulations, the amount of radioactive material administered should result in the subject's receiving as small a dose as possible without compromising the value of the study. Under no circumstances should radiation dose exceed the limits in Table 17.1. The single-dose limits in the table refer to doses from a single study. The annual and total dose commitment refers to limits on the cumulative dose from a number of studies over a 1-year period and include the contribution of all radionuclides (including any contaminants in the administered product) and any contributions to dose from x-ray procedures that are part of the research study (FDA 21 CFR 361 1990). These dose limits apply only to RDRC studies. There are no federal regulations that restrict radiation dose in other research studies.

The dose limits listed in Table 17.1 are comparable to occupational exposure limits for adult radiation workers(U.S. NRC 10 CFR 20 1991b). Accordingly, the risk acceptable for volunteers is comparable to the risk acceptable for radiation workers. Although volunteers derive no direct medical benefit from their participation, they may be satisfied in other ways by accepting payment or other form of compensation as a reward for their participation.

SPECIAL POPULATIONS

In human subjects research, pregnant women and children are considered special populations. Depending upon the purpose and nature of the study, participation by pregnant women or children in research studies may be restricted.

Federal law requires that pregnant women be excluded from participation in clinical trials or human research protocols except (*a*) when the purpose of the study is to meet the health needs of the mother and the fetus will be placed at risk only to the minimum extent necessary to meet such needs or (*b*) when the risk to the fetus is minimal (Department of Health and Human Services 45 CFR 46 1990). In protocols requiring RDRC approval, no pregnant woman may participate. Each female research subject of childbearing potential shall either state in writing that she is not pregnant or on the basis of a pregnancy test be confirmed as not pregnant before entering into any study (FDA 21 CFR 361 1990).

The basis for exclusion of pregnant women from participation in research studies requiring RDRC approval is the enhanced radiation sensitivity of the developing embryo and fetus. The developing embryo/fetus is subject to an increased risk of several kinds of radiation effects, including mental retardation and malformations, that are not seen in radiation exposure of the adult.

Table 17.2 provides the radiation risks for the major biological effects expected at low doses in the adult and the embryo/fetus. Such risks cannot be directly compared because the spectrum of radiation effects are different. However, a comparison of the radiation risk as a percentage of the natural occurrence of effects suggests greater radiation sensitivity of the embryo/fetus.

Federal law also restricts the use of children in research studies. For the purposes of IRB requirements, children are persons who have not attained the legal age for consent to treatments or procedures involved in the studies (Department of Health and Human Services 45 CFR 46, 1990). Children may participate in research studies if (*a*) the research does not involve greater than minimal risk, (*b*) the research involves greater than minimal risk but presents the prospect of direct benefit to the individual subjects, (*c*) the research involves greater than minimal risk and no prospect of direct benefit to individual subjects but is likely to yield generalizable knowledge about the subjects' disorder or condition, or (*d*) the research, not otherwise approvable, presents an opportunity to understand, prevent, or alleviate a serious problem affecting the health or welfare of children (Department of Health and Human Services 45 CFR 46, 1990).

In protocols requiring RDRC approval, research subjects shall be at least 18 years of age. Exceptions are permitted when it can be demonstrated to the RDRC and the IRB that the study presents a unique opportunity to gain information not presently available, requires the use of subjects less than 18 years of age, and is without significant risk to the subject (FDA 21 CFR 361 1990).

The basis for restricting the use of children in research studies involving radiation exposure is the demonstrated enhanced risk, particularly for cancer, in

TABLE 17.2. COMPARISON OF RADIATION RISKS IN THE EMBRYO/FETUS AND ADULT[a]

	Embryo/Fetus	Adult
Principal health effects at low dose	Mental retardation, cancer	Cancer mortality
Normal risk in absence of radiation	6%	18%
Radiation risk (per cGy)	0.2%	0.08%
Radiation risk (per cGy) as percentage of normal risk	3.2%	0.4%

[a] National Research Council, Committee on the Biological Effects of Ionizing Radiations (BEIR V). Health effects of exposure to low levels of ionizing radiation. BEIR V. Washington, DC: National Academy Press, 1990; United Nations Scientific Committee on the Effects of Atomic Radiation. Sources, effects and risks of ionizing radiation. 1988 report to the General Assembly, with annexes. New York: United Nations, 1988.

TABLE 17.3. EXCESS RELATIVE CANCER RISKS PER SIEVERT BY AGE AT EXPOSURE AND FOLLOW-UP TIME[a]

Age at Exposure	Follow-up Time		
	1956–1965	1966–1975	1976–1985
0–19	0.94	0.93	0.69
20–34	0.64	0.44	0.52
>35	0.14	0.22	0.38

[a] From Pierce DA, Vaeth M, Preston DL. Analysis of time and age patterns in cancer risk for A-bomb survivors. Radiat Res 1991;126:171–186.

younger age groups. Table 17.3 provides a recent analysis of temporal patterns in cancer risk estimates in atomic bomb survivors.

Excess relative risks are highest in the youngest age group (at time of exposure) for all follow-up periods. Statistical analysis indicates that for age at exposure groups less than 35 years, the excess relative risk remains constant into the current follow-up period. In the oldest age group at time of exposure (greater than 35 years), there appears to be an increasing trend in excess relative risk. The decrease in relative risk with age at exposure is not fully understood. The large excess risk observed at early ages may be due to the small number of excess deaths during the portion of the lifespan when the normal risk is very small (Pierce et al. 1991).

REFERENCES

Department of Health and Human Services. *Protection of Human Subjects.* Washington, DC, U.S. Government Printing Office, Title 45, Code of Federal Regulations, Part 46 (45 CFR 46), revised as of 1 October 1990.

Food and Drug Administration. *Prescription Drugs for Human Use Generally Recognized as Safe and Effective and Not Misbranded: Drugs Used in Research.* Washington, DC, U.S. Government Printing Office, Title 21, Code of Federal Regulations, Part 361 (21 CFR 361), revised as of 1 April 1990.

Pierce, D.A., Vaeth, M., and Preston, D.L. Analysis of time and age patterns in cancer risk for A-bomb survivors. *Radiat. Res. 126:*171–186, 1991.

U.S. Nuclear Regulatory Commission. *Medical Use of Byproduct Material.* Washington, DC, U.S. Government Printing Office, Title 10, Code of Federal Regulations, Part 35 (10 CFR 35), revised as of 1 January 1991a.

U.S. Nuclear Regulatory Commission. *Standards for Protection against Radiation.* Washington, DC, U.S. Government Printing Office, Title 10, Code of Federal Regulations, Part 20 (10 CFR 20), revised as of 21 May 1991b.

Index

Page numbers in italics denote figures; those followed by "t" or "f" denote tables or footnotes, respectively.

A

Adrenal carcinoma, *79*
Alpha-emitters, cancer risk, 138–153
American Society for Testing and Materials (ASTM), 243
Aneuploidy, 14
Ankylosing spondylitis
  A-bomb survivors, 130t
  radiotherapy for, 92
Antimicrobials, therapy for sepsis, irradiated mice, 212
Atmospheric testing litigation, 159–162
Atomic and thermonuclear weapons
  cancer mortality in survivors, 129t
  cancer risk in survivors, 128t
  exposures from, 97–101, 97t
    Japanese survivors, 97
Auerbach, Charlotte, 7

B

Bacteria, radiation injuries in mice, 204
Bane litigation, 163–164
BEIR IV, 138–153
BEIR V, 15–20, 121–137, 164
Bone cancer, dial painters, *147*
Bone marrow transplantation, 192
Breast cancer
  after radiotherapy, 91
  atomic and thermonuclear weapon exposure, 98
Breast disease, benign, radiotherapy for, 93

C

Cancer, 77–88
  see also specific type
  age at irradiation and, 124t, 127
  atomic and thermonuclear weapon exposure, 98, 132t, 133t, 135t
  childhood, radiotherapy for, 91
  clusters, nuclear installations, 108
  constitutional and physiological factors, 128
  excess cancers, 90t, 92t, 95t, 97t, 101t, 103t
  linear energy transfer, 126–127
  other cofactors, 128
  radiation, latent period, 129
  radiation dose-effect relations, 122–126
  risk coefficients, 130
  risk estimates as compared to ICRP-26, 134–136
  sex and, 128, 129t
Carcinogenesis, 77–88
  human cell systems model, 85–86
  in utero ionizing radiation
    human, 32–39
    rat and mouse, 26–32
  radiation, 86, 89–120
Central nervous system, acute radiation exposure, 32–39
Cervical cancer, radiotherapy for, 90–91
Cesium, radioactive, 55
Chromium, radioactive, 55
Chromosomes, 5–11
  unbalanced rearrangement, 13–14
Cystic fibrosis, 9, 13
Cytogenetics, 7–8

D

DNA, 6–10, 12
Dominant mutant, 13
Down syndrome, 6, 9, 14
Drosophila melanogaster, 5, 7

E

Embryo
  diagnostic radiological exams, 50, 65t
  dose from various radiopharmaceuticals, 65t
  ionizing radiation effects
    human, 32–39, 38t
    rat and mouse, 26–32
    research protocols, 259–261
Epidemiology, radiation carcinogenesis, 89–120
    caution in interpretation, 111
Eye, lens, radiation effects, 178–180

F

Fetus
  diagnostic radiological exams, 50
  radiation effects
    human, 32–39
    rat and mouse, 26–32
    research protocols, 259–261

G

Gastrointestinal radiation syndrome, 188
Genetics, principles of, 3–11
Genetic damage, categories of, 12
Genetic disease
  current instances in human population, 14–15t
  risk estimation, 15, 17t, 18t
    dose-response models, 20
  societal cost, 20–21
Genome, 6
Germ cell stages, 14
Gestation
  acute radiation effects
    human, 32–39
    rat and mouse, *24–26*, 27t, 28t, 29t, 30t, 31–32
Growth retardation, 36

H

Hardy-Weinberg equilibrium, 8
Head and neck cancer, 78–79
Health physics, program content, 240–254, 245t, 246t
Hematological radiation syndrome, 188
Hematopoietic growth factors, 193
Hiroshima-Nagasaki studies, 20–21
Hodgkin's disease, radiotherapy for, 91
Human cell systems, model, carcinogenesis studies, 85–86
Huntington disease, 13

I

Immunomodulators, 204
Informed consent, 256
Institute of Nuclear Power Operations (INPO), 243
International Commission on Radiological Protection (ICRP), 138, 154–157, 230–239
Intrauterine growth retardation, 32
Iodine, radioactive, 52, 55
  cancer risk and, 150
Iodine-131
  diagnostic, 102
  therapeutic, 102

L

Leukemia, 10, 39–43
  risks in specific groups, 41t
  atomic and thermonuclear weapon exposure, 97, *127*
  radiotherapy for, 91
Linear energy transfer, 126–127, 145, 152, 155–156
Luminous dial painters, radiation exposure, 103
Lung cancer, atomic and thermonuclear weapon exposure, 98

M

Marshall Islands, 99–100
Mastitis, radiotherapy for, 92
Mendel, Gregor, 3–11
Mendelism, 4
Mental retardation, 36
Microcephaly, 32–36
Miners, underground, radiation exposure, 106
Morgan, Thomas Hunt, 5
Mortality, modulation of, tissue trauma and sepsis after radiation injury, mice, 202–217
Multifactorial disease class, 14
Mutagenesis
  chemical, 7
  principles of, 3–11

N

National Council on Radiation Protection and Measurements (NCRP), 138, 154–157, 230–239, 242
Neurofibromatosis, 13
Neurovascular radiation syndrome, 188
Neutron quality factor, 230, *232*
Nevada, 100
Non-Hodgkin's lymphoma, radiotherapy for, 91
Nuclear installations, cancer clusters around, 108
Nuclear installation workers, radiation exposure, 105
Nuclear reactor accidents, 110
Nuclear weapons test participants, 100–101
NUREG, 16–20

O

Oncogenes, 83–84
  modulation of radiation response by, 84–86, 85t
Ovaries
  exposure to radiation, 49–51
  radiation effects, 177–178

P

Pelvimetry, 39
Phosphorus, radioactive, 55
Phosphorus-32, 102
Photon energy, *233*
Placenta, radionuclides and, 55
Plutonium, workers exposed to, 106
Polonium, cancer risk and, 144–145
Polyploidy, 6, 9
Potassium, radioactive, 55
Pregnancy
  diagnostic radiation, 43, 46–51, 50t
  exposure to radiation, 43–46, 44t, 45t
    counseling, 43–46

dose assessment, 64–65
health physics considerations, 63–73
legal and policy considerations, 67–71
medical, 68
occupational, 68
risk assessment, 65–67, 66t
internally deposited radionuclides, 51–58
radiation exposure of ovaries, 49–51
radiation exposure of uterus, 49–51
radiation therapy, 48–49
Probability of Causation formula, 164
Psychological stress, radiation emergencies and, 197
Pulmonary radiation syndrome, 189

R

Radiation, 7, 9
accidents, *221–222*, 222t, 223t, 224t
history of, 218–225
long-term medical follow-up, 220t
physician preparedness for, 198
cancer induced by, 86, 89–120
contemporary issues in health, 229–239
diagnostic, 95–97, 95t
dose assessment, 64–65
dose-effect relations, 122–126
dose estimation, 219
dose limits, 237t, 258
environmental exposures, 107–111
exposure limits, 236
external, cancer risk estimates, 121–137
eye, 178–180
genetic damage induced by, 13
genetic effects, 12–22
injuries, 184–201, 189t, 190t
tissue trauma and sepsis in mice, 202–217
ionizing, embryonic and fetal exposure, 23–62
legal and policy considerations, 67–71
medical exposure, 68
natural background, 108
neutron quality factor, 230
occupational exposure, 68, 103–107, 103t, 235t
ovary, 177–178
preconception exposure, 109
prenatal, 95
cancer risks in A-bomb survivors, 128t
oncogenic effects, 39–43
quantities, 230
radiation biology and operational health physics, 240–254
research protocols, human subjects in, 255–261
risk assessment, 65–67
skin, 174
special natural background sources, 234
sensitivity value, *81*
testis, 175
therapy, 48–51
benign disease, 92–94, 92t
cervical cancer, 90–91
childhood cancer, 91
head and neck cancers, 78–79
oncogene modulation of response to, 84–86
radical or palliative, 82–83
sensitivity, resistance, and responsiveness, 79–82
training, 244t
Radiation Effects Research Foundation (RERF), 154
Radiation emergency assistance center, 220
Radiation protection, risk assessment in, 154–157
Radiation review boards, 256–261
Radiation syndromes, 187
diagnosis and triage of, 189–190
gastrointestinal, 188
hematological, 188
neurovascular, 188
psychological stress in, 197
pulmonary, 189
treatment considerations, 190–194
treatment resources, 194
Radioactive fallout, 99–100
Radiobiology
acute exposure, 247
chronic low-level exposure, 248
human, 185–187
operational health physics and, 240–254
special populations, 259
Radioepidemiological tables, radiation litigation, 158–165
Radiography, diagnostic, embryo doses, 65t
Radiologists, radiation exposure, 104
Radionuclides
contamination, 195–197
decorporation methods and agents, 197t
internal
cancer risks from, BEIR IV, 138–153
exposure to, 101–103, 101t
pregnancy and, 51–58
Radiopharmaceuticals
clinical procedures, 53t, 54t, 55t
doses to embryo from, 65t
Radiosensitivity, tissues and organs, 169–183
Radium
cancer risk and, 145–146
ingestion, 148t
Radium-224, 101
Radium aircraft instrument dial litigation, 162–163
Radon
cancer risk and, 140–144, 142t, 143t, 144t
indoor, 107
Recessive mutant, 13
Regulatory control, exemption from, 233
Retinoblastoma, 10
Ringworm, scalp, radiotherapy for, 93
RNA, messenger, 10

S

Scoliosis, diagnostic radiation, 95
Selenium, radioactive, 55
Sepsis, tissue trauma and, radiation injuries, mice, 202–217

Sex-linked mutant, 13
Sickle cell anemia, 9, 13
Skin
  radiation injuries in mice, 204
  radiosensitivity of, 170–174
Sodium, radioactive, 55
Sperm, radiation effects, 174–177
Squamous cell carcinoma, 78–79
S-TDCM, therapy for sepsis, irradiated mice, 212
Strontium, radioactive, 55
  cancer risk and, 150, 151t

T

Tay Sachs disease, 13
Teratogenic agents, humans, 67t
Teratogenesis
  in utero ionizing radiation
    human, 32–39, 32t, 33t, *34–36*
    rat and mouse, 26–32
Testis, radiosensitivity of, 174–177
Tetraploidy, 6, 9
Thallium, radioactive, 55
Thorium, cancer risk and, 146–148
Thorotrast, 103, 148t
Thymus, enlarged, radiotherapy for, 93
Thyroid cancer, atomic and thermonuclear weapon exposure, 98, *126*, 150t
Tissue trauma, radiation, 194
  mice, modulation of mortality in, 202–217
Tonsils, radiotherapy for, 94
Transuranic elements, cancer risk and, 149, 149t
Triploidy, 6, 9
Tuberculosis, diagnostic radiation, 95
Tumor suppressor genes, 83–84

U

United Kingdom, 101
United Nations Scientific Committee on the Effects of Atomic Radiation (UNSCEAR), 15–20, 121, 154, 155t
United States Department of Energy (DEO), 242
United States Nuclear Regulatory Commission (NRC), 241, 256–261
United States Occupational Safety and Health Administration (OSHA), 242
Uranium, cancer risk and, 148–149
Utah, 99–100
Uterus, exposure to radiation, 49–51